WITHDRAWN

I.C.C. LIBRARY

MAPPING *the* WORLD

EASTERN HEMISPHERE
NORTH POLE
ARCTIC OCEAN
EUROPE
RUSSIA
SIBERIA
ASIA
CHINA
INDIA
PERSIA
ARABIA
AFGHANISTAN
TURKEY
SYRIA
IRAQ
SPAIN
FRANCE
MEDITERRANEAN SEA
MOROCCO
ALGERIA
TUNISIA
LIBYA
EGYPT
AFRICA
FRENCH WEST AFRICA
ANGLO EGYPTIAN SUDAN
NIGERIA
CAMEROON
ETHIOPIA
BELGIAN CONGO
PORTUGUESE WEST AFRICA
BRITISH SOUTH AFRICA
MADAGASCAR
NORTH PACIFIC OCEAN
INDIAN OCEAN
EQUATOR
TROPIC OF CANCER
TROPIC OF CAPRICORN
SOUTH ATLANTIC OCEAN
AUSTRALIA
NEW GUINEA
BORNEO
SOUTHERN OCEAN
SOUTH POLE
MAP of DISCOVERY

MAPPING *the* WORLD

An Illustrated History of Cartography

NATIONAL GEOGRAPHIC
WASHINGTON, D.C.

Contents

Ottoman Map of Africa, 1803
In Nizam-ı Cedid, first atlas printed in the Muslim world

اقليم افريقا
بحر سفيد
بحر خزر
مصر
حجاز
جزيرة العرب
يمن
بحر محيط عرب
حبش
عدل
صحراى كبير
بودنو
غصنيه
بنين
ممالك كفريه
قونغو
بحر محيط شرقى
صوفاله
ناطال
بحر محيط
هندستان
اميد برونى
مقياس

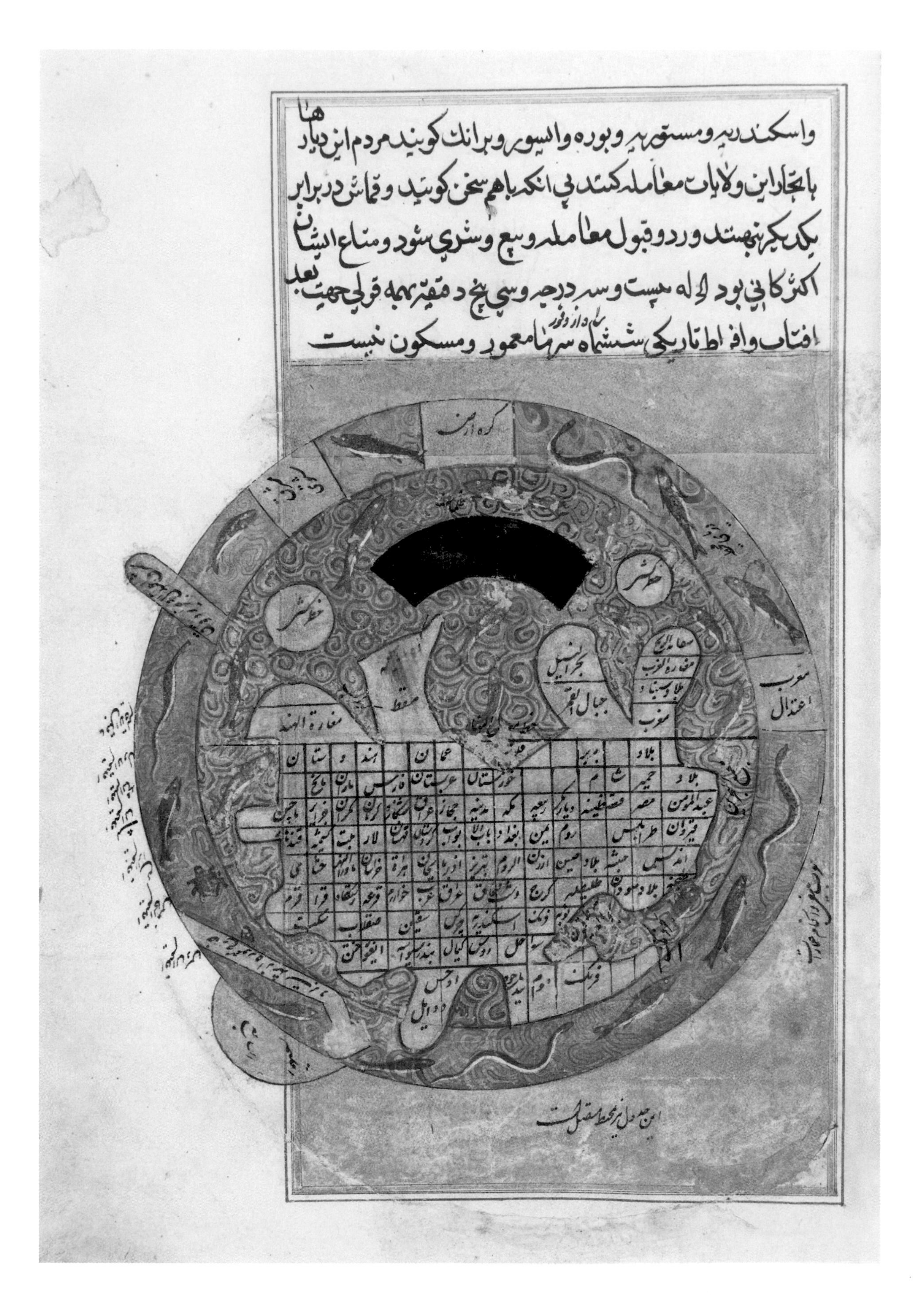

Hand-Colored Persian Map Showing Distribution of Land and Sea, 1565
by Zakariya ibn Mahmud al-Qazwini

Foreword

Casually peruse the maps in this book, and you're sure to have a rewarding experience on a purely aesthetic level. Every page presents colorful and intricate webs of line and pattern that are endlessly pleasing to the eye. Yet a distinctly different (and perhaps deeper) pleasure can be had by thoroughly considering the historical context, the cartographic principles, the geography, the politics, and the complex human aspirations behind each map. As one of the privileged few who have made cartography the focus of my career, I've long since come to the conclusion that it's this rich mix of visceral, left-brain pleasure and intellectual, right-brain engagement that makes maps so alluring. Maps stimulate our whole mind: the emotional and intuitive, the rational and analytical. Creating maps always presents interesting mental challenges: determining scale and projection, weighing the features to depict and how to symbolize them; balancing color, type, form; distilling elements of geography onto a map that tells a story, does a job, and pleases the eye. Much of the pleasure in creating maps, though, lies simply in the quest to create something beautiful.

Reading maps is challenging, too: interpreting patterns, deriving locations and routes, understanding interrelationships, making the not insignificant mental leap from color, label, and symbol back to the original form—the surface of the Earth. But if map reading involved only effort, books of beautiful maps wouldn't be published, and few cartographers would have jobs. Reading maps, fortunately, is fun. Each map is a challenge to explore, imagine, dream, question. And the best maps, such as the ones in this book, thrill us with their sheer beauty.

Maps are perhaps unique in their dual role as artistic creations and as precise, scientific exercises. They're profound expressions of the unique human compulsion to explore the world around us, to measure, analyze, model, and survey. We seek to fully possess the world by projecting its essence onto paper, sphere, vellum, or computer screens via the efforts of our intellects. But they're also a celebration of the wonder of the world, and of our continuing quest to understand our place in it.

Allen Carroll
Chief Cartographer
National Geographic Society

Introduction

During a critical juncture of the Lewis and Clark transcontinental expedition in 1804-06, Capt. Meriwether Lewis asked the brother of Sacagawea, the expedition's female Indian guide, to instruct him "with rispect to the geography of his country. This he undertook very cheerfully, by delineating the rivers on the ground," Lewis noted in his journal. "He place[d] a number of heaps of sand on each side which he informed me represented the vast mountains of rock eternally covered with snow through which he passed." In this dramatic passage, Lewis described the preparation of an Indian map. Using the tools at hand, Cameahwait, a Shoshone headman, gave visual expression to his geographic knowledge of the formidable ranges and deep river valleys of present-day central Idaho. His cartographic creation contributed directly to the success of the Lewis and Clark expedition, which in turn produced the first widely disseminated printed map of the trans-Mississippi West.

Cameahwait was one of many early Indian mapmakers whose work shaped the North American map. His rudimentary sketch of the Bitterroot and Salmon regions was part of a larger process of systematic geographical exploration and mapping. Native American mapmakers such as Cameahwait played crucial roles in the discovery, delineation, and definition of this great continental landmass. Their maps documented geographical discoveries, pro vided the spatial framework for further exploration, and projected the power of empires across the landscape.Mapmaking, as Lewis's story suggests, has a rich and varied history.

The impulse to map one's world dates from the earliest records, and is associated with all cultures. A recent work devoted to the history of the traditional cultures of Africa, the Americas, Australia, and the Pacific islands, edited by David Woodward and G. Malcolm Lewis, reveals that indigenous peoples, before European contact, mapped for reasons similar to those of later mapmakers who left more permanent records. These vary from culture to culture, but all include some form of expressing spatial knowledge. The Pawnee and Lakota of the Northern Great Plains, for example, painted celestial maps on deerskin, which they carried in their medicine bundles. Mesoamerican cartographers favored maps that recorded their communities' history and mythology. The Chukchi of Siberia produced detailed river maps, painted on boards with deer or reindeer blood. Pacific islanders constructed stick charts and sophisticated star compasses from coral, coconut leaves, and banana fibers to teach ocean navigation. And Australian Aborigines communicated maps of their sacred myths through songs and art.

The earliest surviving maps and charts are associated with ancient Babylonia and Egypt. By the third millennium B.C., both empires possessed the necessary mathematical and drafting skills and the bureaucracy for surveying and mapping. Sumerian scribes invented writing in the fourth millennium B.C., and developed the sexagesimal system, a numerical system based on the number 60, which is still used to measure angles, time, and geographic coordinates. They also learned to measure distance and calculate area, knowledge used to prepare world maps, city plans, and property plans complete with appropriate symbols for streets, rivers, and canals beginning in about 2300 B.C.

Unlike the practical cartography of Babylonia, the majority of Egyptian maps were pictorial renderings of mythical lands and imaginary routes to the afterlife. A small number have survived in the form of stone funerary monuments on temple and tomb walls, decorative pottery, wooden tablets, and painted coffins. Also surviving is a topographical map of a gold-bearing region in Nubia, drawn on papyrus in about 1150 B.C. Now housed in the Egyptian Museum in Turin, Italy, it portrays settlements associated with gold extraction and roads linking the Nile with the coast of the Red Sea.

The scientific foundation of the Western cartographic tradition was formed in classical Greek theories relating to the nature and form of the Earth. The Pythagoreans introduced the concept of a spherical Earth in about the fifth century B.C. Aristotle divided the known world into five climatic zones of equally placed parallels, foreshadowing the concept of a geographic coordinate system. Eratosthenes, head of the library at Alexandria, conducted the first measurements of the circumference of the Earth. The astronomer Hipparchus of Nicaea was the first to divide the globe into an imaginary grid of equally placed parallels of latitude and meridians of longitude. He also has been credited with inventing the astrolabe, an instrument later used by mariners to determine latitude, longitude, and the time of day. In the second century A.D., the Greco-Egyptian geographer and astronomer Claudius Ptolemy systematized the work of earlier Greek thinkers in two treatises that influenced Western and Islamic cartography for some 14 centuries. In the *Almagest*, he formulated a geocentric model of the solar system. His *Geography* provided detailed instructions for mathematical mapmaking, including the construction of map projections and the preparation of a world map and regional maps.

While Greek astronomers and geographers provided the theoretical basis for Western cartography, their Roman contemporaries focused more on the practical aspects of surveying and mapmaking. The few surviving Roman maps from this era pertain to large-scale property surveys, town plans, and road maps.

AT THE SAME TIME, ANCIENT CHINESE DYNASTIES were evolving their own, distinct cartographic traditions. One blended quantitative method with historical sources. From the third century A.D. onward, the map elements of distance, direction, and terrain features were described verbally by narrative text on the maps themselves or in accompanying pamphlets, rather than by symbols or numbers as on Western maps. Another cartographic tradition that emerged was the map in the form of wall hangings or mural paintings. Generally associated with the visual arts, these map paintings contained cartographic and geographic elements of particular places and spaces.

While few maps survive from this early period, a small number of surprisingly sophisticated plans and maps with a "modern" look have recently been unearthed from Chinese tombs. Incised on bronze plates, or drawn on silk, wooden boards, and tomb walls, some appear to be prepared to scale. The earliest Chinese maps were used for military planning and state security as well as for education and aesthetic appreciation. With the establishment of a centralized bureaucracy during the Qin and Han dynasties (221 B.C.-A.D. 220), a variety of property, terrain, and boundary maps were prepared for administrative purposes, but few have survived. After the invention of woodblock printing in the eighth century, map production increased. Representative of this later period are the *Lidai dili zhizhang tu* (Easy-to-use maps of geography through the dynasties), printed from woodblocks in 1098-1100, and the *Yu ji tu*, a map of China engraved on stone in 1136. The *Yu ji tu* is noted for its grid, which resembles a modern graticule of latitude and longitude but in fact was used to calculate distance and area.

Korea and Japan developed indigenous mapping traditions of provincial and national maps, but they relied upon China for their world maps, which they copied, adding their countries to the periphery.

In medieval Europe, the scientific advances represented by Ptolemy were lost after the collapse of the Roman Empire, replaced by Christian doctrine and fervor that pervaded scholarship and literature. Guided by Scripture, monastic mapmakers discarded the spherical Earth for a flat, circular one to produce world maps known as *mappaemundi*. These pictorial maps, centered on the holy city of Jerusalem, portrayed Old Testament and classical historical and geographical lore. Alongside this religious cartography, emerged practical and functional maps to aid travelers and seafarers. Unlike the theoretical *mappaemundi*, these itinerary maps and portolan charts were drawn to scale. They provided detailed information on distances and directions.

After the conquest of Egypt by Arab armies and the establishment of Islamic rule throughout the Middle East in the seventh century, Greek scientific works, including Ptolemy's *Almagest* and *Geography*, were translated into Arabic. Two Islamic cartographic traditions subsequently evolved. The oldest was based on Ptolemy and classical Greek traditions, and is represented by the Berber geographer Abu Abdallah Muhammad ibn Muhammad al-Sharif al-Idrisi, who prepared a map of the world in multiple sheets for the Norman king of Sicily in 1154. The other emerged in the early tenth century in reaction to Ptolemaic and Western influences. Known as the Balkhi school after its founder, geographer Abu Zayd Ahmad ibn Sahl al-Balkhi, it drew upon more traditional Islamic works and focused on the Islamic empire. Calligraphers and painters prepared highly stylized maps reflecting the aesthetic values of Islamic society.

Chinese Muslim admiral Zheng He led seven naval expeditions into the Indian Ocean from the east, reaching India, the Arabian Peninsula, and Africa by 1421. His exploits were recorded on a contemporary navigation chart.

THE REDISCOVERY OF PTOLEMY'S *GEOGRAPHY* BY Western Europe following its translation into Latin from a Byzantine copy carried to Italy in about 1400, combined with the invention of the printing press by Johannes Gutenberg, sparked a revival in scientific cartography that spread throughout Europe, and eventually around the world. Ptolemy's work reestablished the concepts of the spherical Earth, latitude and longitude, and north orientation, and provided the geometrical framework for drawing maps with accuracy. Many woodcut and copperplate editions of Ptolemy's atlas were issued over the next three centuries. The European voyages of discovery and exploration to Africa, the Americas, and the

Spice Islands by Spanish and Portuguese seafarers further stimulated European mapmaking and map use. The portolan chart expanded beyond the Mediterranean, and new world maps were added to revised editions of Ptolemy's atlas.

Inspired by Ptolemy, mapmakers in Vienna, Nürnberg, Heidelberg, and Freiberg initiated elementary land surveys of areas in northern Italy and central Germany during the first half of the 16th century, creating the first rudimentary regional topographic maps in the process.

At the same time, Ottoman sultans spurred by the same impulses that motivated European royalty and Chinese mandarins, sought maps and descriptions of new lands they had conquered in southeastern Europe, western Asia, and northern Africa.

IN THE 1560S AND 1570S, THE CENTER OF THE EUROpean map trade shifted from Italy to Amsterdam and Antwerp, inaugurating a golden age of Dutch cartography. During the next century, the Netherlands produced some of the greatest mapmakers of the world: Gerard Mercator, Abraham Ortelius, Jodocus Hondius, Willem Janszoon Blaeu, and Joan Blaeu. They abandoned the Ptolemaic system of cartography and developed a new world projection, by Mercator, which still is used for navigation today. They also devised and introduced the atlas as a physical object and popularized printed sea charts and wall maps. The Dutch maps of this period remain unmatched in terms of artistic and technical quality. Jesuit cartographers introduced Western mapmaking practices to China in the last half of the 16th century, and ultimately through trade to Korea and Japan, where they competed with but did not replace Asia's indigenous cartographic traditions, which continued to thrive until the end of the 19th century.

In the late 17th and 18th centuries, the French Royal Academy of Sciences advanced surveying, geodesy, and cartography. The academy sponsored measurements of an arc of meridian, leading to a more accurate determination of the dimensions and shape of the Earth; perfected methods for determining longitude; and initiated the triangulation of France. The French were the first to establish an official hydrographic office for marine surveying, the Dépôt des Cartes et Plans de la Marine, and the first to conduct an official national land survey based on triangulation. The resulting *Carte Géométrique de la France*, begun in 1747, was completed 40 years later. It consisted of 182 sheets, with maps of uniform size and scale, 1:86,400.

With the emergence of France and Great Britain as dominant world powers, the commercial map trade shifted from the Netherlands to Paris and London, bringing to the forefront the great publishing houses of Nicholas Sanson, the Delisles, Jean-Baptiste Bourguignon d'Anville, and the Robert de Vaugondys in France; and Herman Moll, Thomas Jeffreys, William Faden, Thomas Kitchin, and the Arrowsmiths in England.

THE LATE 18TH AND 19TH CENTURIES WITNESSED three major developments in cartography: the spread of official topographic and hydrographic surveying and mapping, the widespread use of thematic maps, and the introduction of lithography and color printing. Inspired by the French national survey, other countries undertook official large-scale topographic mapping programs. Belgium initiated a national land survey in 1770. Great Britain established its Ordnance Survey in 1784, and the British Hydrographic Office in 1795. By the late 19th century, official large-scale topographic map series had been prepared for most of Europe. The Austrian-Hungarian Empire, for example, was surveyed and mapped at a scale of 1:75,000 by Vienna's Imperial and Royal Military Geographical Institute from 1875 to 1893. The United States established the Survey of the Coast (later Coast & Geodetic Survey) in 1807, the Navy's Hydrographic Office in 1866, and the Geological Survey in 1879.

Isolated by Japan's closed-door policy, surveyor Ino Tadataka nevertheless conducted a national survey of Japan's coastlines from 1800 until his death in 1818. His survey has been preserved in 225 manuscript sheets drawn at three different scales. Between 1895 and about 1924, the Japanese Imperial Land Survey produced a multi-sheet map at a scale of 1:50,000.

European colonial powers introduced large-scale topographic mapping to other parts of the world. British surveyors carried the French model to India in 1802, where they undertook the Great Trigonometrical Survey of India. When the survey was completed near the end of the 19th century, India was probably the best-mapped region in the world. The Dutch Topographic Service began mapping in the Netherlands East Indies (Indonesia) in the 1860s. The French Army's Topographic Bureau prepared the first official topographic maps of Cambodia, Laos, and Vietnam in 1886. Similarly, the Japanese Army mapped Manchuria in the 1930s.

Thematic maps are special purpose maps that illustrate a single theme or particular subject. They were introduced in the 1600s, but their value was not recognized until the 19th century. The emergence of the natural sciences such as geology and meteorology, and the creation of national censuses, beginning with the United States in 1790 and Great Britain in 1801, provided the necessary data. The adaptation of lithography to map printing during the first decade of the 19th century, and the introduction of color printing in the 1840s, provided an inexpensive and versatile method for reproducing the tones, shadings, and colors required to illustrate the qualitative and quantitative data portrayed by thematic maps.

Commercial publishers in Germany and England issued the first thematic maps, but government agencies soon began compiling large-scale thematic maps in multiple sheets. The German state of Saxony, for example, conducted a geologic survey under the direction of geologists Carl Friedrich Nauman and Carl Berhard von Cotta, beginning in 1830. The first national geological survey—the Geological Survey of England and Wales—began issuing maps in 1835. By the second half of the 19th century, statistical maps portrayed people and their activities with a variety of innovative symbols and techniques, such as graduated circles, isolines (lines of equal value), layer tinting, shading, color-coding, and flow lines.

COMMERCIAL MAP PUBLISHERS REMAINED ACTIVE IN Europe and America during the 18th century. One of the new developments in atlas publishing was the production of atlases for specialized markets. The German *Hand-Atlas* was a high-quality scholarly world atlas of general geographic maps published initially for the educated classes, military officers, scholars, and scientists. Constantly updated with the latest contemporary maps, hand atlases were printed in various formats by Adolph Stieler and other German map publishers from 1797 to 1945. A distinctive American atlas that appeared in the last half of the 19th century was the county atlas, which was sold through subscription campaigns. These atlases contained detailed cadastral maps of townships along with city plans, and often lithographic views of homes, farmsteads, and even prized livestock, which were included for an additional fee.

The mapmaking process became more specialized, efficient and accurate with the invention of the aerial camera during World War I, and the related development in the 1920s of instruments of photogrammetry—the science of making reliable measurements by the use of photographs. These plotting instruments reduced the need for ground surveys and helped translate the information on aerial photographs into detailed maps.

DURING THE 1920S AND 1930S NEW FORMS OF COMmercial maps appeared in response to new markets. The illustrated automobile road map and the airline souvenir map were designed to aid motorists and entertain airline passengers. Financed by oil companies and airlines as part of their advertising campaigns, these maps were distributed free of charge in the United States until the 1970s. At the same time, geographical journals, popular magazines, and newspapers introduced millions of readers to thought-provoking maps with each issue. NATIONAL GEOGRAPHIC magazine included page maps with its first publication in 1888, and large, full-color map supplements beginning in 1899. *Fortune* featured the innovative maps of cartographic-artist Richard Edes Harrison during the Great Depression and World War II. The *Chicago Tribune*, the *Los Angeles Times*, and the *San Francisco Examiner* kept their readers informed with full-page pictorial maps, some in color.

Military units mapped much of the world during World War II and the subsequent Cold War with the aid of aerial photographic techniques; improved plotting instruments; and high-speed, multi-color offset printing presses. Preeminent cartographer Arthur H. Robinson, the former head of the Office of Strategic Services' cartographic division, observed that more maps were made and printed during the war "than had been produced in the aggregate up to that time." The U.S. Air Force's Aeronautical Chart Service (now the National Geospatial-Intelligence Agency) published the first international map of the world. Designated the World Aeronautical Chart, or WAC, it was designed for visual navigation at altitudes up to about 17,500 feet. This chart, issued in multiple sheets of uniform size and scale (1:1,000,000), remained the basic aeronautical chart favored by pilots and navigators until official production stopped in the 1990s.

THE LAST QUARTER OF THE 20TH CENTURY CAN BE characterized as one of continuous and rapid innovation in cartography. Remote sensing satellites, global positioning systems, and lunar and extraterrestrial space probes record mappable data at a daily rate unimaginable just a few years ago. This data, combined with the personal computer, geographic information systems (GIS) software packages, and new output devices for hikers, boaters, automobiles, and aircraft have made maps omnipresent in today's society.

大雪山
香酔山
鉢露羅国
信度国

SECTION ONE

Emergence of Mapping Traditions

Detail from Gotenjiku Zu (Map of the Five Indies), 136
by Jukai

Introduction

During the two millennia from the 7th century B.C. to the end of the 15th century A.D., mapping cultures emerged in three major geographical areas: Europe, the Middle East, and Asia. Each developed its own traditions of expressing spatial concepts in graphic form, although some exchange of ideas began near the end of the period. Indigenous mapmaking flourished as well in many other regions, as illustrated by the "stick chart" that Pacific islanders used to navigate from island to island well into the 20th century, but generally these maps did not survive and their contributions are lost to history. The maps selected to illustrate these developments provide a visual history of map development. Each was chosen for its significance either as an individual item or as representative of a class of maps.

The two oldest maps included here illustrate cartographic legacies of ancient Mesopotamia and classical antiquity. A Babylonian world map, dated about 600 B.C., reveals an ancient desire to map a people's worldview. The scientific and theoretical practices introduced by Greek geographers are represented by Claudius Ptolemy's world map, originally prepared about A.D. 150, which influenced Islamic and European cartographers for 14 centuries.

In Europe, several mapping traditions emerged simultaneously during this period. Early developments in "way finding" maps are found in the Peutinger road map of the Roman Empire, initially drawn in about A.D. 335 ; Matthew Paris's 1245 map of Britain; and a *portolan*, or sea, chart of the Mediterranean and Black Seas, which appeared along with the mariner's compass in about the 13th century. Medieval Christianity's influence is revealed in the tripartite *mappamundi,* or world map, by St. Isidore, Bishop of Seville, and the world map accompanying Beatus of Liébana's *Commentary on the Apocalypse of Saint John*. The merger of several mapping traditions is illustrated by Cresques Abraham's map of 1375 and Fra Mauro's map of 1450, both of which incorporate tantalizing images of new worlds beyond Europe and the Mediterranean Basin.

Two Islamic cartographic traditions emerged in the tenth century, one a blend of classical Greek, Persian, and Indian geographical lore, and one derived from religious tenets. The first is illustrated by al-Idrisi's sectional map of North Africa, drawn in 1154 for the Norman King Roger II. Islam's influences are represented in al-Wardi's world map of 1086, and particularly in the *qibla* world map, a prayer map representative of a genre that first appeared about 1250.

Mapping traditions in Asia evolved along similar tracks. The functional and scientific tradition is illustrated by the *Yu ji tu*, a remarkably accurate 12th-century map of China, and by an extraordinary world map by a Korean Confucian scholar, which is dated 1402. The Korean map incorporates Korean, Chinese, Japanese, and Islamic sources. A 1364 Buddhist pilgrimage map of "the Five Indies" illustrates the Asian religious impulse.

Most of the maps reproduced here are etched, drawn, or painted on permanent materials such as clay tablets, stone, animal hide, or silk. These survived the ravages of time because they were more durable than papyrus. Paper, invented in China in the 2nd century A.D., did not reach Europe until the 11th century and was not used as a map medium until near the end of this period of map development. With two exceptions, the maps in this section are one-of-a-kind items—map reproduction was limited to hand-copying until about the 12th century. The *Yu ji tu* map, included here as a rubbing taken from the stone stele dated 1136, represents the earliest example of a map of which multiple copies were mechanically reproduced. The earliest known map printed in Europe is the "T-O" woodcut, published in Augsburg, Germany, in 1472.

DESPITE THE MAPS' UNIQUENESS, COMMON ELEMENTS AND ATTRIBUTES ARE FOUND. MANY MAPS OF THIS era are works of art, reflecting the training of those who prepared them. Painters and calligraphers drew most of the maps, except for sea charts, for which a professional chartmaking trade emerged. Trained in Chinese landscape painting, European illumination, Islamic calligraphy, or Ottoman manuscript illustration, they incorporated their aesthetic values into their maps. Color, script, ornamentation, and structure were used to this end. Even Ptolemy's "scientific" map is embellished with drawings of winds personified as human heads.

Lack of scale continuity and inaccurate renditions of geographic features are common characteristics. Organized mapping programs and surveying instruments were not available to provide raw materials for detailed scientific maps. Some distortion, however, was deliberate. Particular features often were enlarged or highlighted in color to illustrate significance. This was standard procedure on portolan charts. Another example is the enlargement of Korea in relation to China by Korean cartographers in their 15th-century world map. Sometimes features were reduced or reshaped merely because a mapmaker ran out of drawing room. Fra Mauro notes one such example on his 1450 world map relating to information obtained from envoys of the Coptic Church of Abyssinia. A similar comment is found on Matthew Paris's map of Britain.

A number of the maps combine different viewing perspectives. Three-dimensional landscape features such as mountains, towns, and buildings are rendered in profile, as if seen from the side or from an angle slightly above the horizon. At the same time, other features such as roads, rivers, lakes, and political boundaries are depicted as lines of various widths and colors, as if observed from above.

Finally, directional orientation varies. Influenced by Scripture, early European Christian mapmakers oriented their *mappaemundi* eastward, toward Jerusalem. Islamic mapmakers placed south at the top of their maps in respect to the locations of Mecca and Medina. And some maps—the 1350 portolan chart and the 1481 qibla chart—were drawn without regard to orientation.

The Invention That Changed the World

THE INSTRUMENT THAT HAD THE GREATEST IMPACT on the preparation and use of maps through the 15th century was the magnetic compass. Science writer Amir D. Aczel, in his popular book on the compass, called it "the invention that changed the world." Whether or not one agrees with this assertion, there is no argument that the instrument's perfection in Italy at the end of the 13th century, along with a compass card marked with compass points, or directions, heralded a revolution in medieval mapmaking. It led to the development of the nautical sailing chart and, later, to accurate terrestrial maps.

Chinese documents refer to the use of "south pointers" as early as the fourth century B.C., south being the favored Chinese direction. These very first compasses were made from lodestone, a mineral with magnetic properties that attracts iron. Frequently, thin strips of iron were shaped in the form of spoons, fish, and turtles. They were used primarily as divining tools to aid in positioning graves and buildings in accordance with *feng shui*, the Chinese art of orienting objects. Needles began to be used as pointers after the seventh century, when new developments in Chinese steel production made these precision pins possible. In 1088, mathematical cartographer Shen Kuo described how to magnetize the point of a needle with a lodestone so that it faced south, and how to suspend it by "a single cocoon fiber from new silk." By the early 12th century, Chinese sailors were consulting their south-pointing needles in "dark weather," but no compass cards from this period are known. Surviving cards, with 24 directions, date from the Ming dynasty (1368-1644).

From China the magnetic compass made its way separately to Europe and the Near East. In 1190, British scholar Alexander Neckam wrote in his classic work *De Naturis Rerum* that when sailors are lost "they touch the magnet with a needle . . . . This then whirls round in a circle until, when its motion ceases, its point looks direct to the north." By 1242 the compass was reported in the Indian Ocean. Early Chinese, European, and Near Eastern compasses were no more than a needle and a piece of wood floating in a bowl of water.

The modern mariner's compass with compass card, mounted in a square or rounded box with gimbals to reduce vibrations, was perfected in the Italian port of Amalfi near the end of the 13th century, according to some scholars. The compass card, or wind rose, was a major innovation. It evolved from the Mediterranean tradition of associating direction with prevailing winds. The compass card depicts the primary directions, or rhumb lines, with tapered points, giving it a starlike appearance. The north and east points of the compass often received additional decoration. Medieval Christian sailors called the compass their "guiding star," or *stella maris*—star of the sea—because it always points to the North Star.

Circa 1570 Italian Mariner's Compass
Ivory Case with Brass Gimbal Ring and Vellum and Paper Compass Card
National Maritime Museum, London, England

Circa 600 B.C. Babylonian World Map
Incised on Clay Tablet, 4.8 x 3.1 in.
British Museum, London, England

Oldest World Map

CENTURIES BEFORE THE CHRISTIAN ERA, THE Babylonians, thriving in vibrant urban communities scattered along the courses of the Tigris and Euphrates Rivers and on the Persian Gulf, were compiling geographical itineraries, surveying irrigation canals and property lines, and drawing maps.

It was here that Sumerians in southern Babylonia invented writing in the fourth millennium B.C., with cuneiform characters on clay tablets. Scattered among the fragments of thousands of these tablets, collected over the centuries by the great museums of Europe and the Middle East, are a few dozen examples of maps and plans. Three hold particular interest, illustrative of the antiquity of basic cartographic concepts such as scale, orientation, and map type.

A detailed plan of Nippur, a Sumerian religious center, for example, is believed to be one of the earliest town plans drawn to a specific scale based on a measured survey. Inscribed about 1500 B.C., it is now preserved in the Hilprecht Collection at Friedrich Schiller University in Jena, Germany. City walls, canals, buildings, and a temple are clearly shown.

Another clay tablet is inscribed with two elements essential to the development of mapmaking: symbols for depicting cultural and natural landscape features and notations indicating map orientation. Excavated in 1930-31 from the ancient city of Nuzi (near modern Kirkuk, Iraq) and now in the Semitic Museum at Harvard University, the Nuzi map dates from about 2300 B.C. It depicts a river valley, an estate, and three towns bounded by two ranges of mountains (perhaps the Zagros Mountains on the Iran-Iraq border). Surprisingly, the scribe conveyed the information with a combination of symbols and pictures readily understood today. Four thousand years after these symbols were etched in the soft clay, they are still decipherable: Double lines suggest waterways, circles indicate towns, and double rows of bell-shaped images indicate mountains. Cardinal directions are inscribed along three sides of the tablet—east, north, and west—making this the oldest surviving map to show orientation, according to historian A. R. Millard. It is oriented with east at the top, a convention often followed by later cartographers in the Middle East.

Pictured here is the earliest known world map. Dating from about 60 B.C., it portrays a flat Earth encircled by water. Beyond, seven islands were originally inscribed as triangles, each described in some detail by the cuneiform text that accompanies the map. According to Babylonian cosmology, these islands link Earth's oceans with the heavenly ocean and animal constellations that inhabit it. The Euphrates River—vertical parallel lines—bisects the map, extending from mountains at top to marshlands of southern Iraq, portrayed by a set of horizontal parallel lines at the bottom of the circle. An oblong horizontal marker above the center of the disk identifies Babylon, surrounded by eight circles representing nearby cities.

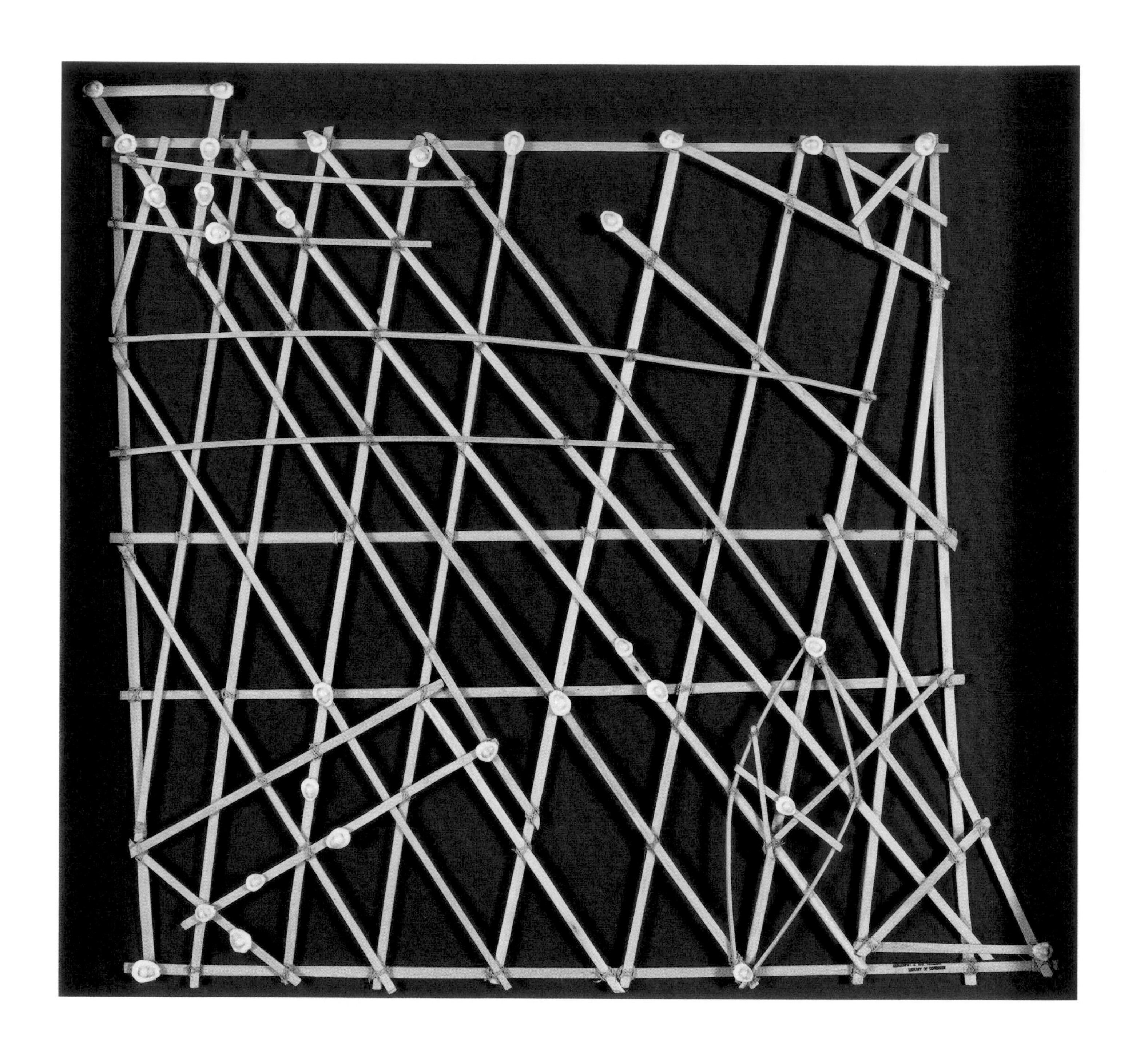

Circa A.D. 300 Marshall Islands Stick Chart
Reconstructed from Coconut Palms or Pandanus Reeds and Cowrie Shells by Marshall Islanders, 1972
Library of Congress, Washington, D.C.

Stone Age Nautical Chart

LONG BEFORE EUROPEANS EXPLORED AND MAPPED the Pacific Ocean, Pacific islanders were navigating and exploring its wide expanse, guided by mental maps and maps in the form of "stick charts." Later European sailors and missionaries were astonished to meet indigenous people able to outline entire island groups in the sand. Jesuit missionary Paul Klein copied a pebble map of the Caroline Islands prepared in 1696 by Carolinians who had drifted by canoe to the Philippines after 70 days at sea. Seventy-three years later, the British maritime explorer Captain James Cook traced a chart of Polynesia sketched by a Tahitian priest named Tupaia that covered an area larger than the continental United States. Tupaia's map, drawn from memory, shows Tahiti and 74 outlying islands. Later, Tupaia piloted Cook's ship, the *Endeavour*, some 300 nautical miles to a small volcanic island named Rurutu by dead reckoning (estimating distances from previously known positions).

Eighteenth- and nineteenth-century observers recorded native voyages of 700 to 1,200 miles lasting 30 days or more. More recently, the ability to navigate similar distances without instruments or nautical charts has been confirmed by the revival of traditional navigational practices by Caroline Islanders and of long-distance sea trials. The trials were conducted by anthropologist Ben Finney and colleagues in a reconstructed twin-hulled Polynesian voyaging canoe powered only by sails.

Most Pacific islanders navigated by forming mental map images of their natural environment, particularly the stars and other celestial bodies, ocean swells, and prevailing winds; and by observing the flights of migratory and land-based birds, and island-influenced cloud cover. Navigators were trained from a young age, using a variety of techniques and memory devices. In the Gilbert Islands, celestial maps were re-created on the ceilings of training houses. Carolinian navigators studied sophisticated star compasses constructed on beach mats from coral, coconut leaves, and banana fibers. These short-lived maps and navigational aids were then memorized through verbal exercises, including songs, chants, and dances.

Marshall Islands navigators also relied upon the stars and dead reckoning for long-distance navigation, but for coastal navigating they devised the stick chart. Stick charts aided navigators in sailing near land and in sighting landfalls. Surviving examples are made of coconut palm or pandanus reeds and cowrie shells. The reeds are arranged to replicate patterns of ocean currents and sea swells. Shells represent islands. Three types of stick charts were used: *rebbelib*, a small-scale chart for major island groups; *meddo*, a larger-scale chart displaying one or two island chains in greater detail; and *mattang*, an abstract chart for instruction. Unlike modern nautical charts, stick charts were consulted prior to a voyage but then were left behind to guide other seafarers.

All Roads Lead to Rome

THE ROAD SYSTEM OF THE ROMAN EMPIRE IS ONE of the wonders of antiquity. Like spokes on a wheel, some 50,000 miles of paved roads radiated from Rome, linking the capital with Roman outposts in Britain, Spain, Germany, Persia, and North Africa. Constructed and maintained over seven centuries, many of the roads and bridges are used today. Lacking maps or compasses, Roman military engineers laid out roads with the *groma*, or surveyor's cross, a line-of-sight instrument used to plot straight property lines and to lay out building foundations.

Primarily built to move Roman legions from post to post, the imperial road network eventually served government officials, the Roman postal service, and, after the formal recognition of Christianity by Rome in 313, Christian pilgrims. Travelers were guided from place to place by the first written road itineraries and the first road maps. Use of these geographical travel aids was facilitated by milestones placed along the main roads. Road itineraries were simple lists of places along a route, with distances recorded in Roman miles and leagues. Maps were described as "painted itineraries." Surviving road itineraries are common, but only one Roman road map has been found.

The Peutinger road map is a 12th-century manuscript, found in 1494 and later given to Konrad Peutinger, a German scholar and collector, for whom the map is named. Historians believe this manuscript was copied from an earlier work, probably drawn shortly after 335. The Peutinger manuscript is a strip of parchment, 22 feet long and 1 foot wide. This format was convenient for travel in horse-drawn carriages and chariots, but it contributed to significant distortion in the map image. The north-south map scale is greatly compressed, as is evident in the segment shown here. From top to bottom, it depicts: Illyricum (the Balkans), the Adriatic Sea, southern Italy, the Mediterranean, and North Africa.

The Peutinger map delineates major roads of the Roman Empire, but like a modern road atlas it also depicts other features. These include stops at staging posts, depicted as sharp turns in the road. Distances between staging posts are given in Roman miles. Popular destination points include villas, illustrated by twin towers; spas or baths, shown as open squares and designated "aqua"; and temples and harbors, portrayed graphically. Attractions along routes include lighthouses, major rivers, mountains, and forests. A drawing of a goddess seated on a throne identifies Rome. Twelve roads radiate from Rome. Its harbor and prominent lighthouse are pictured below the city.

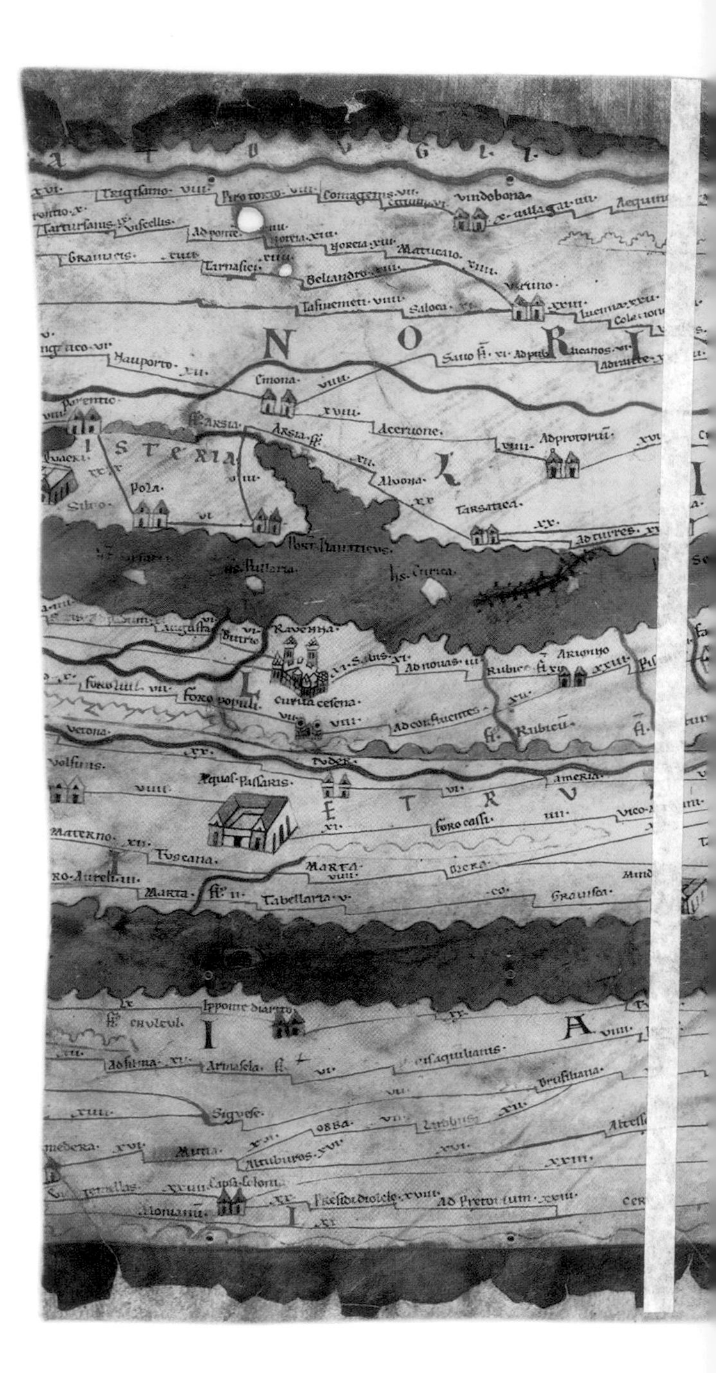

Circa 335 Road Map of Rome and Region, from a Larger Strip Map of the Roman World, Copied from Fourth-Century Map
Twelfth-Century Manuscript on Parchment, 12.8 x 23.1 in.
Österreichische Nationalbibliothek, Vienna, Austria

Circa 630 Ecclesiastical World Map by St. Isidore, Bishop of Seville
Illuminated Manuscript from a French cosmography, Circa 1220, 10.6 x 6.6 in.
Walters Art Museum, Baltimore, Maryland

T-O Mappaemundi

IN THE MIDDLE AGES, WORLD MAPS APPEARED IN the West unlike any seen before. They represented an entirely new worldview. Still known by their Latin name, these *mappaemundi* were devised and promoted by the fathers of the early Christian church for religious, not geographical, instruction. Most were made to illustrate Bibles, psalters, and other religious works. They were generally simple in design, more diagram than traditional map, but became more complex by the late Middle Ages. Some 1,100 mappaemundi have been recorded, including the first map printed in Europe, a 1472 woodcut from Augsburg, Germany (see inset). They were popular for 900 years.

The most common ecclesiastical mappaemundi are of the tripartite type, which first appeared in the manuscript works of St. Isidore, Bishop of Seville, after the year 600. They are popularly called "T-O" maps, where *O* represents a spherical world surrounded by water, and *T*, placed within the *O*, divides the world into the known continents. The stem of the *T* represents the Mediterranean Sea, and separates Europe from Africa. The crossbar represents the Don and Nile Rivers, and separates Asia from Europe and Africa. Asia is the largest region, according to St. Isidore, because St. Augustine said it was "the most blessed." East is usually at the top. The cardinal points are designated in Latin: *Oriens* (east), *Meridies* (south), *Occidens* (west), and *Septentrio* (north).

T-O maps contain Christian symbols commonly understood by the faithful. The *T*, for example, represents the *tau* cross. In use among early Christians, it was less conspicuous than the familiar Latin cross at a time when Christianity was illegal. Mapmakers often placed Jerusalem at the center of the map, particularly after the Crusades of the 13th and 14th centuries, and named continents for Noah's sons said to have populated them: Sem, whose name is associated with Semites of Asia; Ham, with Hamites of Africa; and Japhet, with Japhites of Europe.

Several types of T-O maps evolved over time. The manuscript map shown here comes from a French cosmography dated about 1220. It combines medieval and classical elements. The map portrays a tripartite Earth surrounded by Aristotle's wind system, associated with the 12 points of the compass. The mapmaker named the winds for the directions from which they blew. He also personified them with drawings of human heads, a convention generally associated with later mapmakers. Similar "headwinds" enhanced maps well into the Renaissance.

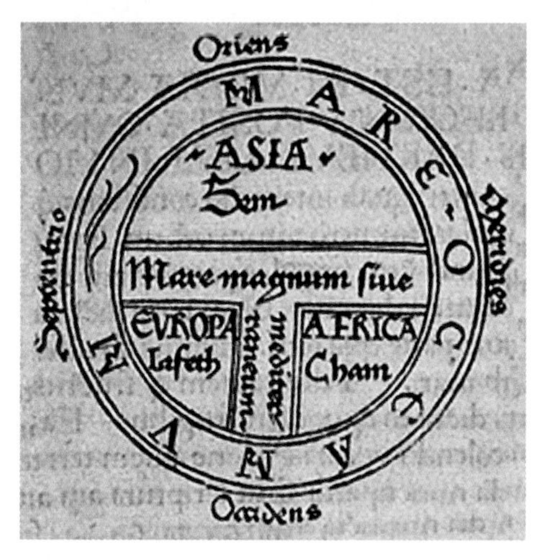

1472 TRIPARTITE WORLD MAP, PRINTED VERSION
WOODCUT, 2 X 2 IN.
LIBRARY OF CONGRESS, WASHINGTON, D.C.

Mapping the Apocalypse

MEDIEVAL MAPPAEMUNDI TOOK MANY FORMS. MOST intriguing is one made to accompany a *Commentary on the Apocalypse of St. John*, prepared in the year 776 by the Spanish Benedictine monk Beatus of Liébana. The *Commentary* was "a somber treatise" that stressed "the tribulations of the end of time and the terror of the final day of reckoning," as one commentator recently wrote. A compilation of biblical stories and illustrations, it was widely copied and circulated for four centuries, exerting a strong influence on Western Christianity and Romanesque art.

To aid the devout in reading the *Commentary*, Beatus, or his illustrator, prepared a map to depict the world into which the Twelve Apostles were sent to proselytize. Fourteen copies have survived. All but one was prepared in Spain. They date from the 10th to the 13th century. Facundus, a Spanish scribe-painter, created one such map in 1047 for his patron, Fernando I of Castile, King of León and Castile. Facundus perceived his map as a visual narrative of the *Commentary* and as a work of art, not a product of geographical science.

The map is drawn in typical T-O fashion, oriented toward the east, but with two new features. It is rectangular—not circular—and divided into four regions, not three. Asia is at the top, separated from Europe and Africa by the Don and Nile Rivers. A vertical Mediterranean Sea at the center, dotted with islands, divides Europe from Africa. The new land area is depicted on the far right of the map, separated from the other three continents by an equatorial sea often identified as Mare Rubrum (Red Sea). According to a legend on another Beatus map, this land is "unknown to us on account of the heat of the sun. Within . . . the antipodeans are . . . said to dwell." It shows, according to new research by Beatus scholar John Williams, the most southerly region of Africa then known, probably Ethiopia.

Reflecting its Spanish-Arab heritage, the map has vibrant colors, creative symbols, captivating sketches, and multiple legends. Fish, an icon of the post-Apostolic Church, populate the ocean encircling the land. Place-names of islands, regions, and towns show the reach of Christianity. Exotic birds' wings and tail feathers depict mountains and ranges such as the Alps and Pyrenees. A gold-and-black city wall and gate at the top of the map represent Jerusalem. Another emblem of Christianity is represented by the inset of the serpent with Adam and Eve in Paradise, "standing in shame," in a world map depicting the missions established by the Apostles as a result of their sins.

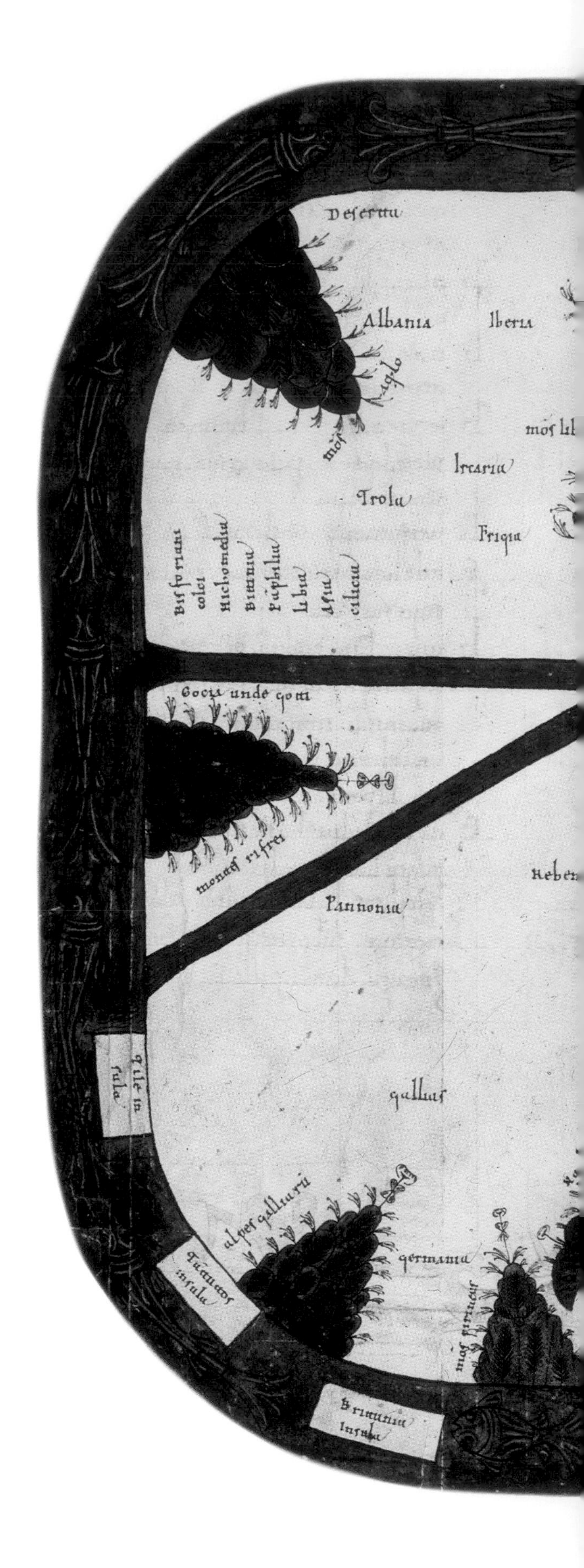

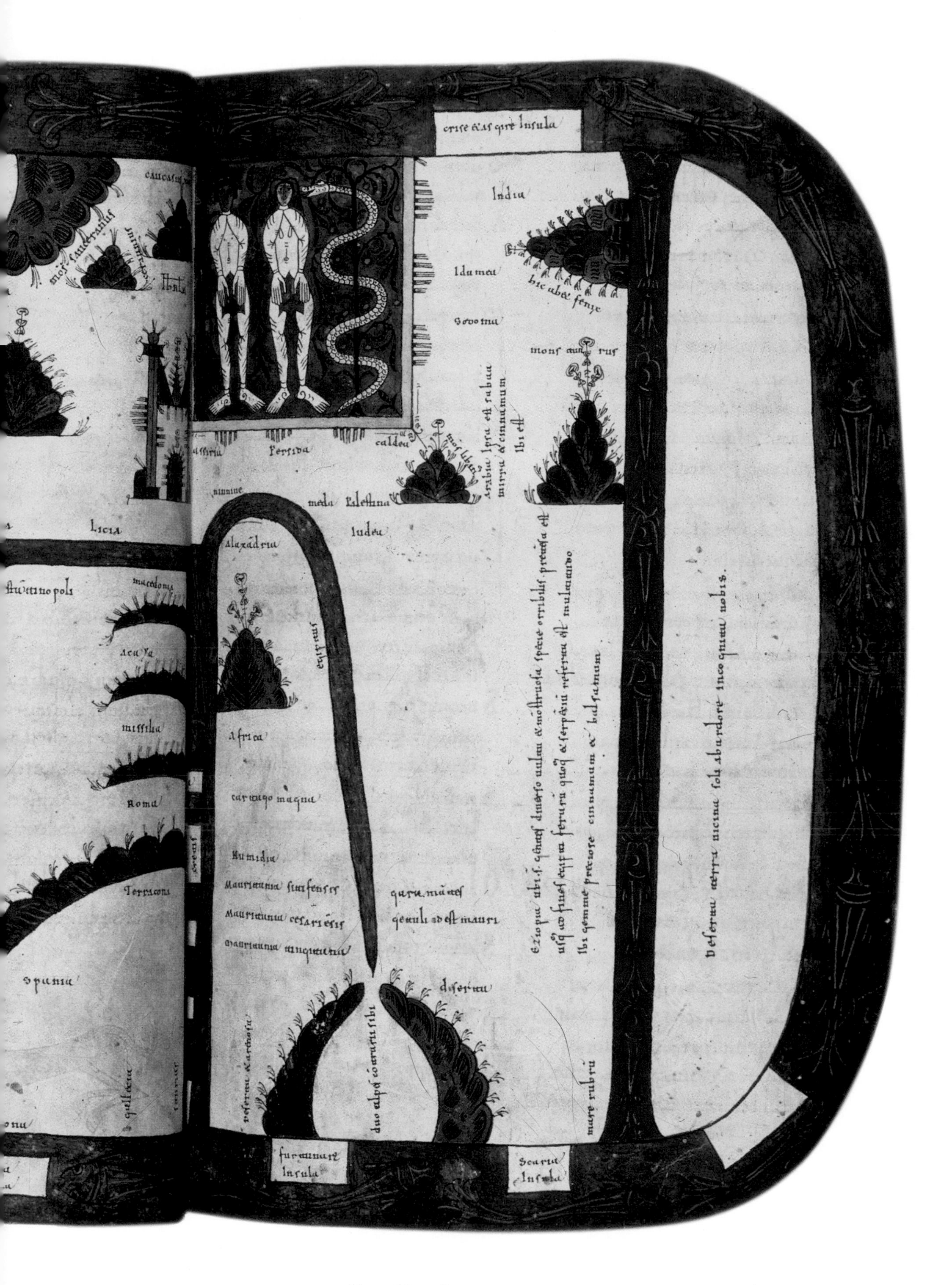

1047 World Map by Facundus copied from one made by Beatus of Liébana in 776
Illuminated Manuscript, 14 x 10.9 in.
British Library, London, England

Circa 1086 World Map from Ibn al-Wardi's Kharidat al-Ajaib wa-faridat al-gharaib
(The Unbored Pearl of Wonders and the Precious Gem of Marvels)
Illuminated Manuscript, 1481, 7.75 x 5.7 in.
Library of Congress, Washington, D.C.

Picture of the Earth

ABU ZAYD AHMAD IBN SAHL AL-BALKHI, AN EARLY TENTH-century Iranian geographer working in Baghdad, introduced new maps of the Islamic caliphate based on traditional Islamic tenets. His original treatise, *Suwar al-aqalim (Picture of the Climates)*, has not survived, but his cartographic commentary was expanded initially by Abu Ishaq Ibrahim ibn Muhammad al-Farisi al-Istakhri in about A.D. 951, and more definitively by Abu al-Qasim Muhammad ibn Hawqal in 1086. Ibn Hawqal traveled throughout much of Persia, Turkistan, and North Africa before settling in Sicily. His text *Kitab surat al-ard (Picture of the Earth)*, preserved in the Topkapi Sarayi Museum in Istanbul, Turkey, was popular throughout the Middle East.

Twenty-one maps usually accompanied these geographies, including a world map; maps of the Mediterranean, Persian (Indian Ocean), and Caspian Seas; and provincial or regional maps. They were prepared as memory aids "rather than for any other practical use," concludes Gerald R. Tibbetts in his study of the Balkhi school of geographers. The provincial maps are basically maps of camel routes, but staging areas, villages, and oases also are marked. Information for these maps came from road books that began to appear in the ninth century. They list pilgrimage and post routes and provide brief descriptions of towns along the way with their products and revenues. The maps provide coverage of the tenth-century Islamic empire administered by Baghdad and the Abbasid caliphate. Maps of Iranian provinces are more detailed than western areas recently conquered from the Byzantines, a reflection perhaps that al-Balkhi's original patron was a Samanid ruler of Persia.

Most intriguing is the world map, which looks more like a work of modern art than an ancient map. It portrays Ptolemy's inhabitable world as a circle surrounded by mountains and an enclosed sea. Fifteenth-century geographer Siraj al-Din Abu Hafs Umar ibn al-Wardi prepared the version of Ibn Hawqal's map presented here. Reflecting the political and religious views of tenth-century Islam, the world map is centered on Mecca, indicated by a symbol of the Kaaba, and oriented with south at the top "because of reverence for the cities of Mecca and Medina in Arabia, beyond which there is no land," according to Islamic scholar S. Maqbul Ahmad.

In line with Islamic aesthetic styles, topographic and water features are geometric and highly stylized. Mountains are green and red; water is blue, except for the Nile River and a major tributary, which are red. Red circles indicate cities. The map is divided in half by the Indian Ocean (at left with the Arabian Peninsula pointing upward) and the Mediterranean. Africa occupies the top of the map, bisected by the Nile. Europe is at lower right, with Greece and Italy extended up into the Mediterranean Sea. A red crescent moon is Constantinople. Two blue ribbon-like symbols at lower left indicate the Aral and Caspian Seas.

Stone Maps of the Ancient Chinese Empire

MAPS ENGRAVED ON STONE REPRESENT A MAPPING genre unique to China. Their purpose was education. They were commissioned by school directors during the Song dynasty (960-1280) and erected in schoolyards for students to study and copy. Copying was done by ink rubbings, the earliest method of mass-producing map images. China has a long tradition of preserving Scriptures and sacred texts on stone. With the invention of paper in the second century, it became possible to transfer images engraved on stone to this new, portable medium. The process involved pressing a sheet of paper firmly against the stone surface and then rubbing ink on the paper. The rubbing reproduced here was made in 1935 from the original *Yu ji tu* stele by Prof. W. B. Pettus of the College of Chinese Studies in Beijing.

The *Yu ji tu* is modern in appearance and surprisingly accurate. China's two major rivers, the Huang (Yellow) and Chang (Yangtze), are properly placed and aligned. The delineations of the coastline, the Shandong Peninsula, and Hainan Island compare favorably with current maps. Place-names abound. These include administrative names and names of principal cities and towns, rivers, lakes, and mountains. Several other surviving stone maps contain descriptions of adjacent lands, with one listing "several hundred countries."

A feature that has intrigued generations of map historians is the grid system of squares overlaying the map, which serves as a scale indicator. Each side of a square represents 100 *li*, or about 33 miles.

Chinese stone maps represent a cartographic sophistication unknown in the West at that time. The specific source maps and surveys that were used in the construction of the *Yu ji tu* remain unknown, but Chinese mapmakers had mastered the basic cartographic concepts and had the surveying instruments to conduct the necessary local and regional land measurements by the time of the Song dynasty. As early as the Jin dynasty (265-420), cartographer Pei Xiu laid out a set of six basic mapping principles, which accompanied a map that he drew for the emperor. These related to scale, relative positioning of points on maps, distances along routes, elevation, diagonal distances, and curved lines. Instruments available to Song cartographers included magnetic compasses for determining directions and a variety of sighting boards, water levels, plumb lines, and graduated rods for determining distances and heights.

Pei Xiu's maps were known for their accuracy. Commenting on a world map drawn about 630 in accordance with Pei Xiu's principles, Yu Shina wrote, "Rulers, without descending from their halls, could comprehend the four quarters of the world." One historian believes that the *Yu ji tu* was based on a world map by Shen Kuo, an 11th-century cartographer who followed similar principles.

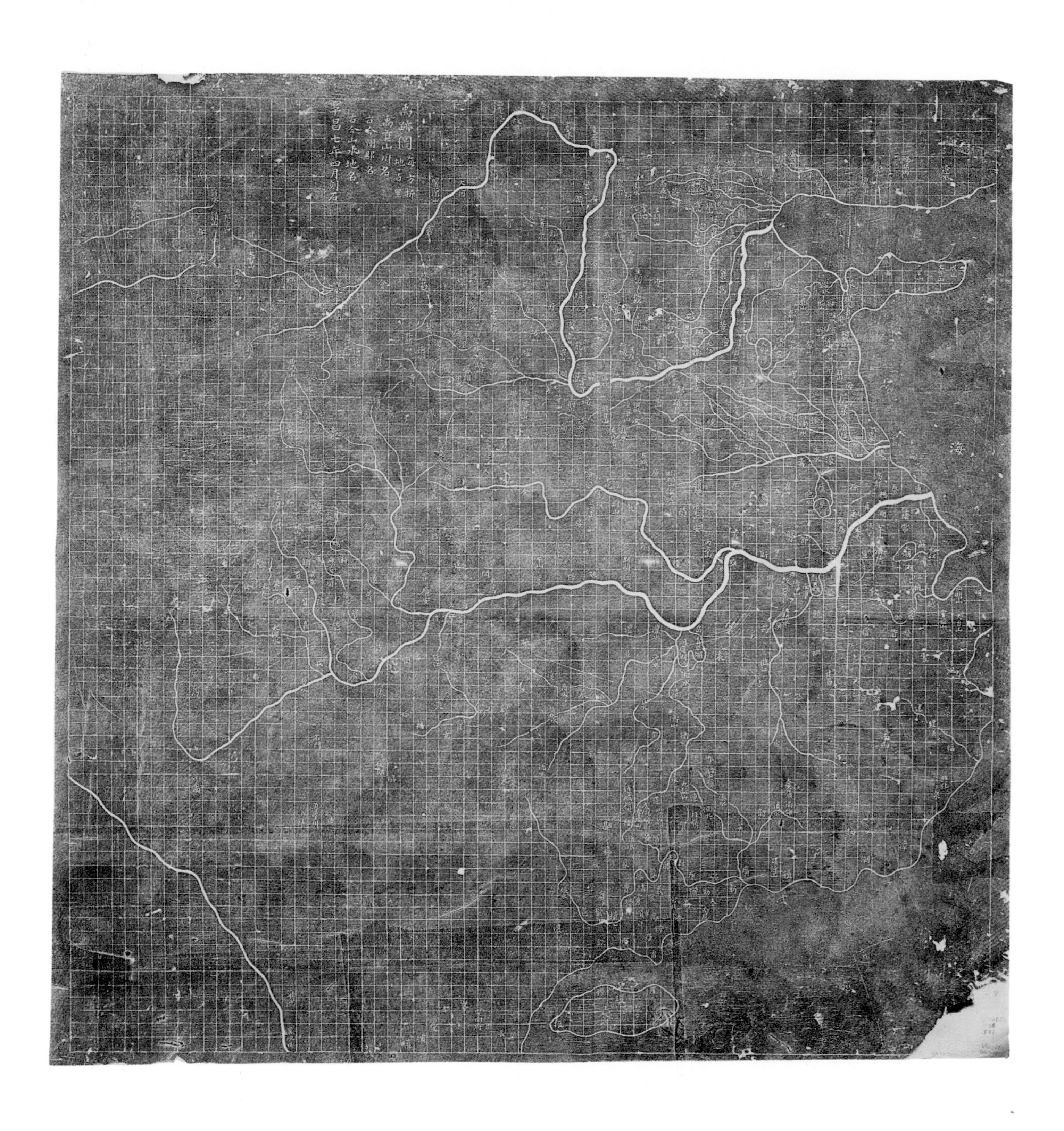

1136 Map of China, Yu ji tu (Map of the Tracks of Yu)
Rubbing by W. B. Pettus from Original Stele, 1935, 31.2 x 30.8 in.
Library of Congress, Washington, D.C.

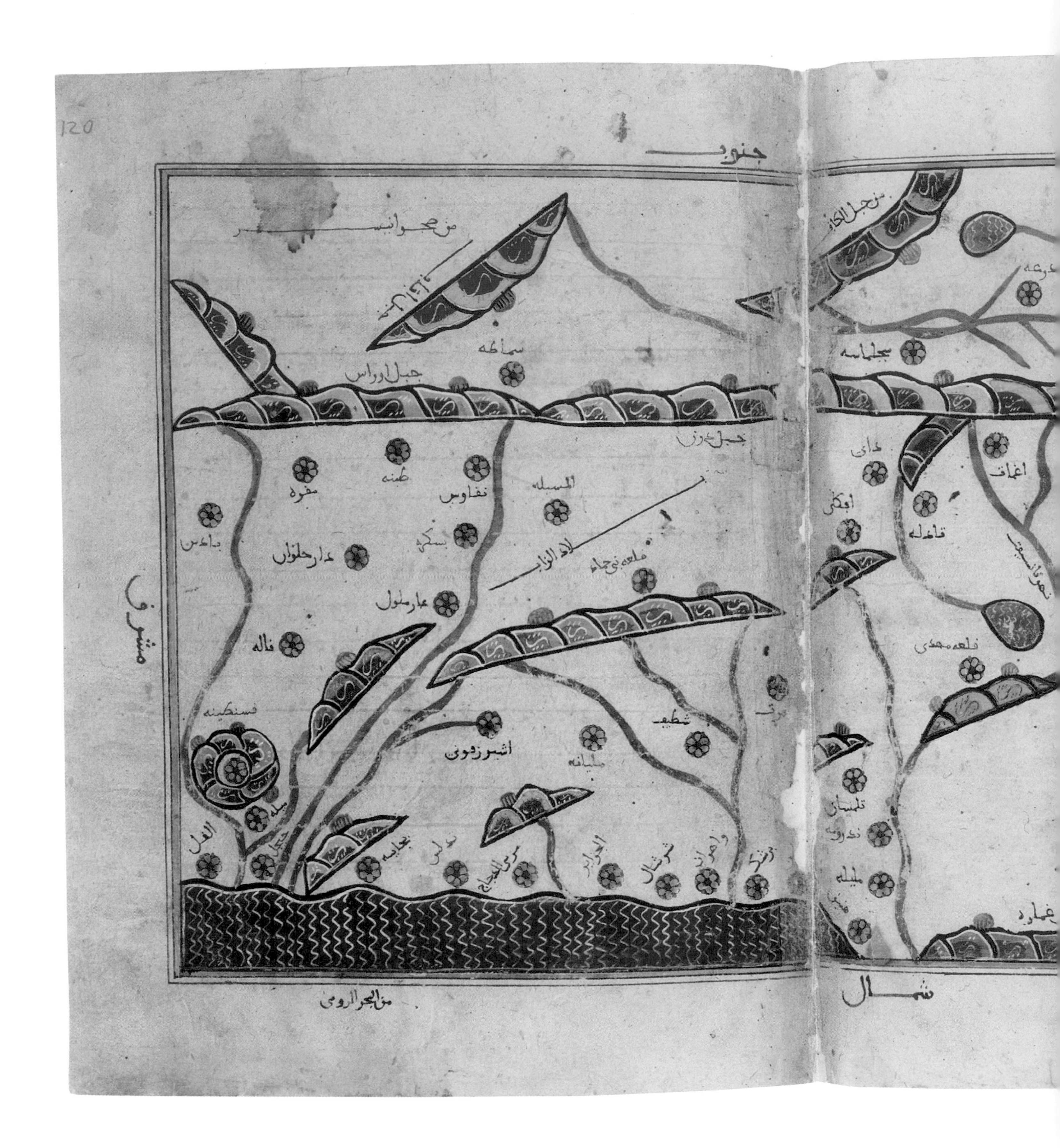

1154 Map of North Africa (Climate 3, Section 1) by Abu Abdallah Muhammad ibn Muhammad al-Sharif al-Idrisi
Illuminated Manuscript Copy by Unknown Hand in late 14th century, 12.5 x 18.7 in.
Bodleian Library, University of Oxford, England

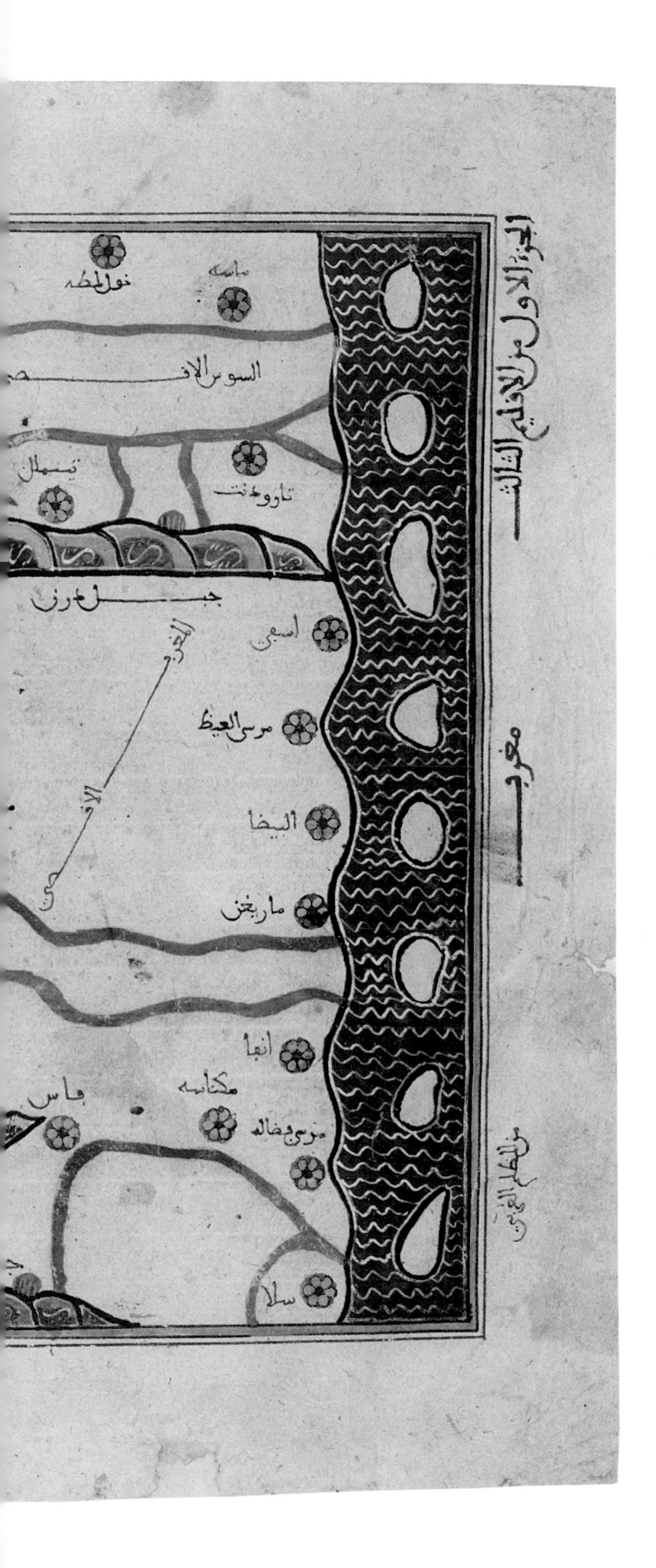

Islamic-Norman Map of North Africa

Among the prizes of the seventh-century Arab conquest of Egypt were copies of Claudius Ptolemy's *Geography*, which were then translated by Islamic scholars. Geographers in Baghdad prepared a world map derived in part from Ptolemy in about 820 for Caliph al-Mamun. It portrayed "the universe with spheres, the stars, land and the seas, inhabited and barren, settlements of peoples, cities." Shortly thereafter, Muhammad B. Musa al-Khwarizmi penned a geographical treatise based on the *Geography*, with maps. Although the original is lost, Arab scholars prepared subsequent translations and versions.

A beautiful example of this Islamic cartographic tradition is an atlas of illuminated maps compiled in 1154 by Abu Abdallah Muhammad ibn Muhammad al-Sharif al-Idrisi, a prince of the Muslim Berber dynasty in Morocco and one of the leading geographers and mapmakers in medieval Europe. Entitled *Nuzhat al-mushtaq fikhtiraq al-afaq* (*The Book of Pleasant Journeys into Faraway Lands*), it was prepared in Palermo for al-Idrisi's patron, Roger II, the Norman king of Sicily who had military designs on Islamic Spain and North Africa. Also known as *The Book of Roger*, the atlas consists of a world map and 70 sectional maps, with a detailed narrative. The original has been lost, but eight illuminated manuscript copies dating from 1300 to 1556 are preserved in libraries around the world. Portions of the book also were printed. An Arabic edition was issued in 1592, and a Latin edition in 1619.

Like Ptolemy, al-Idrisi divided the inhabited world into seven climates. Each extended horizontally from the Equator northward, and ten sections extended vertically from the Canary Islands eastward. The segment reproduced here, Climate 3, Section 1, portrays the northwest corner of Africa. As with all Islamic maps, south is at the top. The Atlantic coast is on the right, the Mediterranean on the bottom. If joined, the 70 sectional maps would measure 7 by 16 feet.

Al-Idrisi used a variety of sources to compile his map at Roger's court, from Ptolemy's commentary to information from travelers and merchants. Educated in Córdoba, he had traveled through Spain and North Africa with visits to England, France, and Asia Minor.

Although this map may appear crude to modern eyes, it was made for 12th-century Islamic sensibilities and for use with accompanying text. Exquisitely rendered symbols, patterns, and colors—mountains (purple and ocher), rivers (green), the sea (blue), cities and towns (gold rosettes with red centers)—combine with the flowing style of Arabic writing.

Medieval Map of Britain

MAPS OF REGIONS AND COUNTRIES DURING THE Middle Ages were rarely compiled, and just a handful have survived. It did not occur to most medieval Europeans to draw maps to guide them from place to place or to record land holdings. They were more likely to rely on written descriptions of routes, field boundaries, and urban property lines.

One of the few active mapmakers of the period was Matthew Paris, an English monk and chronicler born in about 1200. Paris is best known for his *Chronica majora* (*Major Chronicles*), an unrivaled account of events in Europe between 1235 and his death in 1259.

An artist of some skill, Paris illustrated his manuscripts with drawings of animals and people, architectural renderings, scientific diagrams, and maps. Each volume of his *Chronica majora* was prefaced with a picture map, beginning with an itinerary strip map of the route from London to Apulia, a region of southeastern Italy bordering the Adriatic Sea. This map provided context for the accompanying written itinerary, prepared from information obtained from crusaders and pilgrims traveling to Rome and Palestine. Map strips portray towns, with distances noted. Each is identified by profiles of features such as castle towers, church spires, and city walls.

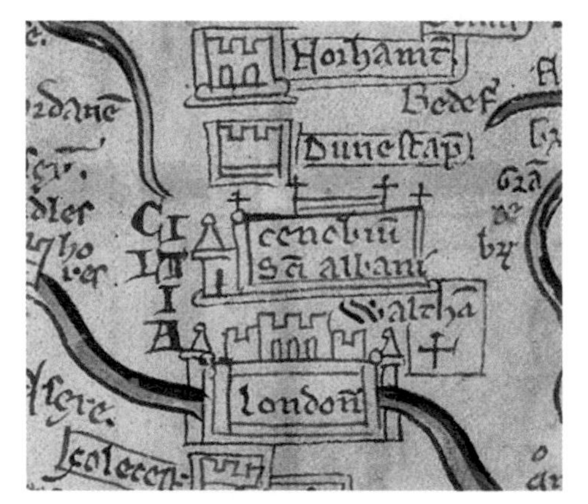

DETAIL OF MAP OF BRITAIN

Volume two of the *Chronica majora* opens with a map of the Holy Land, a popular subject of artists, writers, and mapmakers during the Crusades of 1096 to 1270. The three surviving copies were prepared about the year 1252. The artist highlighted the city of Acre, still in the hands of the crusaders, by drawing it much larger in relation to Jerusalem and Bethlehem, places that had fallen to the Muslims. Illustrations of buildings, animals (including, inexplicably, a central Asian two-humped Bactrian camel), and ships are scattered throughout the map, which is oriented with east at the top. Inscriptions in French and Latin provide further information.

The final volume includes this map (opposite) of England and Scotland. A landmark work, it represents one of the earliest efforts to map a region and the first to place north at the top of the map. Most noteworthy are the depictions of Hadrian's Wall and the main route from Dover to Main Castle upon Tyne via London and St. Albans, indicated by picture symbols of castles and churches. Although small, and not to scale, it is a remarkable document limited only by contemporary materials and technologies. "If the page had allowed it," Paris noted in a legend, "this whole island would have been longer."

Circa 1245 Map of Britain by Matthew Paris
Illuminated manuscript, 13 x 9 in.
British Library, London, England

1481 Qibla World Prayer Map, in Turkish Translation of Ibn al-Wardi's The Unbored Pearl of Wonders and the Precious Gem of Marvels
Manuscript, Hand-Colored, Istanbul, 1481, 7.75 x 5.7 in.
Library of Congress, Washington, D.C.

Mapping Sacred Directions

MAPPING SACRED GEOGRAPHIES IS A GENERAL THEME of medieval cartography. Holy places and pilgrimage sites were the focus of maps in many of the major religions. European *mappaemundi* were centered on Jerusalem. Chinese Buddhist maps focused on Mount Meru. Only in Islam, however, did maps play a central role in the belief system. Practicing Muslims must daily face the holy Kaaba for prayers and other religious observances. "Turn your face toward the sacred mosque, wherever you may be, turn your faces toward it," the Koran commands. The holy Kaaba is a shrine within the Great Mosque in Mecca, Saudi Arabia. Muslims consider it their most sacred site. It symbolizes the presence of God, or Allah. As Islam spread throughout the Middle East and beyond, following its founding by the prophet Muhammad in Arabia, maps began to appear that aided believers in observing the sacred direction, or *qibla*.

The first to appear portrayed a circular world focused on the Kaaba, according to qibla scholar David A. King. In this tradition, which evolved from folk astronomy, the world was geographically segmented into four or more regions in accordance with the orientation of the Kaaba. The four corners of the rectangular shrine are aligned with the four cardinal points, providing the basis for the first division of the world. With time, the four walls of the holy monument also were used to subdivide the world, since Muslims can choose which wall to face during prayers. Later, additional architectural details of the Kaaba were used to further segment the world map. Muhammad Ibn Suraqah al-Amiri, an 11th-century Yemeni legal scholar, devised 11- and 12-sector schemes with detailed instructions for finding local and regional qibla using the stars and winds. Eventually even more complex divisions became common.

Qibla world maps have a long history. The earliest surviving copies date from the 13th century, although textual references are found as early as the ninth century. Most are in manuscript form, but a few have been published. They are usually page-size, and found in religious tracts, geographies, encyclopedias, and historical texts. The 11-segment Ottoman qibla world map reproduced here is representative. The focal point of the circular map is the plan of the Kaaba, highlighted in gold and black and surrounded by a red circle in a red square. The cardinal points are written in the corners of the square. Eleven encircling arched-segments, pointed toward the Kaaba, represent geographical regions. Names of each segment are written in red ink, for example, "Mihrab of Ethiopia." A *mihrab* is a niche in the wall of a mosque indicating the direction toward Mecca. Narrative text within each segment provides astronomical directions for finding the qibla and facing the Kaaba from that region, and information about the part of the Kaaba associated with that direction. Qibla world maps centered on the Kaaba continued to be produced.

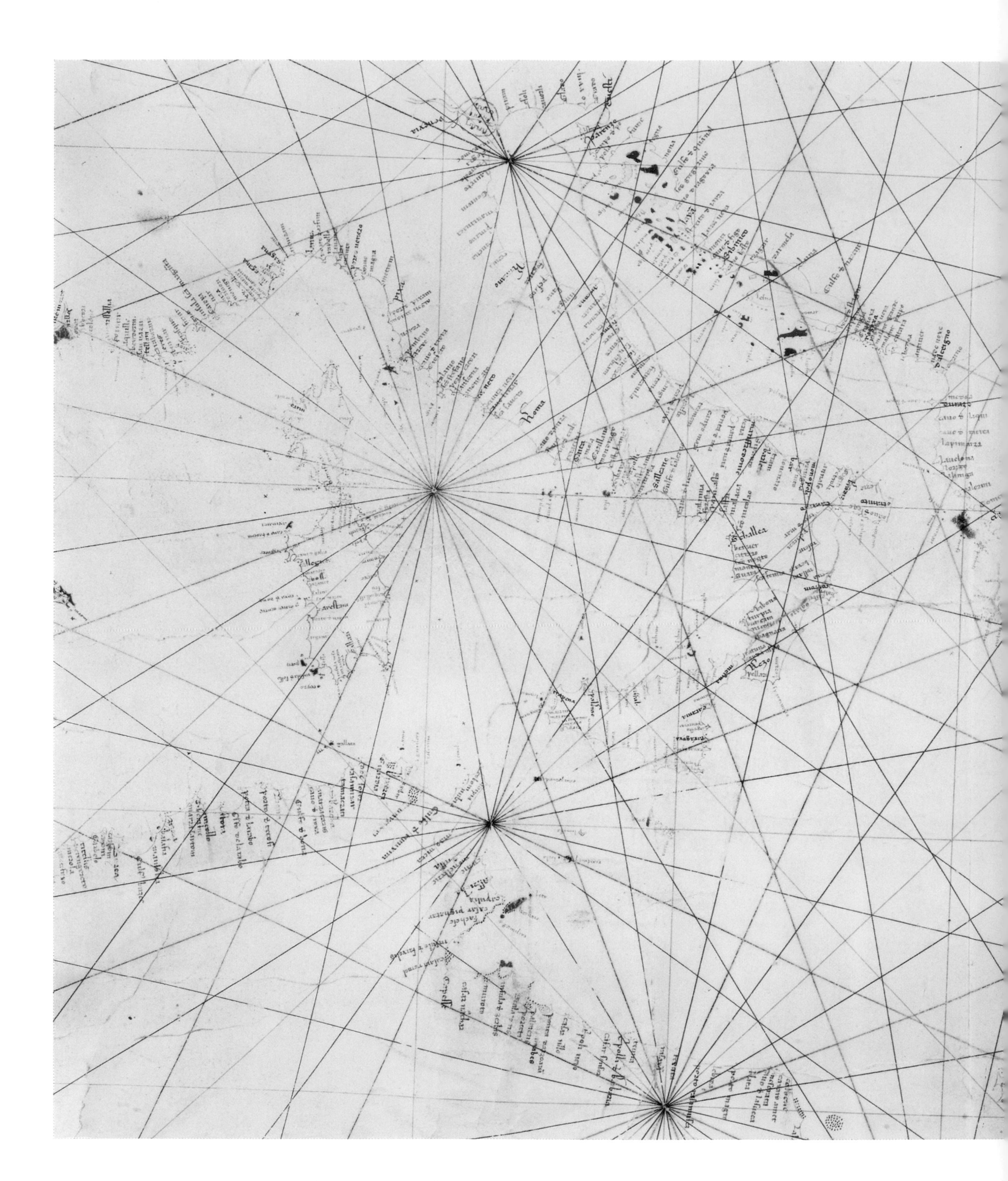

Circa 1350 Portolan Chart of Mediterranean and Black Seas
Illuminated Manuscript on Vellum, 16.3 x 22.4 in.
Library of Congress, Washington, D.C.

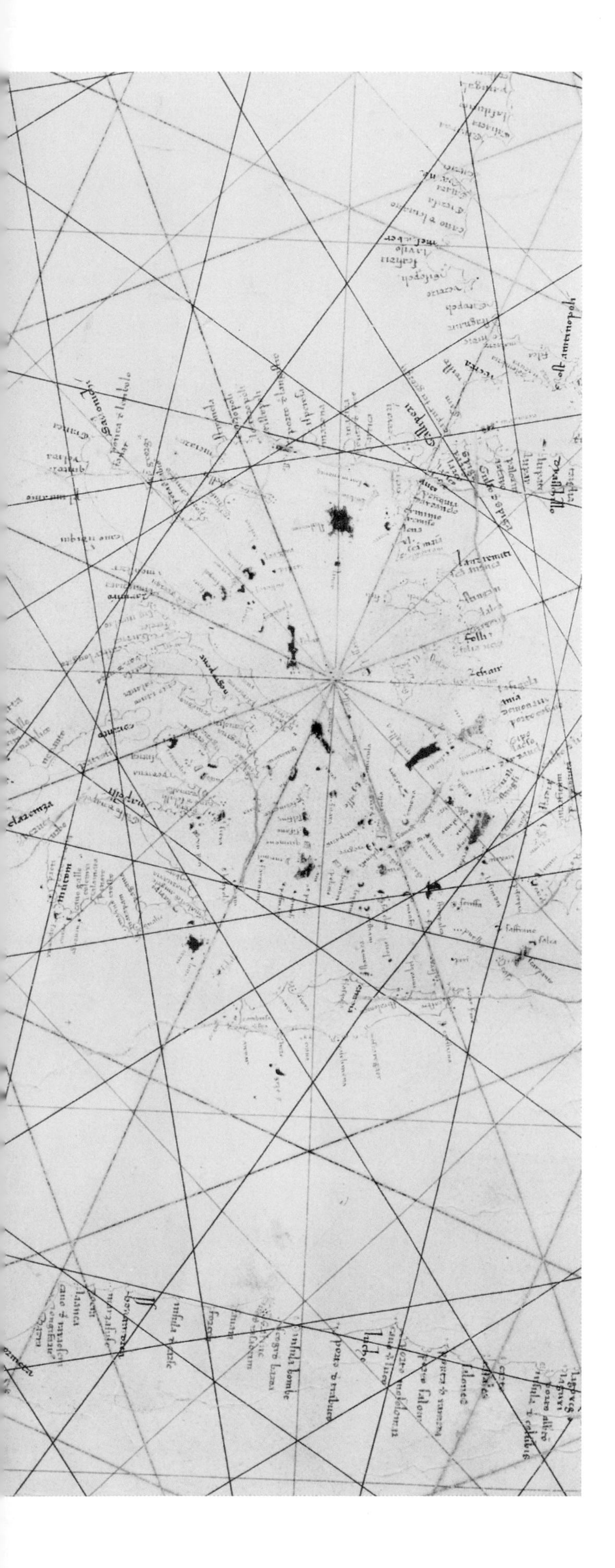

Medieval Sea Charts

PORTOLAN CHARTS BURST, FULLY FORMED, UPON THE medieval world in the later decades of the 13th century. The portolan chart, from the Italian *portolano,* a book of sailing directions, is a map used for navigating at sea. These charts appeared in the Mediterranean area simultaneously with the mariner's compass. When used together with a log line for estimating distance and an hourglass for telling time, they provide the basic aids for navigating by dead reckoning.

Produced primarily in the Italian seaports of Genoa and Venice and the Catalan ports of Majorca and Barcelona, portolan charts changed little over the next few centuries. The earliest charts, such as the one shown here of the Mediterranean and Black Seas, drawn in about 1350, are similar to 17th-century portolans but lack the decorative features.

Although each portolan chart is unique, drawn or illuminated by hand on vellum or animal skin, production of the charts followed standard conventions. The most recognizable feature is the network of rhumb lines (lines of constant geographic direction) crisscrossing the chart. Rhumb lines radiate from a central point, or compass star—located in the Aegean Sea in the chart displayed—through 16 intersecting compass stars. Each compass star provided the navigator with 32 rhumb lines that he used to lay courses. Later, compass points were decorated with elaborate compass roses, and rhumb lines were color-coded to aid navigators. Black or brown represented the principal wind directions, such as north and south; green, the eight half winds, such as north northeast; and red, the 16 quarter winds, such as northeast by east. Lines were extended to the edge of the map sheet, which was usually untitled and without borders. Another innovation introduced by portolan chartmakers was the inclusion of graphical scales, measured in *miglia,* a unit of length still undefined. Coastlines were shown in detail and with remarkable precision, but some features were deliberately distorted: Islands, capes, headlands, and estuaries—serving as landmarks, presenting hazards, or providing sources of supplies—were highlighted and enlarged.

Names of towns and landmarks are provided, written at right angles to the coast on the inland side to avoid obscuring landing sites or hazards. This made it easy for a navigator sailing parallel to the coast to follow the place-names in proper sequence, literally around the Mediterranean coast. The charts, meant to be rotated, have no top or bottom. Place-names are coded. Uppercase lettering identifies regions; lowercase denotes cities. Red lettering indicates seaports.

Road Map for the Soul

ANCIENT FAITHS OFTEN EXPRESSED THEIR COSMOGraphies in spatial terms that could be visibly represented in map form. Southeast Asian specialist Joseph Schwartzberg calls these maps "road maps for the soul." Buddhism was no exception. Originating in northern India in the fifth century, it rapidly spread throughout Southeast Asia, China, and Korea. By the sixth century it had reached Japan. One of the earliest surviving world maps to express Buddhist doctrine is the *Gotenjiku Zu (Map of the Five Indies)*, painted and mounted on a scroll by a Japanese priest named Jukai in 1364. Jukai copied an earlier Chinese version of this map, which is no longer extant. The purpose of the *Gotenjiku Zu* was to record the pilgrimage of Xuanzhuang, a Chinese Buddhist priest who traveled to India in 629-45. Because of the map's association with Xuanzhuang's pilgrimage, surviving copies are considered objects of worship, according to cartographic historian Unno Kazutaka, and most are preserved in Buddhist temples.

Jukai's Earth is portrayed as an egg-shaped, flat disk surrounded by water. It represents Jambudvipa, the Rose Apple Island, from the *jambu*, or rose apple tree, which came to symbolize the Indian subcontinent. Jambudvipa was the site of the mythical Mount Meru, the mountain that linked the universe, the terrestrial world, and the world below. Mount Meru, according to Buddhist belief, is encircleded by seven rings of mountains and oceans, capped by a golden palace, home to Indra, king of Hindu gods. Today, Mount Meru is associated with Mount Kailas in Tibet.

Jambudvipa is divided into five major geographical areas, each outlined by a brown double line. Northern India covers the top half of the map, while the southern portion is subdivided into west, central, east, and south India. Labels locating China, Nepal, and Persia are found near the edges of the map. Several countries are shown outside Jambudvipa, suggesting their island status. These include Japan, Sri Lanka, "Western Woman's Country in the Western Sea," and "Golden Country."

Symbols, pictorial drawings, and labels adorn the map, conveying the Buddhist worldview in the form of a Chinese landscape painting. The pictorial renderings of mountains and sea-tossed waves are distinctive to Chinese and Japanese maps. Color is used to add perspective and realism. Several mountains are colored white to indicate snow cover, suggestive of great height. Roads and rivers are depicted as if viewed from above. Four unraveling concentric circles in the upper center of the map identify Mount Meru, in Buddhist belief, as the source of the world's four great rivers: the Indus, the Ganges, the Sutlej, and the Brahmaputra. Depicted also is Indra's golden temple. A red line identifies Xuanzhuang's pilgrimage route. Labels provide information concerning pilgrimage sites and distances and the number of travel days between them.

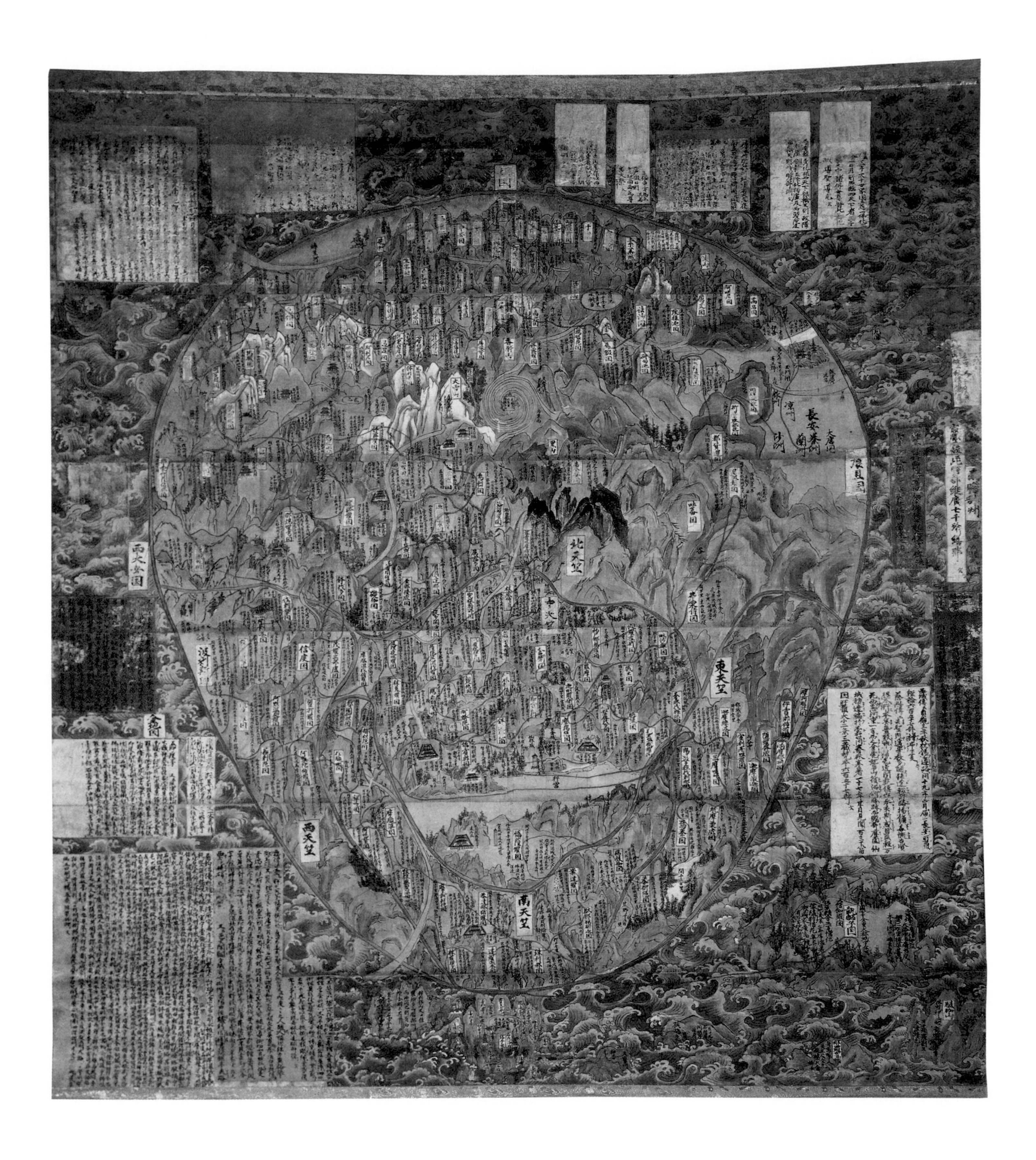

1364 GOTENJIKU ZU (MAP OF THE FIVE INDIES) BY JUKAI
PAINTING ON SCROLL, 69 X 65.6 IN.
HORYUJI TEMPLE, NARA, JAPAN

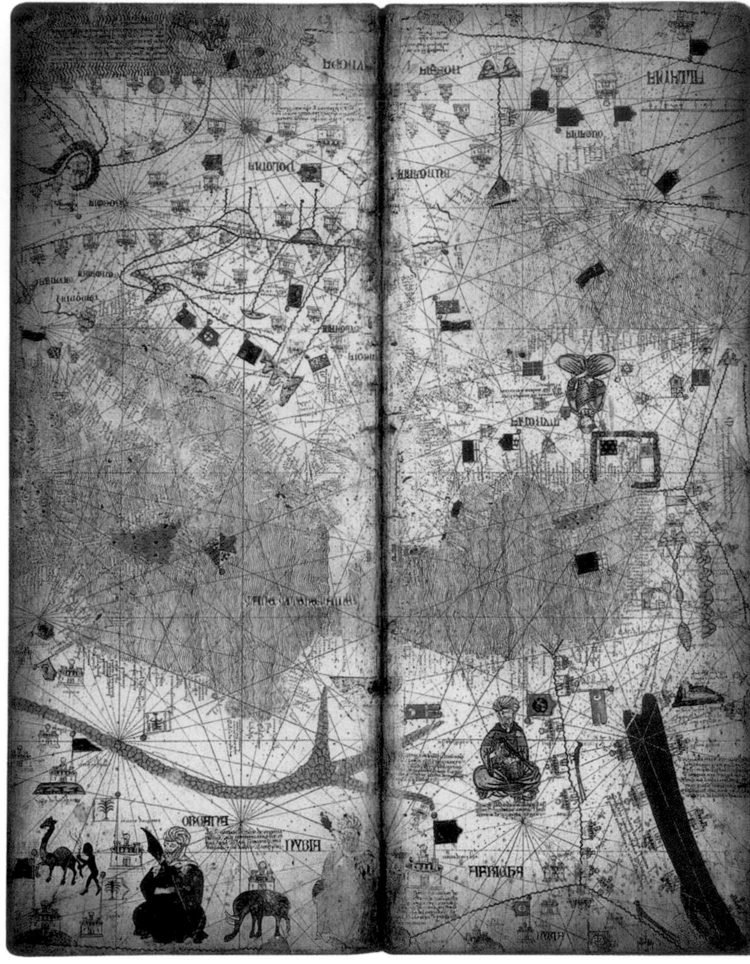

Circa 1375 Mappamundi by Cresques Abraham
Illuminated manuscript in 8 panels, 25.5 x 118 in.
Bibliothèque Nationale, Paris, France

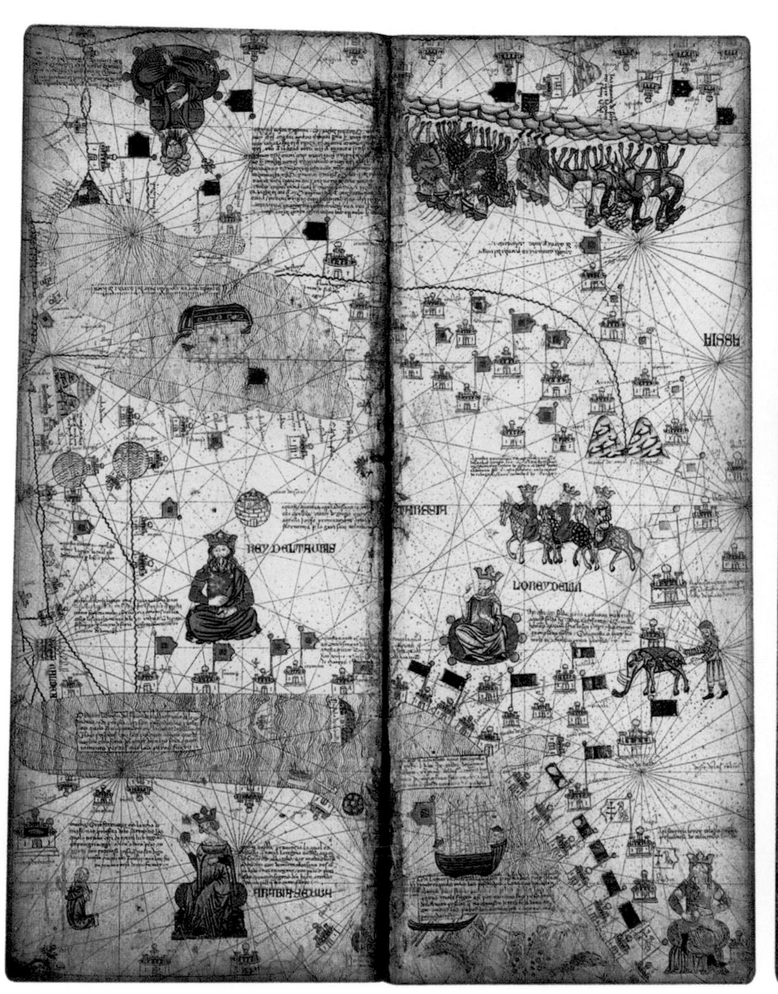

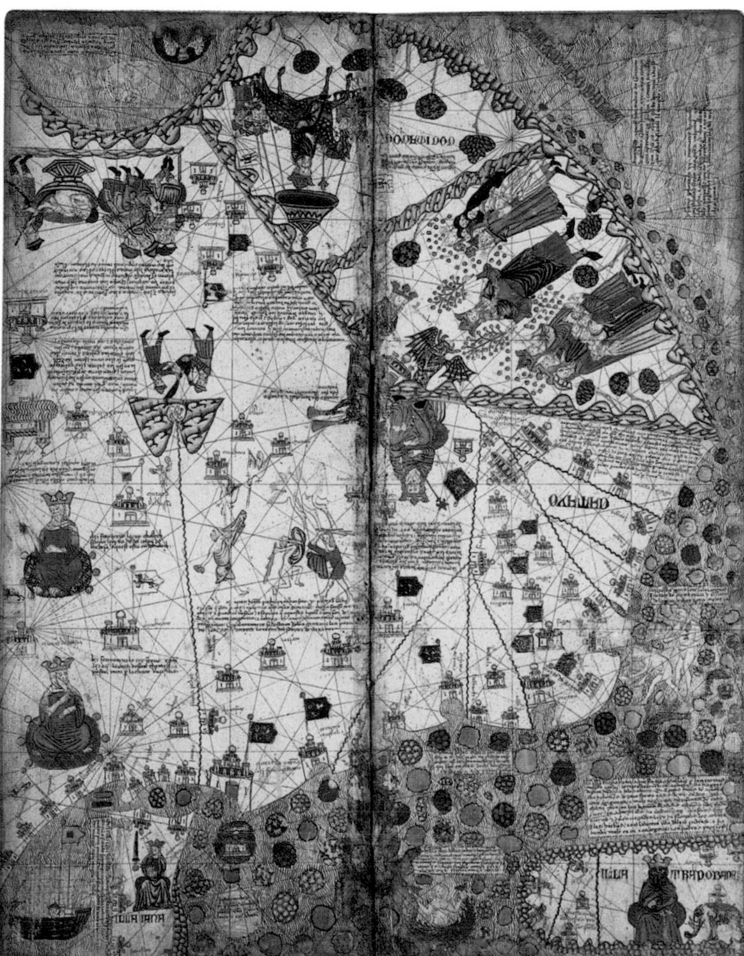

Image of the World

THE CATALAN ATLAS, A MASTERPIECE OF ART AND science, was prepared in about 1375, probably by Cresques Abraham, a book illuminator and "master of maps of the world" from the Jewish quarter of Palma, Majorca. Catalan chartmaking evolved alongside Genoese and Venetian traditions but differed considerably in design and style. Italian chartmakers focused on function, displaying only navigational elements. Their frame of reference was the Mediterranean and Black Seas. Catalan practitioners, however, influenced by their Arab and Jewish heritage and the aesthetics of medieval *mappaemundi*, decorated their distinctive charts with illustrations of topographical and maritime details and descriptive notes. They expanded their coverage to portray the interior of Europe and North Africa.

Cresques Abraham's 1375 world map is the finest example of the Catalan style. Prepared on six sheets of parchment, stretched on wood panels, it was later divided into 12 half-sheets and bound as a book. The first four contain discourses on cosmography, astronomy, and astrology, and include the earliest surviving lunar tide tables, critical data for mariners. The others are devoted to a world map oriented with south at the top.

The European panels are done in true Catalan portolan fashion. Seacoasts are accurately outlined, place-names written perpendicular to the coasts in red and black, and islands and seaports emphasized by size or color. Christian cities are designated by the spire and cross, but Cresques began the custom of identifying Muslim urban centers with a dome. The chart also provides the earliest example of a compass rose—an embellished wind rose, or intersection point of rhumb lines—with an eight-pointed polestar indicating north.

The panels encompassing Asia are drawn with information from written accounts of Marco Polo's travels. For the first time, a recognizable image of Asia is presented, and India appears as a peninsula. Symbols and motifs are derived from mappaemundi, biblical stories, and the classics. In China, Christ as king is pictured with his disciples, and Alexander the Great with a black-winged Satan. In Central Asia, a camel caravan travels the Silk Road and three magi on horseback follow the star of Bethlehem. Noah's ark rests on Mount Ararat. The Queen of Sheba and Tower of Babel represent Arabia. A Muslim at prayer symbolizes Mecca. Chinese ply the Indian Ocean in ships larger than any in the West. In Africa, Mansa Musa, Emperor of Mali, holds a gold nugget, seated next to Tenbuch (Timbuktu), site of his Great Mosque.

DETAIL OF MAPPAMUNDI BY CRESQUES ABRAHAM

Map of the Inhabited World

CLAUDIUS PTOLEMY IS THE MOST IMPORTANT FIGURE in the history of Islamic, Byzantine, and Renaissance European cartographic traditions. He dominated geographic and cartographic thought in Western Europe and the Near East for more than 14 centuries through two major works—the *Almagest*, a 13-volume treatise on mathematics and astronomy, and the *Geography*, in eight volumes. None of his maps have survived. Of Greek descent, he was born in Egypt and lived from about 90 to 168, working mostly in Alexandria.

Ptolemy's major contribution was to synthesize Greek theoretical cartography and Roman practices in surveying and mapping, laying the foundation for mathematical cartography. His *Geography* provided instructions and necessary geographical information for preparing a world atlas, including a world map and 26 regional maps. Of greatest importance, he revised and improved several map projections, or mathematical frames of reference, that reduced distortions associated with representing a round globe on a flat plane. A second contribution was a list of some 8,000 towns and places along with their geographic coordinates, gathered from travelers.

The *Almagest* and *Geography* were neglected, however, until ninth-century Islamic scholars in Baghdad rediscovered them and translated them into Arabic. Four centuries later, Byzantine geographers in Constantinople resurrected the *Geography* and reconstructed the maps following Ptolemy's instructions. The German monk Donnus Nicolaus drew the world map on pages 48-49 about 1466 for the Latin version of the *Geography* by Jacopo Angelo de Scarperia.

The map is projected on a conical grid where meridians of longitude appear as converging straight lines, and parallels of latitude as arcs of circles. Establishing a convention still followed, Ptolemy expressed latitude in degrees north or south of the Equator, but longitude was determined by degrees east of a prime meridian—at that time drawn through the Fortunate Isles (Canary Islands). Ptolemy's world map extends 180 degrees west to east. The Roman Empire is recognizable. The Mediterranean, parts of Western Europe, and the Near East are similar in outline to modern maps. Some of Ptolemy's information was adopted from a map prepared about A.D. 100 by Marinus, a Phoenician and probably the first cartographer to incorporate information from the Romans.

A major feature of the map is the Nile River. Its source, Ptolemy believed, was snowmelt from the Mountains of the Moon (Ruwenzori Mountains in Uganda near the actual source, Lake Victoria). Later mapmakers copied his depiction of six or more rivers flowing from the mountains into two lakes, and then from these lakes into one lake and then one stream. Ptolemy's portrayal of Asia is more mysterious. The Indian Ocean is drawn as an enclosed sea, and Taprobana (Sri Lanka) is depicted as a large island overwhelming the subcontinent of India.

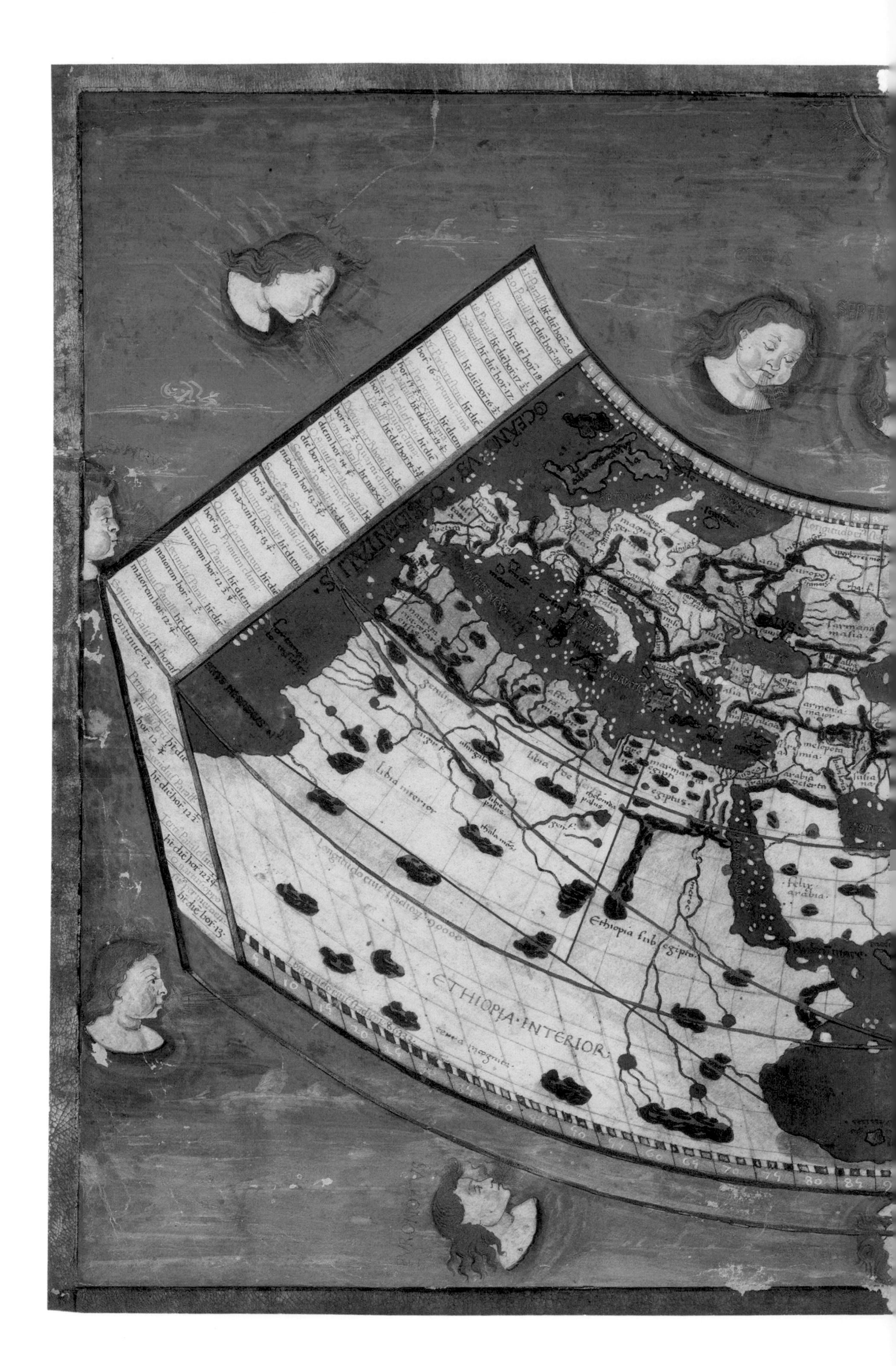

Circa 1466 World Map by Donnus Nicolaus, from Claudius Ptolemy's Geography
Illuminated Manuscript on Parchment, 17.2 x 11.5 in.
Biblioteca Estense Universitaria, Modena, Italy

Serica Rego
Sathia intra imaum montem.
Sathia extra Imaum montem
Sacha regio
INDIA EX TRA·GAN GEM.
Sina regio
Sogdiana
ARIA
INDIA
INTRA GANGEM.
gedrosia
Taprobana insula
INDICVM
M A R E

The World Is Very Wide

"THE WORLD IS VERY WIDE," CONFUCIAN SCHOLAR Kwon Kun asserted in a note on his 1402 world map. It extends, he wrote, "from China in the center to the four seas at the outer limits." World maps with a Chinese perspective began appearing in the 14th century. China rather than Jerusalem became the focus, with Europe, the Mediterranean, and Africa reduced in size and moved to the periphery. The most intriguing surviving example of this cartographic tradition is a Korean world map painted on silk in 1402 by Yi Hoe, directed by Kwon Kun. The original has disappeared but copies exist in libraries in China and Japan, such as this one dated 1470, in Kyoto.

The *Kangnido*, as the map is known, was created during a period of dynastic change in Korea that brought Confucian reformers to power. Kwon Kun was one of these reformers. A scholar and government official, he proposed two maps to "demonstrate the new dynasty's cosmic legitimacy," according to Korean historian Gari Ledyard. The first, a star map based on an A.D. 670 rubbing, was engraved in 1395 on a huge block of black marble that is now on exhibit at Seoul's Royal Museum. The second was the *Kangnido*. Little is known of Yi Hoe's training, but he was a cartographic painter of some skill. His map of Korea, the *P'altodo (Map of the Eight Provinces)*, prepared in about 1400, was the standard for nearly 50 years.

Yi Hoe's map encompasses a very wide world indeed, one stretching from the Korean Peninsula and Japan west to the coast of Europe, and from the interior of the Eurasian continent south beyond the tip of Africa. But the relative size of countries decreases as their distance from Korea increases. The Korean Peninsula is greatly exaggerated compared with China and Japan (depicted below Korea with a northwest orientation). Yi Hoe's own map of Korea provided the model for his world map. Japan's image was taken from a copy of a Japanese manuscript map obtained by a Korean military attaché, according to Ledyard. China was copied from a standard historical map of the Middle Kingdom prepared by a Buddhist monk of the Ming dynasty (1368-1644). The most notable feature is the Great Wall, forming China's northern border.

More speculative is the area west of China. Recent research by Ledyard indicates that the geography of the non-Chinese areas was obtained from a Chinese map derived from Islamic maps. The peninsulas on the left are believed to be Africa and the Arabian Peninsula. Thirty-five Chinese place-names transcribed from Persian Arabic, including Alexandria and Egypt, support this thesis. The Arabic symbol depicting the source of the Nile is found near the tip of Africa. Another version of the *Kangnido* identifies this symbol with a Chinese transcription identified with Djebel al-Qamar, the Persian Arabic name for the Mountains of the Moon (Ruwenzori Mountains), once thought to be the river's source.

1470 Honil Kangni Yoktae Kukto Chi To (Map of Integrated Lands and Regions of Historical Countries and Capitals)
by Yi Hoe and Kwon Kun
Painted on Silk, 64 x 67 in.
Ryukoku University Library, Kyoto, Japan

Circa 1450 World Map by Fra Mauro
Illuminated Manuscript on Ox Hide, 78 x 77 in.
Biblioteca Nazionale Marciana, Venice, Italy

Visions of New Worlds

FRA MAURO, A MONK AND EXPERIENCED CARTOGRAPHER working in Venice, provided tantalizing glimpses of new worlds and oceans with this manuscript map prepared circa 1450 with the assistance of portolan chartmaker Andrea Bianco. This large wall map was the last of the medieval *mappaemundi* and one of the first of a new generation of world maps. A copy was sent to Afonso V, king of Portugal, through the efforts of Venetian aristocrat Stefano Trevisan in 1459.

The map is oriented to the south, in Arabic fashion. Africa is drawn as a freestanding continent and the Indian Ocean as an open sea. These revolutionary concepts suggested the possibility of seagoing trade with the Orient nearly four decades before Bartholomeu Dias rounded the Cape of Good Hope.

Fra Mauro's map embodies three cartographic traditions. The circular format and the pictorial representation of cities, towns, and castles reflect its mappaemundi lineage. Jerusalem has been moved off-center to accommodate newer regions, and elaborate decorations have been banished, except for the rendering of Paradise at lower left. The coastal outlines of the Mediterranean and Black Seas are derived directly from portolan charts. Although Fra Mauro pays homage to Ptolemy's *Geography*, he does not follow its use of meridians, parallels, or degrees, because of the regions unknown to Ptolemy. The Venetian filled Ptolemy's *terrae incognitae*.

This new map data was obtained from a variety of sources, including firsthand accounts based on rumor and speculation. Portuguese portolan charts were consulted for the delineation of the west coast of Africa, which had been charted south to Cape Verde by 1445. Information about Africa's northeastern interior probably was acquired from emissaries of the Coptic Church of Abyssinia "who with their own hands," Fra Mauro noted, "had drawn for me all these provinces and cities, rivers and mountains, with their names—all of which I have not been able to set down in proper order for lack of space." The representation of Africa's east coast and the island at its tip, likely Madagascar, was probably from Arab sources.

The travel accounts of two Venetian merchants, Marco Polo and Niccolo de' Conti, provided details of China and South Asia. China is filled with names of towns taken from Polo's narrative, and for the first time, its main rivers are shown with relative accuracy. Conti may have been the source for Fra Mauro's belief that the Spice Islands could be reached by sailing around Africa.

DETAIL FROM MAP: ASIA

CLAVDII PTHOLOMEI ALLEX
ANDRINI COSMOGRAPHI
APARCTIAS
ASIA
EVROPA
AFFRICA
CIRCVLVS
80
90
SEPTENTRIO
CIRCVLVS
ARCTICVS
MARE GLACIALE SIVE CONGELA
MARE GLACIALE
TARTARIA
MARE HIRCANV
SIVE CASPIVM
PARTHIA
ARIA
MEDIA
CARMANIA
SINVS
PERSICVS
ARABIA FELIX
MARE RVBRV
GETVLIA
LIBIA INTER
IOR
EQVINOCTI

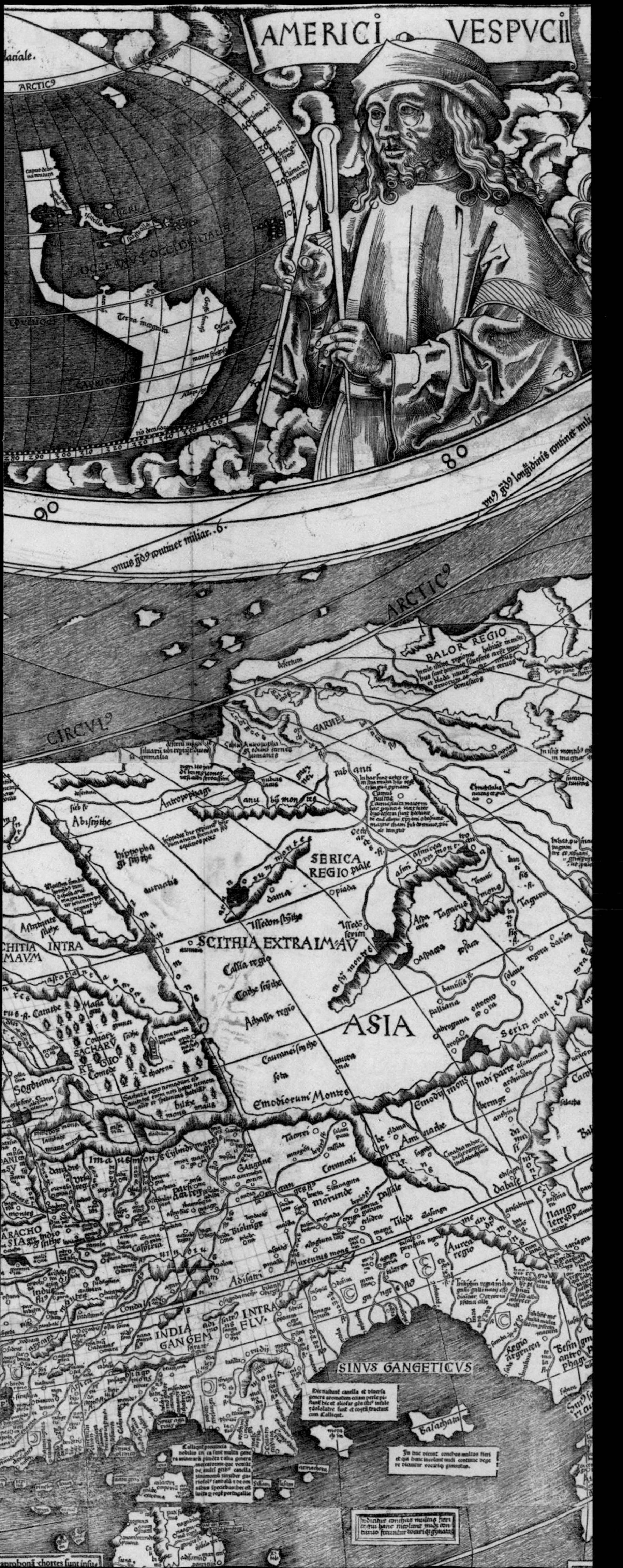

SECTION TWO

Charting the Age of Discovery and Exploration

First World Map to show America as a Continent, 1507 by Martin Waldseemüller

Introduction

Interest in maps expanded greatly during the 15th and 16th centuries, spurred by geographical discoveries and the subjugation of new lands and peoples. Portuguese, Spanish, and Italian mariners initiated the exploration and charting of the Atlantic and Pacific Oceans, beginning with Prince Henry the Navigator along the coast of Africa, from 1418 onward. At the same time, Ottoman sultans sought maps and descriptions of lands conquered in southeastern Europe, western Asia, and northern Africa, and Chinese mandarins continued to map their country, its inland waterways and coastal areas.

In Europe, the rediscovery of Claudius Ptolemy's *Geography* combined with major advances in navigation and surveying instruments and techniques, and the invention of printing from movable type by Johannes Gutenberg revolutionized mapmaking. Ptolemy's *Geography* introduced Renaissance mapmakers to mathematical cartography—the use of map projections to construct maps and the inclusion of latitude and longitude to frame map content. The first maps to be printed were those by Ptolemy described here, or new maps included in revised editions of Ptolemy's *Geography*, such as Johannes Ruysch's 1507 world map. Ruysch's and Martin Waldseemüller's 1507 world maps were based on map projections described by the second-century Greek geographer. Waldseemüller and Antonio Florian (1555 world map) each honored Ptolemy by including his portrait in their map borders. Caspar Vopell's armillary sphere of 1543 illustrates the Ptolemaic universe.

New instruments such as the seaman's astrolabe, quadrant, and cross-staff, as well as books of navigation tables such as *Regiment of the North Star*, aided mariners in charting their discoveries. Seamen adapted the Mediterranean portolan chart to ocean sailing by adding double latitudes to compensate for magnetic variation, which is shown on the 1547 Dieppe chart of the east coast of North America. The appearance of scaled sea charts based on the magnetic compass and measured distances in 13th-century Italy was matched by developments in topographic maps in 15th-century Germany and Austria, reflected in Erhard Etzlaub's 1500 map of Central Europe.

Printing had a profound impact on the appearance of maps, the process of map production, and map distribution and circulation. The earliest printed maps were produced from woodcuts. The wood was cut away, leaving the printing surface in relief. Woodcut printing was particularly popular north of the Alps, a region with a tradition of wood carving. Because woodcut maps could be printed with letterpress text, it was the printing process of choice for the German humanist geographers such as Sebastian Münster. Chinese mapmakers, with a long history of woodcut maps, attained a high level of perfection, illustrated in the 1555 Chinese woodblock entitled *Gujin Xingsheng Zhi Tu*. Metal type with lettering was inserted into Waldseemüller's woodcut to produce letterpress legends, which could easily be changed for different editions.

Copperplate map engravings, generally superior to woodcuts in size, detail, precision, and appearance, were favored in Italy. An intaglio process, the printing image is incised into a copper (or other metal) plate by an engraver.

Copperplate maps could be enhanced with tones and textures, created through stippling the plate with special tools. Hieronymous Cock, one of Europe's premier engravers, etched Diego Gutiérrez's rare 1562 map of the Western Hemisphere on six plates with exquisite skill.

A PROFESSIONAL CLASS OF MAPMAKERS BEGAN TO EMERGE DURING THIS PERIOD, BUT FOR THE MAJORITY cartography remained a secondary interest. Diego Gutiérrez was a licensed chartmaker for the Spanish hydrographic office, but Waldseemüller, whose map first gave America its name, was a cleric. Oronce Fine and Caspar Vopell were mathematicians and astronomers. Münster, the most prolific mapmaker of the era, was a noted Hebraic scholar who translated the Hebrew Bible into Latin. Erhard Etzlaub, compiler of the first road map of Central Europe in 1500, was renowned throughout Europe as a compass maker. Several were artists and illuminators, including Antonio Florian, whose self-portrait, the first of its kind, graces the border of his 1555 engraving. Piri Reis, author of the 1513 map of the New World, was a Barbary pirate, later beheaded by Süleyman the Magnificent. With the introduction of printing, map creation became a team effort rather than the work of individuals. Names of engravers and printers begin to join the names of the original cartographer. Cock prominently etched his name next to that of "author" Gutiérrez.

Maps and charts of this period were produced primarily for an emerging commercial market, but the 1540 Aztec estate map was prepared for a land litigation dispute, and the two Chinese maps were probably drawn for official purposes. Etzlaub's map of Central Europe, apparently the first map to be mass-marketed, was created to guide Catholic pilgrims to Rome during the Christian Holy Year of 1500, and perhaps to advertise his compasses. Waldseemüller's large wall map, printed from 12 woodblocks, was designed for display in personal libraries and public courthouses. One thousand copies were printed, according to Waldseemüller, but the one surviving copy was preserved in an atlas in Wolfegg Castle in southern Germany for 350 years. The ornately painted manuscript portolan atlases and charts by Battista Agnese and an unknown Dieppe chartmaker were most likely prepared on consignment for wealthy patrons. Some maps became extremely popular. Seventy copies of Agnese's atlases survive. Münster's maps of the four continents were reissued in 36 editions of his atlases over eight decades.

The focus of many of these maps was geographical discovery and exploration. Portuguese expeditions along the coast of Africa and in the Indian Ocean are reflected on several world maps beginning with Henricus Martellus's map, dated 1489. Similarly, Spanish discoveries of the New World were reported to the wider public in world maps drawn by Ruysch and Waldseemüller in 1507. The confusion concerning the exact nature of the New World is reflected in several world maps that depict the present North American coastline as an extension of Asia. Astonishingly, the Muslim mapmaker Piri Reis recorded the most accurate map of Columbus's view of his discoveries, basing it on an original Columbus map taken from a captured Spanish slave.

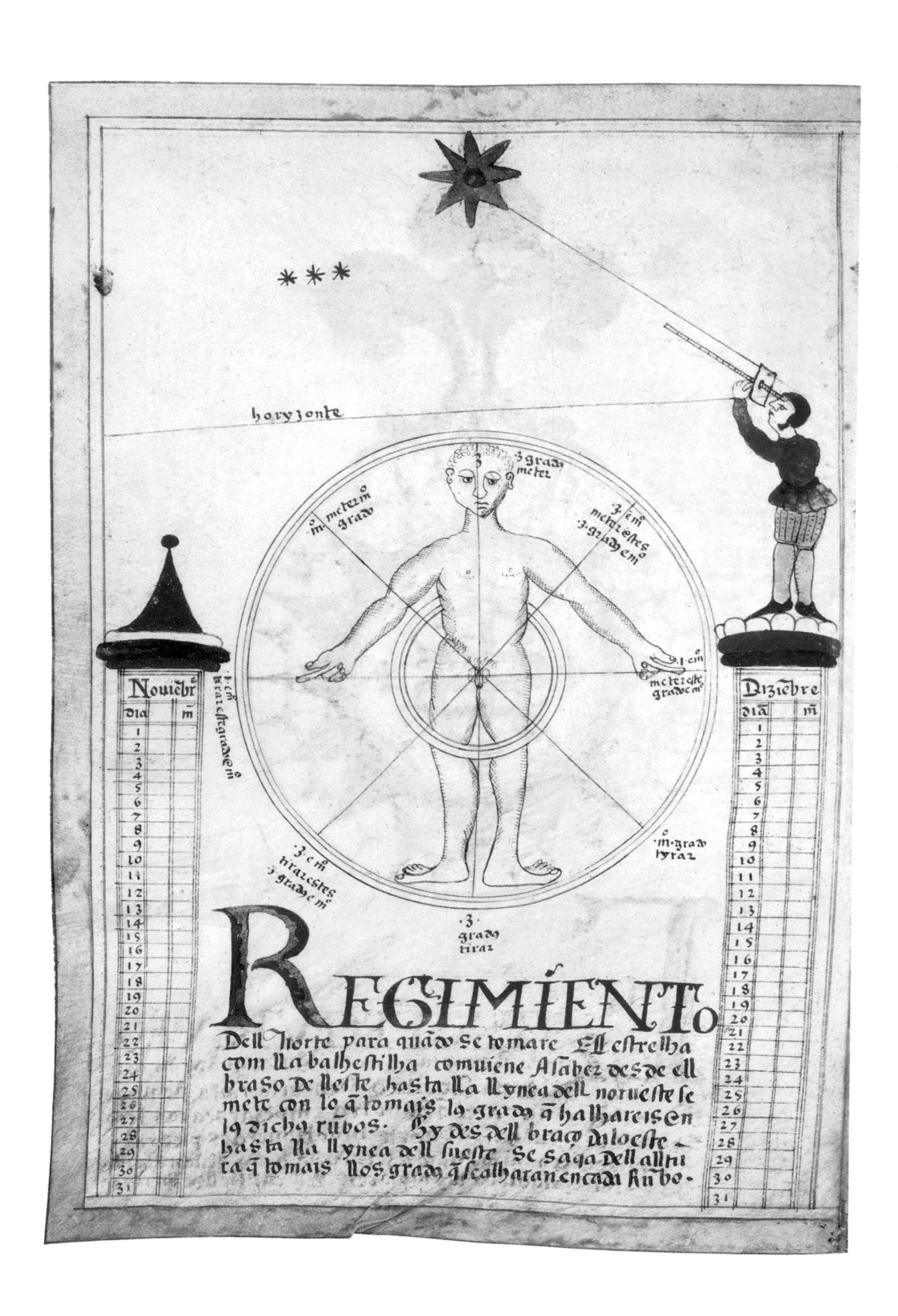

Circa 1555 Portuguese Chart Book Entitled Regiment of the North Star, Author Unknown
Illustrated Frontispiece
National Maritime Museum, London, England

Regiment of the North Star

EUROPEAN MARINERS IN THE AGE OF EXPLORATION charted their courses by a combination of dead reckoning and celestial navigation. Dead reckoning involved deducing a position from a previously recorded position based on direction and distance. Direction was determined by compass; distance was determined by estimating a ship's speed, usually by tossing overboard a small log attached to a knotted rope and then timing with a sandglass as the rope played out. The daily route was tracked on a traverse board, a wooden panel marked with the 32 points of the compass rose, each row of points containing eight holes to accommodate pegs. At half-hour intervals during an eight-hour watch, a helmsman would add a peg indicating the location. Another series of holes recorded the ship's speed. The course information was later transferred to a slate, and then recorded at the end of the day.

Celestial navigation became more important when sailing in unknown seas and for fixing the location of newly discovered lands so that sea routes could be retraced. The preferred method of navigation involved "running down the latitude," sailing north or south to the desired latitude and then following it east or west. Sailors determined latitude by measuring the angle of the North Star above the horizon or observing the meridian of the sun by noon sightings. While the polestar's altitude could be estimated by eye—one Venetian sailor off the coast of Africa in 1454 reported it was "about the third of a lance above the horizon"—several instruments evolved to aid mariners. Columbus carried a seaman's astrolabe, a disk with a rim calibrated into 360 degrees with sighting bar, and a quadrant, a quarter circle with plumb line and twin peep sights; however, both were difficult to use on pitching decks. He preferred dead reckoning. Easier to use was the cross-staff, a wooden pole with a transom, a crosspiece moved back and forth to read the altitude of the sun or a star from a scale on the side of the staff. It was adapted from an Arab navigational instrument, the *kamal*.

In the 1480s, Portuguese officials began providing seamen with navigational tables and the *Regiment of the North Star*, a set of rules for finding latitude. Mariners also kept track of elapsed time with the *Regiment*. As a memory aid, 15th-century sailors envisioned the figure of a man centered over Polaris with eight imaginary compass bearing lines crossing him, each representing three hours, or 45 degrees. Two "guard stars" of the Little Dipper, Beta Ursae Minoris and Gamma Ursae Minoris, revolved around this figure counterclockwise. Columbus, in reference to the *Regiment*, wrote in his journal on September 30, 1492, "I am surprised to find that the Guards are near the arm on the west at night but at day break appear below the arm to the east. If this observation is correct, it appears that I only proceeded three lines last night or nine astronomical hours."

World Map on Eve of Columbus Voyage

PLEADING WITH QUEEN ISABELLA OF SPAIN TO finance his voyage to the Indies, Christopher Columbus asserted that his proposed route to the Indies, westward rather than around Africa and the Cape of Good Hope, was the shortest and safest. In support of his theory, Columbus may have submitted a world map similar to one prepared by Henricus Martellus in 1489. Heinrich Hammer was a talented German map illuminator known to history by his Latinized name, and one of a group of mapmakers working in Florence, Italy, then Europe's leading center of cartography. They were actively reviving Ptolemy's atlas, supplemented with "modern" maps.

Two of Martellus's world maps survive. The largest, a tempera painting on paper, now at Yale University, measures nearly four by six feet. Some historians believe it was copied from a prototype prepared by Columbus's brother Bartolomeo, a chartmaker who worked in Lisbon and later in Fontainebleau for the sister of King Charles VIII. The smaller map, now in the British Library and shown here, apparently was copied from the Yale version.

The first of the Renaissance landmark maps of discovery, these world maps display voyages of exploration within the frame of Ptolemy's worldview. Constructed with curved meridians and parallels, they are the first to depict both latitude and longitude. Martellus (or maybe Bartolomeo) derived his outlines and place-names of the northwest coast of Europe and the Arctic from the Fra Mauro map of 1450; of England and the Mediterranean and Black Seas from portolan charts; and of East Asia from Marco Polo's book, a copy of which was heavily annotated by the Columbus brothers. The delineation of the African continent was based on a series of portolan charts by Portuguese mariners culminating with Bartholomeu Dias's voyage of 1488 around the Cape of Good Hope.

Most intriguing is how closely these maps reflect Columbus's worldview on the eve of his voyage in 1492. The combined width of Europe and Asia is exaggerated by almost 100 degrees (7,000 miles), reducing the Atlantic by a similar amount. The tip of Africa is shown at 45 degrees south—11 degrees farther than recorded by Dias—adding 1,000 miles to the African sea route. Geographer Arthur Davies believes the map originally showed the Cape of Good Hope correctly at 34 degrees south, but that Columbus revised it to support the idea that the African route was too long. The map also extends the Southeast Asia Peninsula. This "Horn of Asia," Davies suggests, was inserted to further illustrate the futility of a southern sea route.

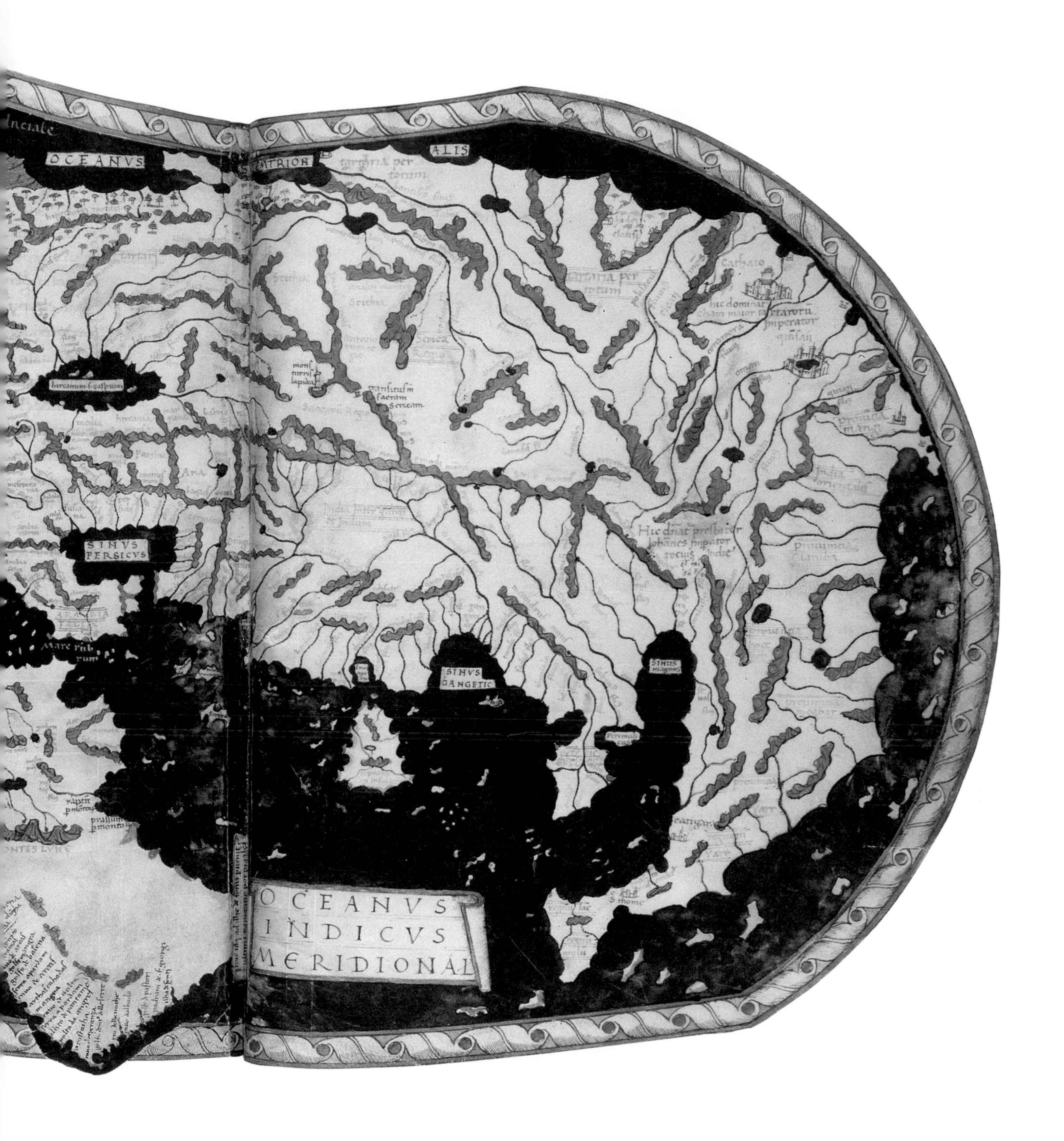

1489 World Map by Henricus Martellus Germanus
Illuminated Manuscript on Paper, 48 x 72 in.
British Library, London, England

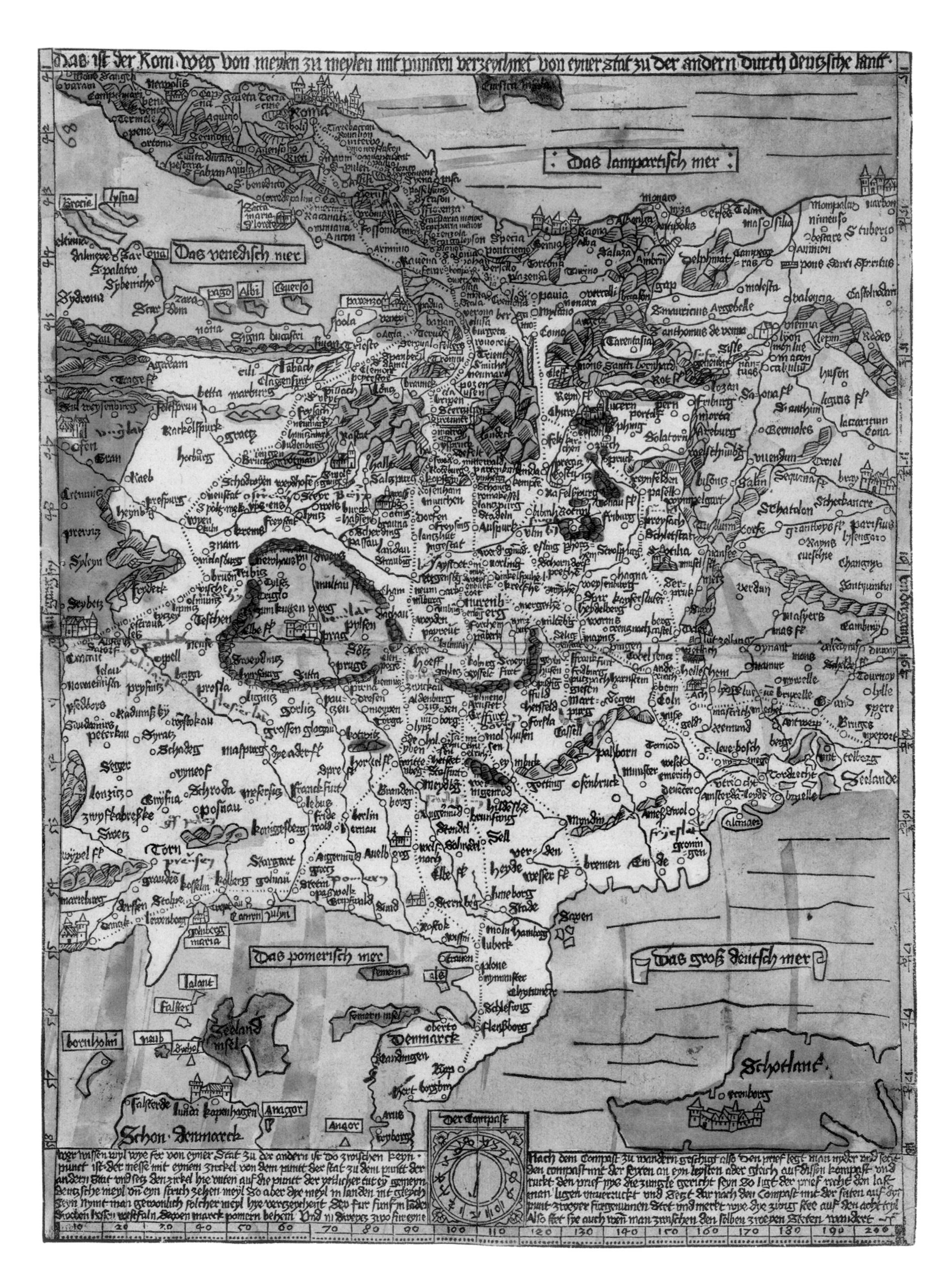

Circa 1500 Central Europe Map entitled Rom Weg (Route to Rome) by Erhard Etzlaub
Woodcut broadsheet, Hand-Colored, 16 x 11.5 in.
Royal Library of Copenhagen, Denmark

First Printed Road Map

THE CHRISTIAN HOLY YEAR OF 1500 WAS THE inspiration for the first great map of Central Europe, a road map designed to lead pilgrims to Rome. First proclaimed in the year 1300 by Pope Boniface VIII, jubilee years drew thousands of the pious to St. Peter's and other churches in Rome. The 1,500-year anniversary of the birth of Christ promised to attract the largest number of pilgrims to date, prompting a Nürnberg compass maker, Erhard Etzlaub, to prepare the first mass-produced guide map. Although primarily a craftsman — his compasses made in the form of pocket sundials with compass points were valued throughout Europe — Etzlaub was "admirably learned in the principles of geography and astronomy." He was also a competent land surveyor and, during the last years of his life, served the poor as a physician. He died in 1532 in his seventh decade.

Etzlaub prepared at least three woodcut maps, all oriented with south at the top. His *Rom Weg* map is the most notable, covering an area extending from Scotland and Denmark south to central Italy, and from Paris east to Budapest. Eight main routes from Danzig, Rostock, Ribe, Lübeck, Bremen, Utrecht, Nieuport, and Marburg, indicated by dotted lines, converge on Rome. More than 800 towns are shown, located by open circles or, in the case of pilgrimage sites, by miniature pictures of churches. Intended for everyday travelers rather than officials and scholars, the map was printed in German rather than Latin and included a variety of innovative travel aids similar to those found in modern road atlases. Distance was determined by counting the dots in the routes, with each dot equaling a German mile. An analysis of the "mileage dots" by historian Brigitte Englisch reveals that the distances between towns were quite accurate. A further aid to determining travel time, along the right-hand border, lists the number of maximum daylight hours by latitude in quarter-hour units. Latitude is listed in the left-hand border. Etzlaub's map was designed for use with a sundial compass. Cardinal points are indicated in the borders, and a drawing of a compass rose and instructions for determining directions by compass appear in the bottom margin.

Another helpful feature was the use of coloring to indicate the language areas or kingdoms through which the pilgrims traveled. For the fist time, specific colors were assigned to a map. Etzlaub listed them on his accompanying information sheet as a guide for the colorist as well as the map user. The "white area in the center is German country," Etzlaub noted, but the bordering regions each had their own colors: Italy, "vegetable or light green"; France, red; the Netherlands, light yellow; Scotland, green; Denmark, "thick stone color"; Poland, brown; Hungary, "grass green"; and "the lands of the Wends," deep yellow. The conspicuous yellow circle in the center of the map is Bohemia, "surrounded by a green forest." Islands were colored in "red oxide of lead."

Land of the Holy Cross or The New World

IMAGES OF A NEW WORLD BURST UPON EUROPEAN consciousness in 1506 and 1507 with the publication of three revolutionary world maps, drawn by Giovanni Contarini, Martin Waldseemüller, and Johannes Ruysch. Only Ruysch's map, however, appears to have been widely circulated in its original form. Until the rediscovery of Waldseemüller's map in 1901, Ruysch's map was believed to be the oldest printed map of the New World.

Drawn to illustrate a description of the New World by Italian monk Marcus Beneventanus, Ruysch's fan-shaped conical projection, centered on four fictitious islands near the North Pole, was one of six new maps added to a reprinting of the 1490 Rome edition of Ptolemy's *Geography*. Little is known about Ruysch, a native of Utrecht. He was, wrote Beneventanus, the "most learned of geographers and well skilled in depicting the globe."

Reflecting the consensus of the time, Ruysch depicted Greenland and Terra Nova (Newfoundland), discovered by John Cabot in 1497, as an extension of Asia. Ruysch himself may have sailed with Bristol seamen to the great cod-fishing areas off Newfoundland, according to Beneventanus and later historians.

Three prominent islands inhabit the region south of Newfoundland. From east to west, these are the mythical Antilia, the original home of the "Seven Cities of Gold," whose virtues Ruysch expounds in nine lines of text; Spagnola (Haiti/Dominican Republic); and an unnamed Florida-like landmass that still draws interest, since its appearance on this map predates the first documented sighting of the peninsula by six years. A scroll device marking its western boundary notes that Spanish ships reached this point. Ruysch himself, in an extended note adjoining the island, suggests that Marco Polo's Japan was located in this area.

The concept of a New World is most fully expressed in Ruysch's delineation and descriptions of the continental landmass portrayed south of these islands. In a legend Ruysch refers to this region, derived primarily from Portuguese maps and reports, as "Land of the Holy Cross or the New World." "This country," he notes, ". . . generally considered another continent, is inhabited [by people who] live to be over 150 years."

Portuguese discoveries in the Indian Ocean are portrayed for the first time, including the islands of Madagascar, Sri Lanka, and Sumatra, here called Taprobana. India and Africa are displayed close to their modern forms. The delineation of the Nile River and its source in the Mountains of the Moon reflect Arabic mapping tradition.

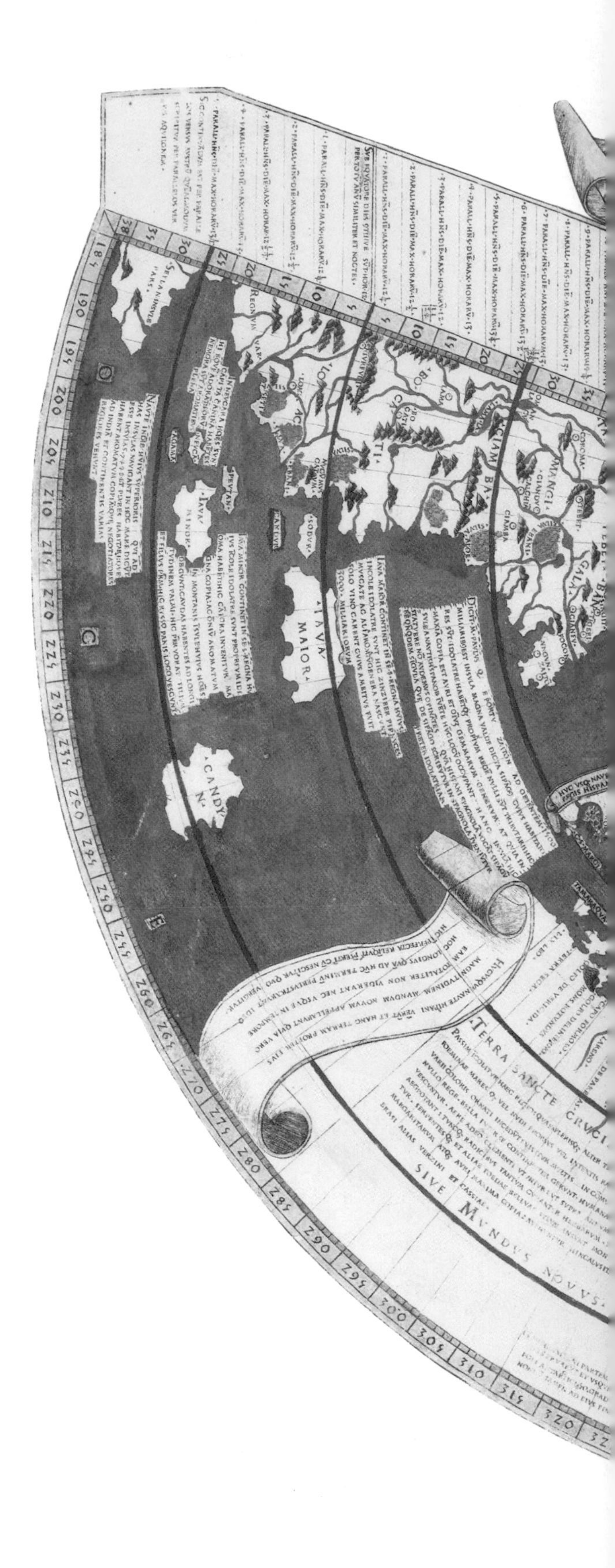

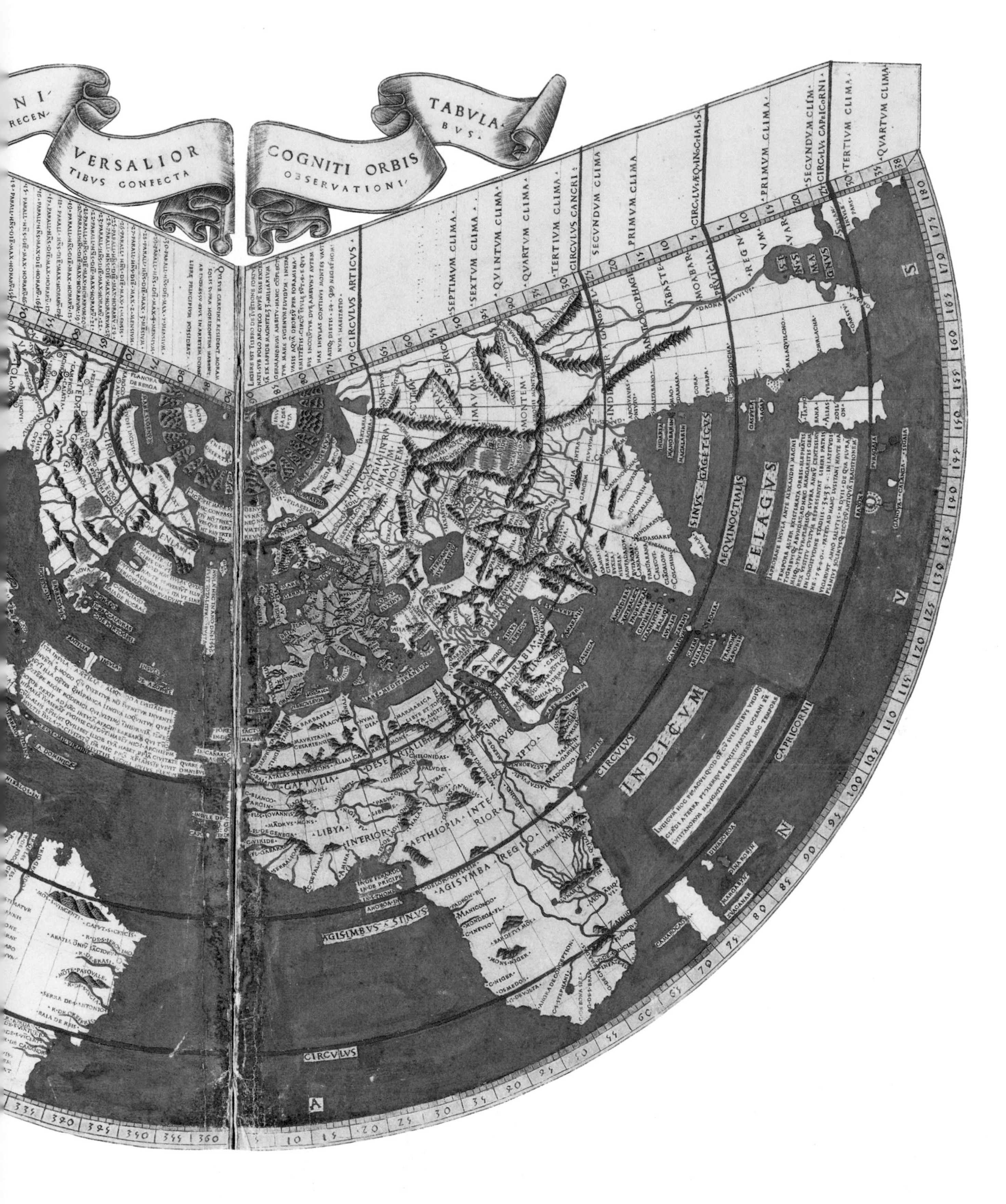

1507 World Map Entitled Universalior Cogniti Orbis Tabula Ex Recentibus Confecta Observationibus
by Johannes Ruysch From Claudius Ptolemy's Geography
Copperplate Engraving, Hand-Colored, 16 x 21.5 in.
Library of Congress, Washington, D.C.

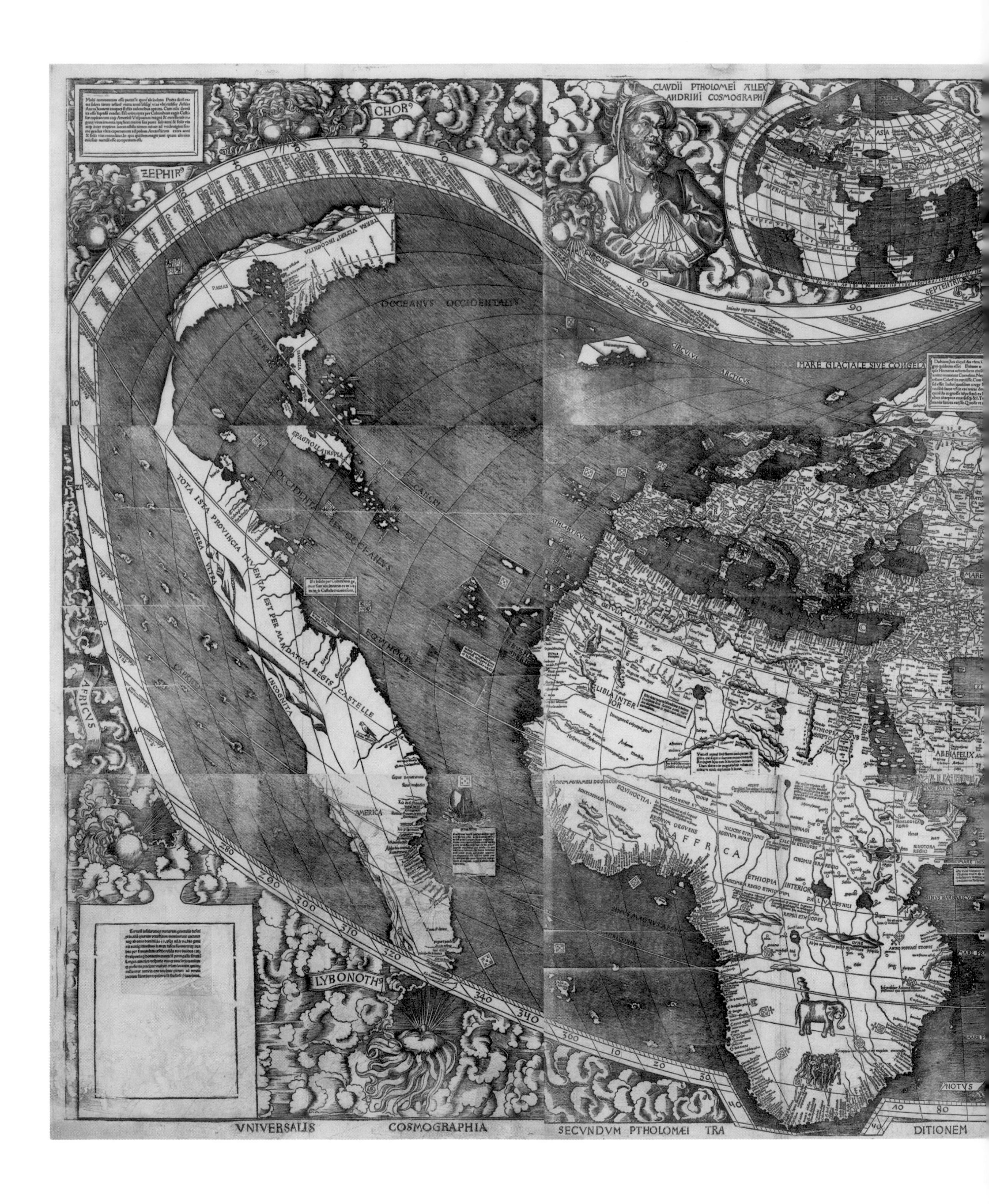

1507 Universalis Cosmographia secundum Ptholomaei Traditionem et Americi Vespucii alioruque lustrationes
(A Drawing of the Whole Earth According to the Tradition of Ptolemy and the Voyages of Amerigo Vespucci and Others)

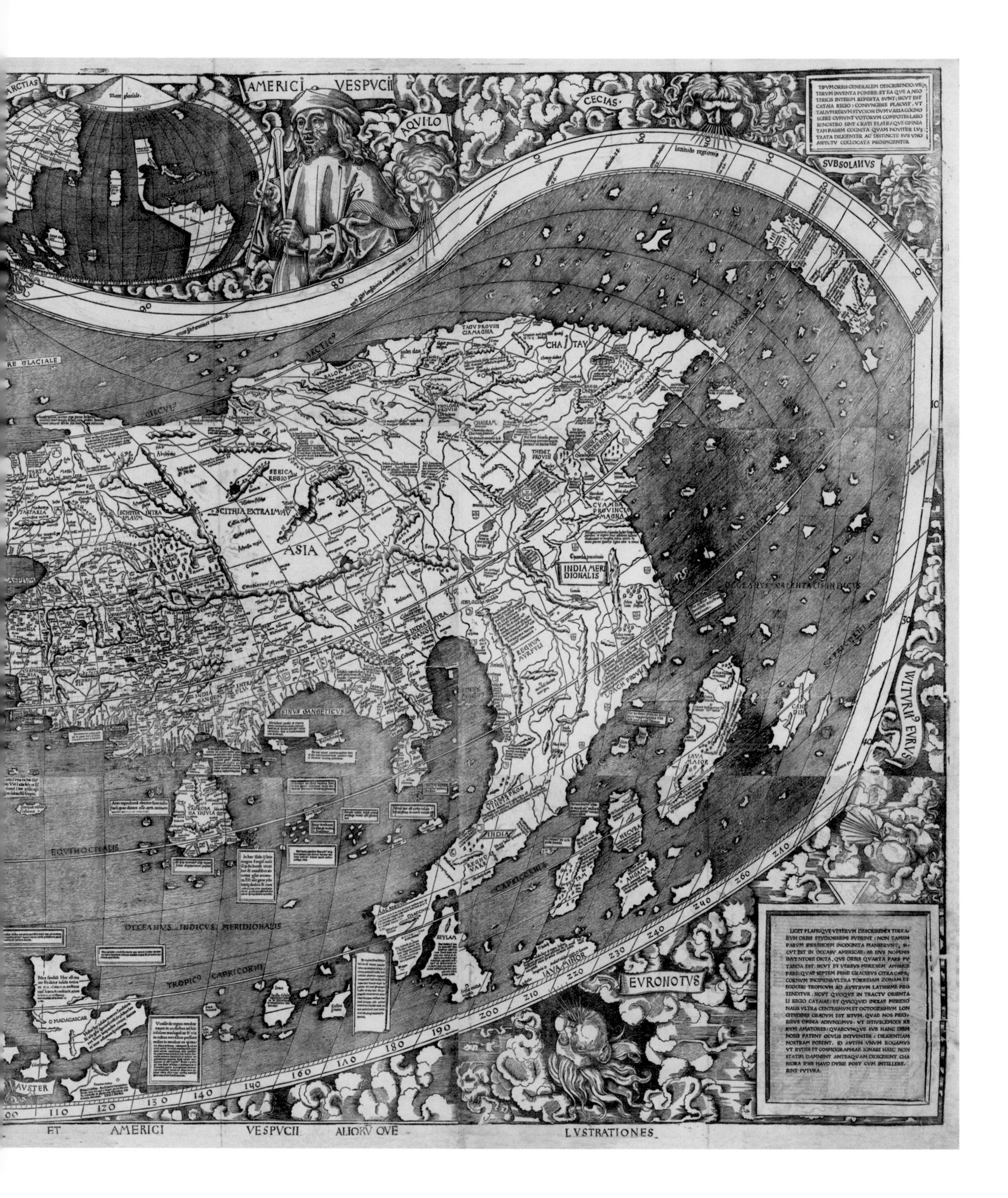

Attributed to Martin Waldseemüller
Woodcut in 12 Sheets, 51 x 92 in.
Library of Congress, Washington, D.C.

The Map That Named America

THE NAME "AMERICA" APPEARED FOR THE FIRST TIME on Martin Waldseemüller's revolutionary world map, the first large world map to be printed. Waldseemüller was part of a small team of German clerics and humanists working to produce a new edition of Ptolemy's *Geography* that would include recent Spanish and Portuguese discoveries. Waldseemüller, a skilled draftsman and printer, was chosen to prepare the maps. The group had begun revising Ptolemy's atlas when they obtained a copy of Amerigo Vespucci's printed report describing his 1501-02 voyage in which the Italian navigator stated for the first time that he had sighted a new continent. After following the South American coast southward as far as Patagonia, Vespucci wrote, "We arrived at a new land which . . . we observed to be a continent." Inspired by Vespucci's claim, the group prepared two maps and a geographical treatise, *Cosmographiae Introductio*, incorporating this new information.

Waldseemüller's monumental map, measuring 4 by 7.5 feet, was printed in Strasbourg from 12 woodblocks in 1507. Towering above the map are engravings of Waldseemüller's mentors: Ptolemy and Vespucci flanking small-scale maps of the Old and New Worlds. The main map, centered on the Gulf of Arabia, is drawn in a heart-shaped form known as a pseudo-conical projection, first described by Ptolemy. Europe, the Middle East, and Asia are copied from Martellus's world map of 1489 or a prototype. Waldseemüller's Africa and New World are derived in part from secret Portuguese sea charts smuggled out of Lisbon. In the Caribbean, the Florida-like peninsula is similar to the one on the Ruysch map. It remains one of the great puzzles of the age of discovery. If it represents Florida, it predates the peninsula's official discovery by Juan Ponce de León by six years.

More significantly, Waldseemüller placed a distinctive landmass and two oceans between Europe and Asia—a radical step in mapping the world—and inserted the name "America" in the center of the South American continent. This name first appeared in the *Cosmographiae Introductio*: "I do not see why anyone would rightly forbid calling it (after the discoverer Americus . . .) 'Amerige,' that is, land of Americus, or 'America,'" wrote Waldseemüller.

Waldseemüller's map was mounted on library and countinghouse walls throughout Europe, providing Europeans with their first image of a new continent named America, yet it was almost lost to history. Of the thousand copies printed, none could be found in 1828 when Washington Irving wrote a book on the life of Columbus. German geographer Alexander von Humboldt later looked for it in vain. It was known only by a description in *Cosmographiae Introductio*. Then in 1901, Jesuit scholar Josef Fischer found a copy in Wolfegg Castle in Germany while researching Norse settlement. In 2003 the Library of Congress purchased the map from Count Johannes Waldburg-Wolfegg for ten million dollars.

DETAIL OF FIRST MAP CONTAINING NAME "AMERICA"

Maps of the Four Continents

MOST RENAISSANCE EUROPEANS FIRST VIEWED their world and new discoveries in America, Africa, and Asia through the eyes of the 16th-century German humanist Sebastian Münster and his major illustrated books, *Geographia Universalis* and *Cosmographia Universalis*. From 1529 until his death in 1552, Münster taught at the University of Basel, where he was best known as a Hebraic scholar, but his other passions were geography and cartography. At Heidelberg University, he had persuaded fellow students to map their hometowns using a quadrant and magnetic compass. In 1528 Münster published an appeal to "scholars and artists" for assistance in preparing a new atlas of Germany. "Let everyone lend a hand to complete a work in which shall be reflected . . . the entire land of Germany with all its peoples, its cities, its customs." He traveled through southern Germany conducting surveys and collecting data. His maps of central and eastern Europe made notable contributions, including the first accurate depiction of the source of the Danube River in the Black Forest.

Münster's revision of Ptolemy's *Geography* is considered the culmination of Ptolemaic revisions. Twenty-one new maps were added to the standard 27 Ptolemaic maps, some from personal surveys and others prepared by geographers from as far away as England and Greece. First printed at Basel in 1540, it was followed by three later printings.

While Münster's *Geographia* represents the end of an era devoted to classical geography in northern Europe, his *Cosmographia* introduced comprehensive world geography, a new genre designed to appeal to the general public as well as the scholar. Extremely popular, it went through 36 editions and revisions between 1544 and 1628. Enlarged with successive printings, the 1550 edition numbered 68 maps, 910 woodcuts of town views, topographical illustrations, and portraits of celebrities and animals drawn by well-known artists. More than 120 persons submitted maps and essays. It was published in Latin and German, with later editions in French, Italian, and Czech. The maps were printed as double- or single-page woodcuts, each followed by a brief descriptive text. Map titles, legends, and most place-names were printed from metal type inserted into the wooden blocks, a technique first used for woodcut maps in 1475.

Münster was the first to introduce the practice of displaying the four continents as separate maps. His map of Europe, oriented with south at the top, is the most accurate of the four, except for exaggerated depictions of England and Ireland. Africa and Asia differ little from Waldseemüller's great map of 1507, but the "Horn of Asia" has been reduced in size to a more recognizable Malay Peninsula, in deference to Magellan's circumnavigation. And confusion reigns with three places named India. Africa is inhabited by a cyclops, parrots, and an elephant.

Münster's map of the New World first appeared in his 1540 *Geographia*. It was a remarkable achievement, reflecting the latest information available. The Western Hemisphere is portrayed as a separate landmass for the first time on a printed map, and Waldseemüller's name for the region, America, is reinforced by its inclusion in South America.

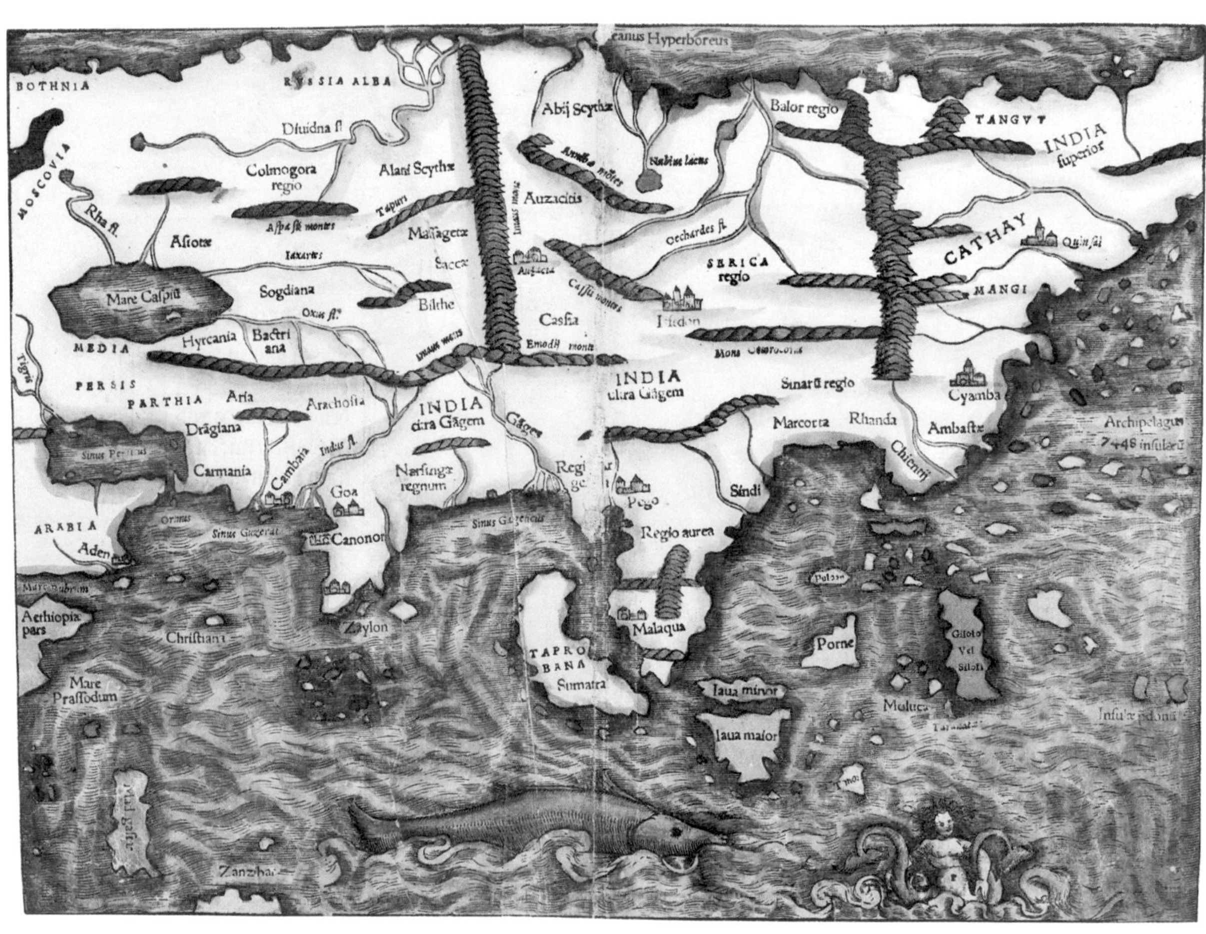

1550 Maps of the Four Continents by Sebastian Münster from Cosmographia universalis
Woodcut with Original Watercolor Wash, Each 10 x 13.5 in.
Library of Congress, Washington, D.C.

Hispania
Granat
Gades
Marc mediter. ancum
Sardinia
Sicilia
Malta
Candia
Cyprus
SYRIA
Libanus
Asia
Regnum Fes
Barbaria
Mauri
Hippa
AFRICA
Regnū Tunis
Marmarica
Ammon
AEGYPTUS
Syene
Cyrene
Alexādria
ARABIA deserta
ARABIA Petrea
ARABIA Felix
Mare Rubrū
Insule Canarie
Budezor
Nigritæ
Dara fl.
Garamantes
Lacus Libyæ
Hodeni
Regnum Senegæ
LIBYA Interior
LIBYAE desertum
Morland
NUBIAE Regnum
Nubia
REGNUM Orguene
REGNUM Habesth
Hamarich Sedes Pre te Iohan
REGNUM de Seylam
REGNUM Melinde
Monoculi
Caput Album
Caput Viride
Senega fl.
Gambræ
Calamansa regio
Regnum MELLI
Ginega
Imperiū Nebeorum
Sinus Rubrus
Vmb diß gros landt/das ein dritten
theil der weldt begreifft/ist zů vnsern zeit
ein schiffung erfunden/die auß Hispania
zů den Canarien inseln/vnd darnach fur
baß biß zům Caput bone spei / das ist ein
tröstlich schifflendung im aussersten spitz
Africe/ vnd von dannen biß gen Callicut
ghat / do her mann allerlei specerei vnnd
gewürtz bringt.
Tellus Psitacorum
QUIOLA Regnum
Azania regio
AFRICAE extremitas
Caput bone spei

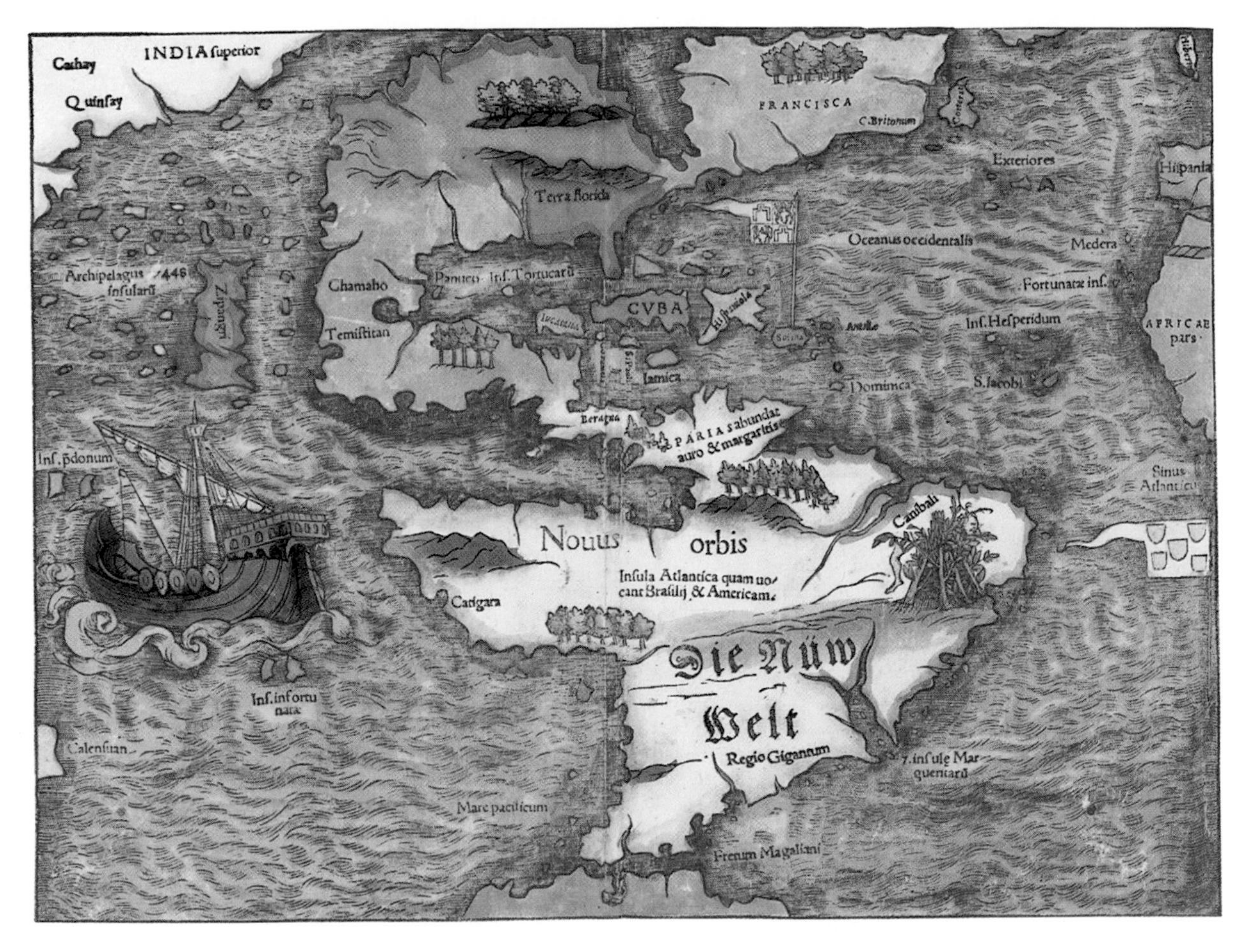
INDIA superior
Cathay
Quinsay
FRANCISCA
C. Britonum
Exteriores
Hispania
Terra florida
Oceanus occidentalis
Medera
Fortunatæ insu.
Archipelagus 7448 insularū
Zipangri
Chamaho
Temistitan
CUBA
Hispaniola
Inf. Hesperidum
AFRICAE pars
Iamica
Dominica
S. Iacobi
PARIAS abundat auro & margaritis
Inf. 3donum
Sinus Atlanticus
Canibali
Nouus orbis
Insula Atlantica quam uocant Brasilij & Americam.
Catigara
Die Nüw Welt
Regio Gigantum
Inf. infortunatæ
7. insule Marqueritard
Calensuan
Mare pacificum
Frenum Magaliani

"No One Has Ever Possessed Such a Map"

A FRAGMENT OF AN OTTOMAN WORLD MAP PROVIDES the only known view of the New World derived directly from a map prepared by Christopher Columbus. Muhyiddin Piri, known to history as Piri Reis, drew it in 1513. A native of Gallipoli, he began his career as a Barbary pirate at age 12, rising to the rank of admiral in the Ottoman Navy. He participated in numerous sea battles for Turkish supremacy of the eastern Mediterranean and the Arabian Sea during a naval career spanning seven decades, fighting "the enemies of our religion mercilessly." Piri Reis was also a keen observer of the seascape and a skilled cartographer who sketched and charted many of the Mediterranean's seaports, islands, and landmarks. He was beheaded at the age of 84 on orders from Süleyman the Magnificent after an unsuccessful campaign against the Portuguese.

During his long career, he compiled at least two world maps and a volume of sailing directions with maps and charts, entitled *Kitab-i bahriye* (*Book of Maritime Matters*). Only a small segment of his most famous work survives, but it reflects the region as envisioned by Columbus. "In this age, no one has ever possessed such a map," Piri Reis noted in one map legend. His world map was compiled from 20 Muslim and Western maps, with the Columbus segment copied from a map taken from a captured Spanish sailor, who claimed to have sailed with Columbus on three voyages to America.

Drawn in the style of a portolan chart with compass roses and rhumb lines, the surviving sheet depicts Central and South America along the left margins of the map and Spain and northwest Africa along the right. Two scale bars, graduated in leagues, are shown in the Atlantic Ocean off the coasts of Spain and Africa. Based on careful place-name and map analysis, some Piri Reis scholars have concluded that the top-left corner of this map was derived from a lost map made by Columbus after his second voyage (1493-96), when he circumnavigated Hispaniola. They suggest that the large rectangular island outlined in red depicts Hispaniola, which Columbus believed was Cipango (Japan), and that the peninsula jutting from the mainland northwest of Hispaniola represents Cuba, also called Mangi, a large peninsula on the Chinese mainland. Columbus believed until his death in 1506 that he had reached Asia.

Compiled during the golden age of Ottoman miniature painting, the map is embellished with vivid images. At the top, the mythical St. Brendan and his crew are depicted lighting a fire on a whale's back, mistaking it for an island. Atlantic sailing ships with square sails and sleek Mediterranean galleys with triangular sails called "lateen" are shown gliding through Atlantic waters. For the first time many animals associated with Central and South America are portrayed, including parrots perching on islands, monkeys, and perhaps a horned llama and a jaguar.

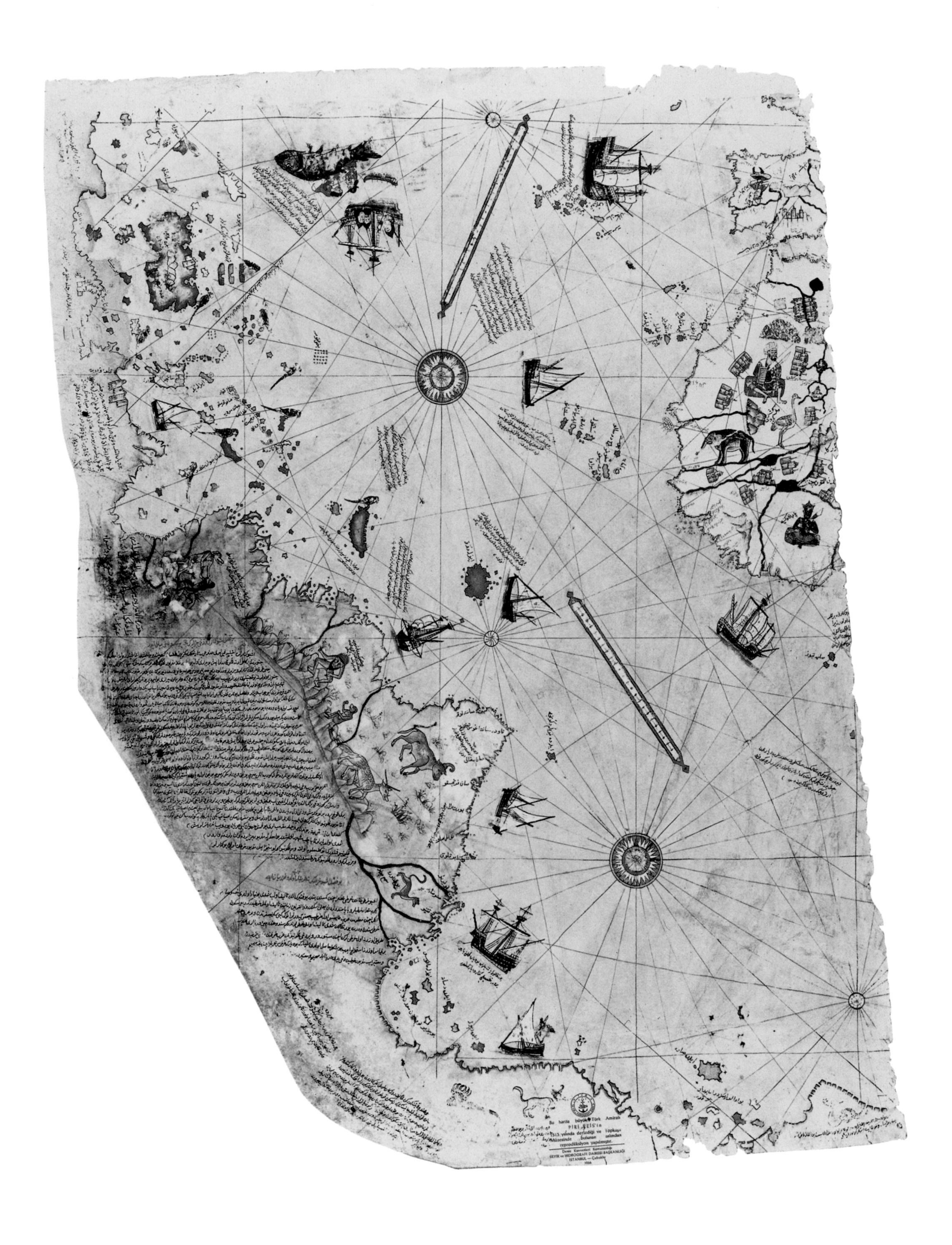

1513 Ottoman Chart of the New World by Piri Reis
Illuminated Manuscript on Deerskin or Gazelle Hide, 35.4 x 24.8 in.
Topkapi Palace Museum, Istanbul, Turkey

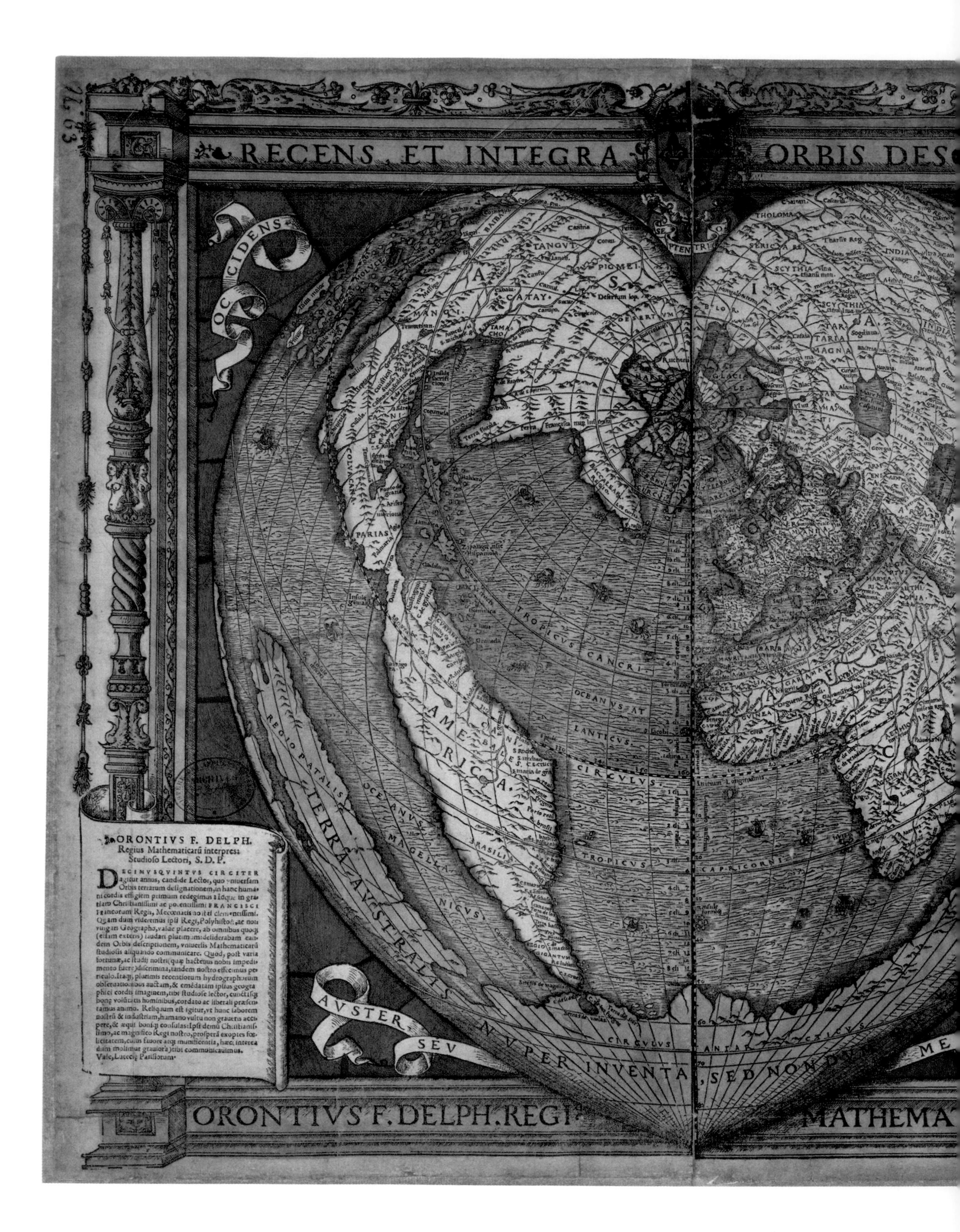

1534 Recens et Integra Orbis Descriptio (A New and Complete Description of the World) by Oronce Fine
Woodcut, Hand-Colored, 20.5 x 23.5 in.
Bibliothèque Nationale, Paris, France

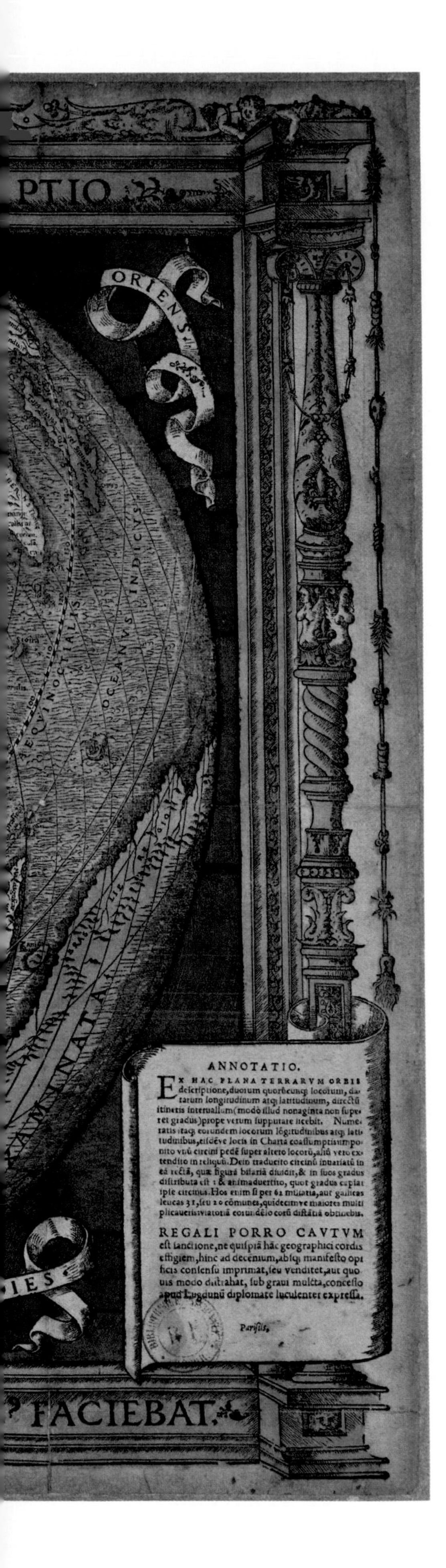

Cosmographic Heart

INTEREST IN MAP PROJECTIONS EXPANDED GREATLY during the Renaissance, spurred by advancements in mathematics and astronomy and an increased awareness of geographical discovery and exploration. The transformation of the surface of the Earth onto a flat or plane surface was the work of a few men who hoped to reduce map distortion or wanted to project a new type of information. Twenty-three map projections are known to have been in use during the Renaissance, including 14 invented between 1426 and 1600.

One of the most intriguing was the heart-shaped, or cordiform, projection, now named the Werner projection after its inventor, Johannes Werner. Sixteenth-century geographer Johannes Schöner called it the "cosmographic heart." Werner described three cordiform projections in a work he published in Nürnberg in 1514. "Vaguely aware of the fact that the bulk of the land mass of the earth was located north of the Equator," historian George Kish observed, Werner devised projections "to emphasize the northern hemisphere at the expense of the southern one."

Oronce Fine chose the cordiform projection for his 1519 map, which he dedicated to the French king. "I designed this map in the form of a human heart for the celebrated savant King Francis I," he noted in a map legend. Fine may have selected the projection because the "heart" motif reflected a theme in vogue in France at the time, a period that included the marriage of the king and his coronation. Fine was a Renaissance man in the broadest sense. For more than 40 years, he combined careers in mathematics, astronomy, astrology, mapmaking, book illustration, and engraving. He was even imprisoned for an extended period after compiling this map, apparently for unfavorable astrological forecasts concerning the king.

Although drawn in 1519, Fine's map was not published until 1534. It is framed with ornate columns and a portico with the title and French coat of arms. Above are four dolphins, Fine's signature device, which represented both the dauphin—the recently crowned Francis I—and Dauphiné, Fine's birthplace. The map reflects a merger of Ptolemaic traditions with recent voyages of discovery. Seemingly unaware of Waldseemüller's world map, Fine portrays North America as an extension of Asia—with Marco Polo's "Mangi" and "Cathay" immediately west of the Gulf of Mexico—but incorporates French explorations in North America and Spanish expeditions in the Pacific. The most dramatic feature is the portrayal of a southern continent named "Terra Australis," a term Fine introduced for the first time on a map three years earlier.

Aztec Cartographic History Map

WHEN SPANISH CONQUISTADOR HERNÁN CORTÉS fought his way across Mexico, he found a mapmaking society. Díaz del Castillo, a Cortés lieutenant, recorded that the Spaniards were guided by a "henequen cloth . . . on which all the pueblos we should pass on the way were marked." In addition to itinerary maps, the conquerors found Aztec maps of communities with extensive narrative histories and genealogies of local people and places; cadastral (or property line) plats; urban plans; trading and war maps; and celestial maps. The community history maps were unique to the Valley of Mexico and the Maya region. In her study of Mesoamerican maps, scholar Barbara Mundy located more than 80 surviving "cartographic histories." After the Spanish conquest and the depopulation of Mesoamerica by disease and migration in the early 16th century, production of indigenous maps was soon replaced by European-style maps introduced by the Spanish.

A fascinating example of a post-conquest Aztec community history map is preserved in the Library of Congress. One or more Aztec mapmakers drew the untitled map in about 1540 on *amatl*, a paper made from the inner bark of the fig tree. The map documents the land holdings and orchards of Don Carlos Chichimecatecotl, a Texcoco nobleman. The holdings were litigated following his public execution in 1539. The Texcoco nobility ruled a fertile region east of Lake Texcoco and the Aztec capital of Tenochtitlan (Mexico City). At the time of Spanish contact, members of the nobility were at war with Montezuma, the Aztec leader, and sided with the invaders to defeat their common enemy. Chichimecatecotl's family was part of this group. Two half-brothers campaigned with the conquistadores, one dying in the process. Cortés himself had helped raise Chichimecatecotl, taking the boy into his household.

The map portrays several estate maps and land plats, drawn at different scales. At the top center, for example, is an urban plan of part of Texcoco depicting the Oztoticpac Palace, which housed the ruler of Texcoco and his court. A description in the local language of Nahuatl by Zacarías Tlacocoua, a converted Aztec, provides a history of the Oztoticpac area and Chichimecatecotl's association with it. Aztec pictographs (bones, foot, fingers) and symbols (black dots) indicate the dimensions of each lot.

The large map at upper right, bounded by four stylized trees, depicts Chichimecatecotl's Oztoticpac estate, including plots given to the nobleman by Cortés. The drawings of small heads represent the renters or commoners who held the adjoining lots, identified by various hieroglyphs. A unique feature, and focus of the litigation, is an inventory of grafted trees from several orchards depicted at lower left by pictographs and hieroglyphs. Five stone fruits (apple, quince, pear, pomegranate, and peach) and grapes are shown, many grafted by Chichimecatecotl.

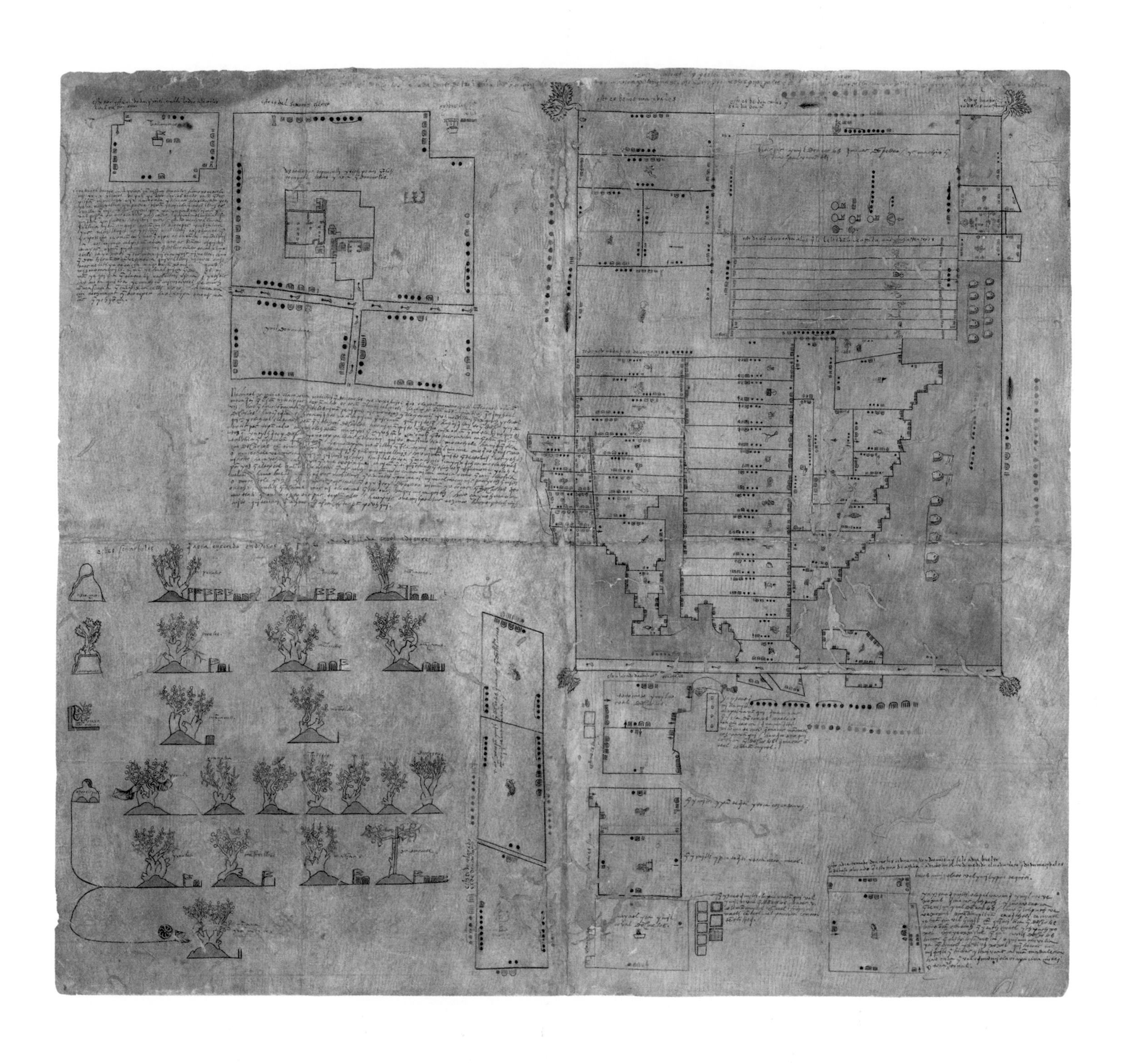

Circa 1540 Oztoticpac Land Litigation Map of a Royal Aztec Estate in the City of Texcoco by an Unidentified Aztec
Manuscript on Amatl Paper, 30 x 33 in.
Library of Congress, Washington, D.C.

1543 Nova ac Generalis Orbis Descriptio (Terrestrial Globe with Armillary Sphere) by Caspar Vopell
Brass, with Manuscript Gores on Wooden Ball, 5.9 in. diameter, Sphere 3 in. diameter
Library of Congress, Washington, D.C.

Ptolemy's Miniature Universe

PTOLEMY'S LEGACY TO RENAISSANCE CARTOgraphers included globes and armillary spheres as well as maps. A globe is a spherical structure that can represent the Earth (terrestrial globe) or the heavens (celestial globe). Ancient Greek geographers and astronomers constructed both types, as did their Chinese counterparts. A celestial marble globe dating from the second century can be viewed in the Museo Archeologico Nazionale in Naples, Italy, where Atlas, of Greek myth, carries it upon his shoulders in the sculpture known as the "Farnese Atlas." An armillary sphere is a scientific device used to demonstrate movement of celestial bodies around Earth or the sun. Consisting of a small globe within a series of movable, metal rings representing celestial orbits, the armillary sphere also appeared early in Greece and China.

Arab scientists continued the art of globe and armillary sphere production when it lost favor in Europe in the Middle Ages. The Istituto di Studi Superiori at Florence, Italy, has a celestial globe constructed in 1085 by Ibrahim Ibn Said-as-Sahli, a Spanish Moor.

Globes and armillary spheres found new European audiences beginning in the late 15th century, when interest in terrestrial globes was sparked by reports of new lands and places. Martin Behaim, a Nürnberg geographer, made the oldest surviving terrestrial globe in 1492; it is now held by the National Museum of German Art and Culture in Nürnberg. Behaim's *Erdapfel* (earth apple) is also significant because its portrayal of Earth reflects Columbus's worldview. It is similar to Henricus Martellus's world map of 1489.

Caspar Vopell, a teacher of mathematics in Cologne, Germany, constructed one of the earliest surviving examples of a terrestrial globe with an armillary sphere. Six are known to exist. In addition, Vopell prepared several globes and maps, including a world map in 12 sheets. He died in 1561 at the age of 50. Vopell's map was hand-drawn on the globe ball in black ink with place-names in red. While South America is portrayed correctly, the mapmaker showed North America as an extension of Asia, but he remained skeptical of this configuration. When Vopell compiled his world map two years later, he sought an audience with Emperor Charles V, who confirmed to him that no sea separated the two. "The Spaniards had again and again gone out West," explained the Spanish ruler, "exploring from Themixtitan, the famous city of the Great Khan, otherwise called Montezuma, and had not been able to find an end of the huge landmass."

Vopell's 1543 armillary sphere, illustrating the Ptolemaic Earth-centered cosmos, was constructed in the same year that Nicolaus Copernicus published his revolutionary treatise transforming the astronomical system by placing the sun at the center. Ptolemaic armillary spheres contiued to be made after the introduction of the heliocentric Copernican system, but the two were displayed together to illustrate differences between Earth- and sun-centered universes.

Verrazano's Sea

SPANISH DISCOVERIES ALONG THE CALIFORNIA COAST were first transmitted to European readers through the manuscript maps and charts of Battista Agnese. A Venetian chartmaker of Genoese birth, Agnese was one of the most prolific draftsmen of his era, compiling at least a hundred manuscript portolan atlases for wealthy patrons from 1536 to 1564. More than 70 survive in private and public libraries. Each atlas typically has a world map drawn on an oval projection and 12 to 14 regional charts in the portolan style. His maps were artfully inked on vellum and brightly painted to compete with published versions of Ptolemy's *Geography* and island books.

Agnese's world map was his tour de force, combining Spanish and Portuguese discoveries with Ptolemaic elements. Several distinctive features found on his world maps after 1539 set them apart. First, they are decorated with colorful cherubs, or wind heads, representing compass directions. Next, they display two historic sea routes. The longest, traced in tarnished silver, is the spice-trading route from Lisbon to the Moluccas through the Strait of Magellan and return via the Cape of Good Hope, a course that followed Ferdinand Magellan's voyage, 1519-21. The shorter route tracks the Spanish treasure galleons from Cádiz to Peru via the Isthmus of Panama. It is inscribed in pure gold.

The strange outline of North America, shaped with a narrow neck at center, can be traced to a voyage led by Florentine navigator Giovanni da Verrazano. Commissioned by the French king in 1524 to find a northwest passage to China, Verrazano mistook Pamlico Sound and the Outer Banks of North Carolina for a strait dividing the Atlantic and Pacific. A map by his brother perpetuated the idea of the "Verrazano Sea." Dots off the east coast, painted green, suggest islands, shoals, or a misplaced Great Bahamas Bank.

Farther west, Agnese portrays for the first time the Baja Peninsula and Gulf of California, based on discoveries of Francisco de Ulloa in 1539-40. The gulf is red to reflect the appearance of the silt-laden waters from the Colorado River. His name for the Gulf was the "Vermillion Sea," to distinguish it from the Red Sea, also red on his map. The blue circle with red buildings represent Tenochtitlan, the Aztec capital, already becoming the Spanish seat of power in the New World.

The interior of South America is more detailed. It depicts six towns, including Tombez, the Inca stronghold visited by Pizarro in 1527, and the Amazon River and Rio de la Plata.

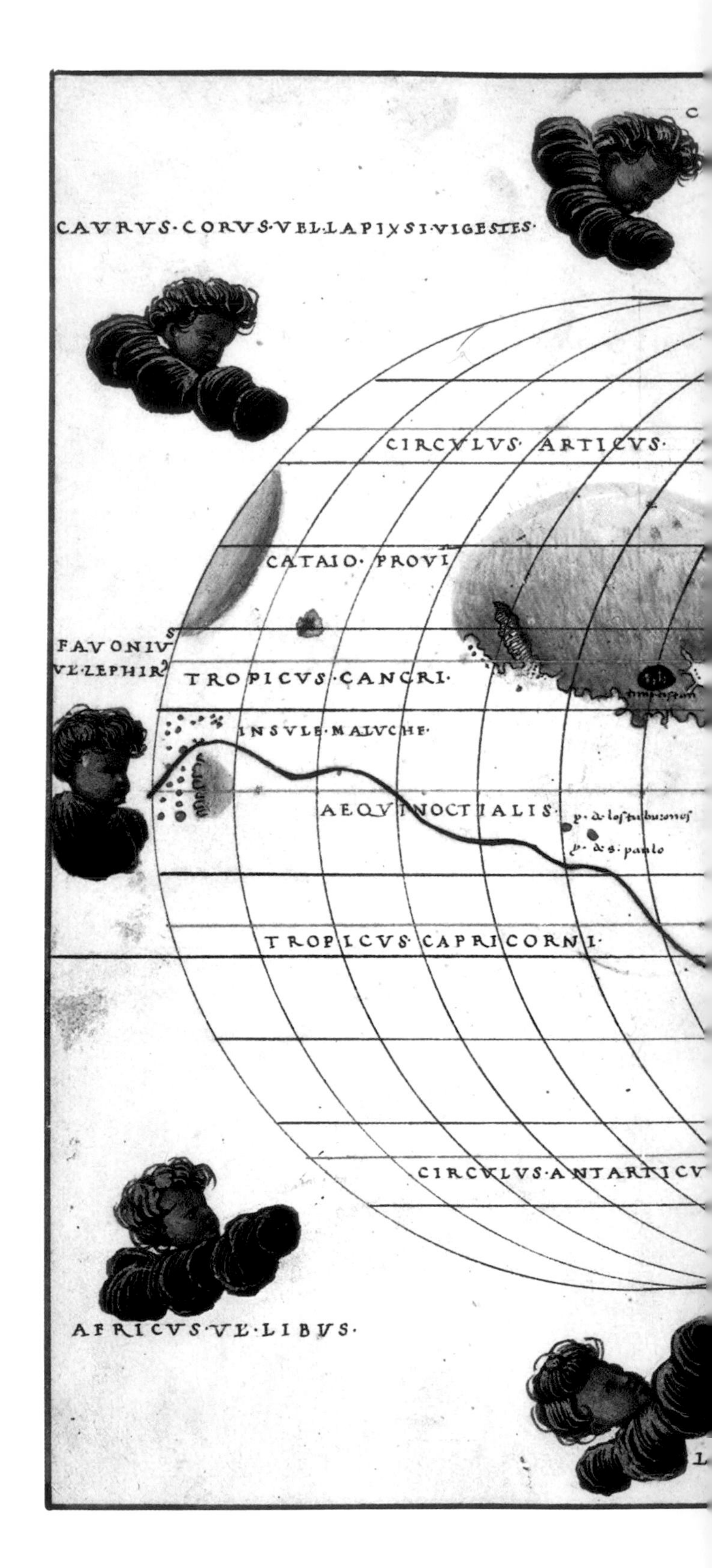

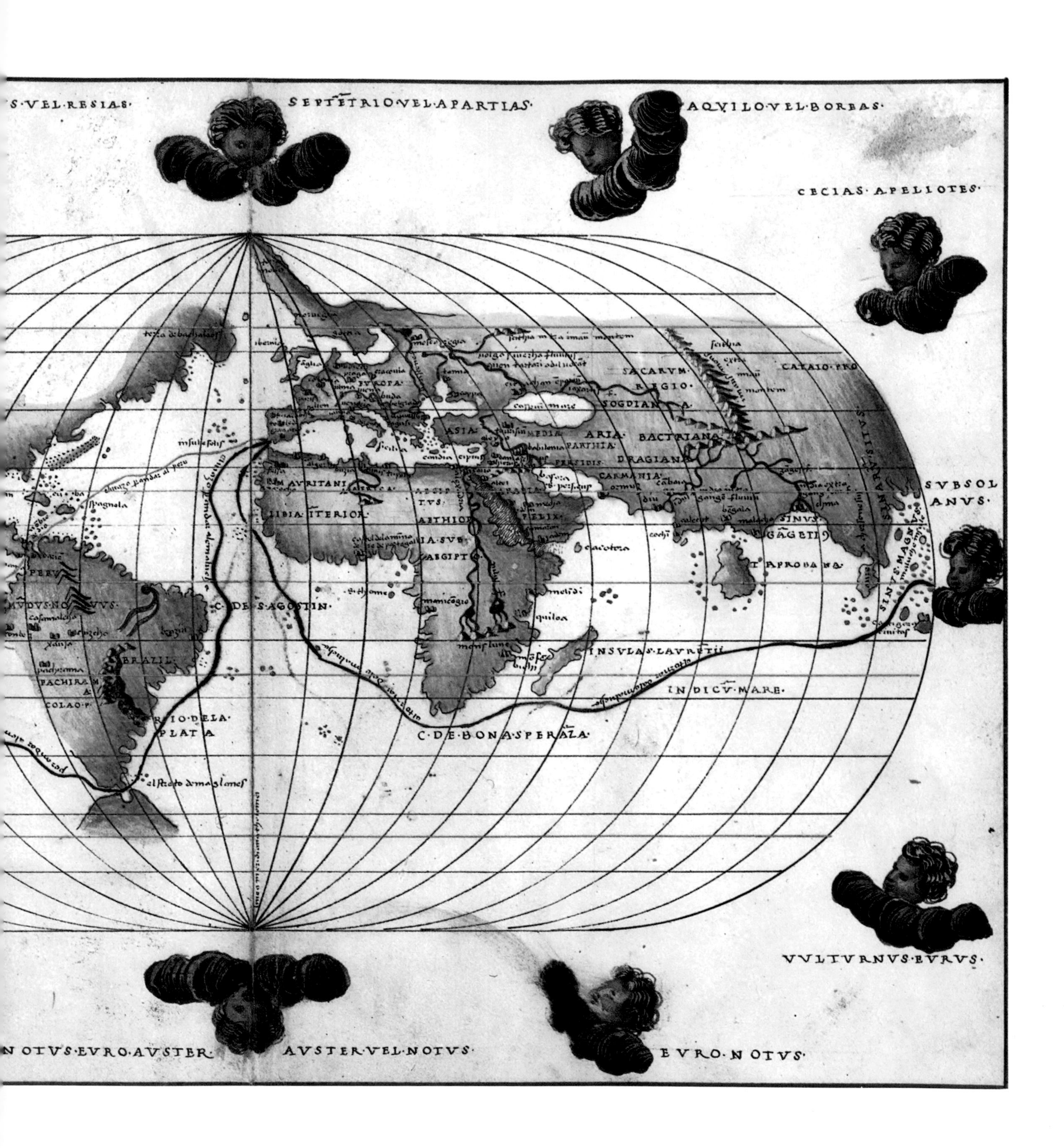

Circa 1544 World Map by Battista Agnese
Illuminated Manuscript on Vellum, 10.2 x 7 in.
Library of Congress, Washington, D.C.

1547 Portolan Chart of East Coast of North America in the Style of the Dieppe School
Illuminated Manuscript on Vellum, 15.7 x 22.8 in.
Huntington Library, San Marino, California

The Dieppe School

A UNIQUE STYLE OF PORTOLAN CHART, CELEBRATED as the Dieppe school, flourished briefly in mid-16th-century France. Dieppe, a seaport and shipbuilding center on the English Channel, provided an ideal environment for chartmakers and artists to collaborate in the production of charts embellished with topographic views and contemporary scenes. Dieppe charts have several characteristics that set them apart from other portolans. These include an orientation toward the Equator; two different latitude scales to compensate for compass variations; and elaborate vignettes filling interior regions. The geography of the coastlines is Portuguese in character, including many place-names.

Most surviving portolan charts in the Dieppe style are unsigned and undated, but three Dieppe chartmakers are recognized: Jean Rotz, the son of a Scottish nobleman, whose atlas dedicated to King Henry VIII is now in the British Library; Pierre Desceliers, a Catholic priest, who was the first to teach nautical science in Dieppe; and Guillaume Le Testu, killed in 1572 while attacking a Spanish treasure ship with his friend Sir Francis Drake.

A fourth person associated with the Dieppe school is Nicolas Vallard, whose name appears on the title page of an illuminated atlas of 15 charts in the Huntington Library in San Marino, California. Historians disagree whether Vallard was the author of the atlas or its owner. No chartmaker by that name is known. Nevertheless, the Vallard map, drawn in typical Dieppe portolan style, is a major work documenting Jacques Cartier's early Canada explorations for France, three voyages from 1434 to 1442.

The map portrays eastern North America, from Labrador to Florida (left to upper right), but its primary focus is the area of the St. Lawrence River, identified as Rio de Canada. Cartier explored the St. Lawrence upstream to present-day Montreal, opening up the interior of North America. The island highlighted with gold leaf at the mouth of the Gulf of St. Lawrence is Newfoundland. The painting is one of the earliest images of French and Indian contact in North America. The company of French gentry and soldiers shown was led by Jean-François de la Rogue, Sieur de Roberval, commander of the third Cartier expedition. The bearded colonist holding a spear and pointing to the Hurons, dressed in furs, may represent Roberval. After a winter near present-day Quebec City, the survivors returned to France, finding it "impossible to trade with the people of that country because of their austerity, the intemperate climate of said country, and the slight profit."

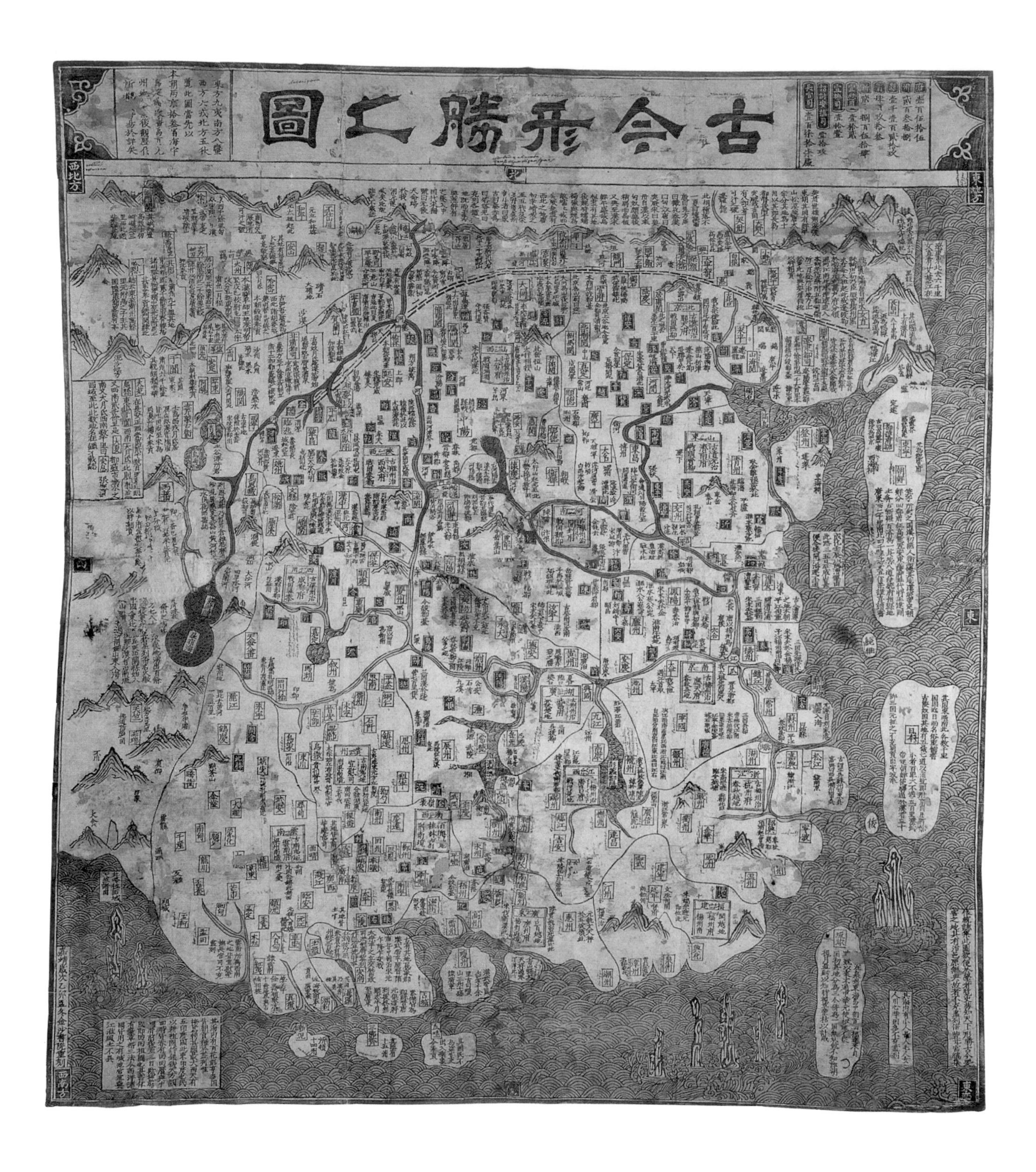

1555 Map of China Entitled Gujin Xingsheng Zhi Tu (Map of Advantageous Terrain Past and Present)
Woodblock, Hand-Colored, 45 x 39.4 in.
Archivo General de Indias, Seville, Spain

Ming Dynasty Map of Asia

One of the treasures sent to Seville by the Spanish viceroy in the Philippines in 1874 was a Ming dynasty map of China entitled *Gujin Xingsheng Zhi Tu* (*Map of Advantageous Terrain Past and Present*). A large hand-colored woodcut, it portrays China and Central Asia as well as border regions, including Mongolia, Korea, and the islands of Japan, Java, and Sumatra.

Map use in early Chinese culture was limited to a relatively small number of scholar-officials and imperial officers who aided the ruling class, but they used maps for a variety of purposes. These included determining geographical distances and space; projecting political power; teaching and historical scholarship; and aesthetic appreciation. The *Gujin Xingsheng Zhi Tu*, for example, served an educational function, describing changes in place-names and administrative boundaries. It provides a kind of historical cartography of the Ming dynasty through its narrative text.

The map is generalized and abstract rather than drawn to a precise scale. China's two major river systems—the Huang (Yellow) and Chang (Yangtze), enhanced with yellow and blue coloring—provide a geographic frame of reference. A major focal point is the red double disk in the left center, representing the source of the Yellow River. The historic Great Wall of China is another principal feature, depicted by a double line of battlements curving across the top of the map. This fortified installation, guarding China's northern approaches for some 1,500 miles, reached its modern form during the Ming and is found on most subsequent maps of China. Single black lines mark provincial and prefecture boundaries. Mountains are represented by pictorial symbols, although a contemporary general map of the Chinese Empire, drawn by Luo Hongxian, introduced abstract symbols for mountains as well as for roads and political units. The depiction of mountains surrounding China may reflect historic, geographic, and political ideas promoted by Yi Xing, a monk and astronomer active in the seventh century. Yi Xing reasoned that mountain systems, reinforced by the Yellow and Yangtze Rivers, formed natural lines of defense against foreign aggressors.

Unique to this map is the portrayal in the China Sea of Korea, Japan, and Taiwan as narrow, elongated, featureless landmasses. Japan is not only misshaped as a single island but misplaced south of Korea. Perhaps this is another instance of limited page size determining the shape and placement of map features.

Finally, the map is supplemented with text, an ancient Chinese tradition. "On a map, one cannot completely draw prefectures, armies, mountains, and rivers," Jia Dan wrote in the eighth century. "For reliability, one must depend on notes." Notes cover the map, providing details such as distances and directions generally found graphically on Western maps. Chinese map narrative provided a "textual scholarship" that included reconstructing topographic and cultural details using geographical and historical information.

Globe Gores

THE SURFACE OF A GLOBE IS COMPOSED OF MAP segments called globe gores, originally printed as triangular or elongated areas with tapered points. The earliest globe balls were generally small in size, constructed of marble, metal, or wood, with etched or painted surfaces. In 1492 the Nürnberg mapmaker Martin Behaim introduced the idea of fashioning the ball from a mold using wood strips, plaster, and fiber, a kind of papier-mâché. Printed globe gores soon followed. One of the earliest was a 1507 woodcut by Martin Waldseemüller. Even the artist Albrecht Dürer got involved, publishing rules for the preparation of globe gores in 1525. A more formal, illustrated guide appeared two years later in a treatise published in Basel by Henricus Glareanus. He proposed using 12 globe segments, each representing 30 degrees of longitude and extending from Pole to Pole.

Antonio Florian's world map of 1555 portrays the Northern and Southern Hemispheres, each subdivided into 36 gores of 10 degrees. It apparently was designed to be used either as a plane map or as the surface cover for a globe. Elements of the plane map are found in the artistic legend and portrait boxes that frame it. Florian's self-portrait, upper right, is one of the earliest of its kind. A portrait of Ptolemy, upper left, symbolically affirms Florian, an artist and architect who actually lacked geographic training. For globe use, each segment was cut from the map, moistened, stretched, and pasted over the sphere. Unlike standard gores described by Glareanus, Florian's map segments extend from Equator to Pole. The sphere covered would have had a diameter of 10 inches.

Florian copied the geographical content of his map from an earlier work by Roman mapmaker Antonio Salamanca, who had reproduced, without acknowledgment, a double-heart-shaped world map published by the Dutch cartographer Gerardus Mercator in 1538. This chain of events, not uncommon in the map-publishing world of the 16th century, occurred when Mercator failed to obtain a privilege for printing his map in Italy. Florian probably was unaware of Mercator's work. The Mercator-Salamanca-Florian polar view provides a unique image of the world. Mercator added the name "America" to the northern continent for the first time, which was continued by later mapmakers, including Florian. Though Florian basically plagiarized his map, he has "the merit," according to 19th-century geographer M. Fiorini, "of having made a world map in gores rather different than that of [Mercator] and [of having] tried a new manner for projecting the globe."

1555 Double-Hemisphere World Map Viewed From the Poles by Antonio Florian
Copperplate Engraving, 18 x 33 in.
Library of Congress, Washington, D.C.

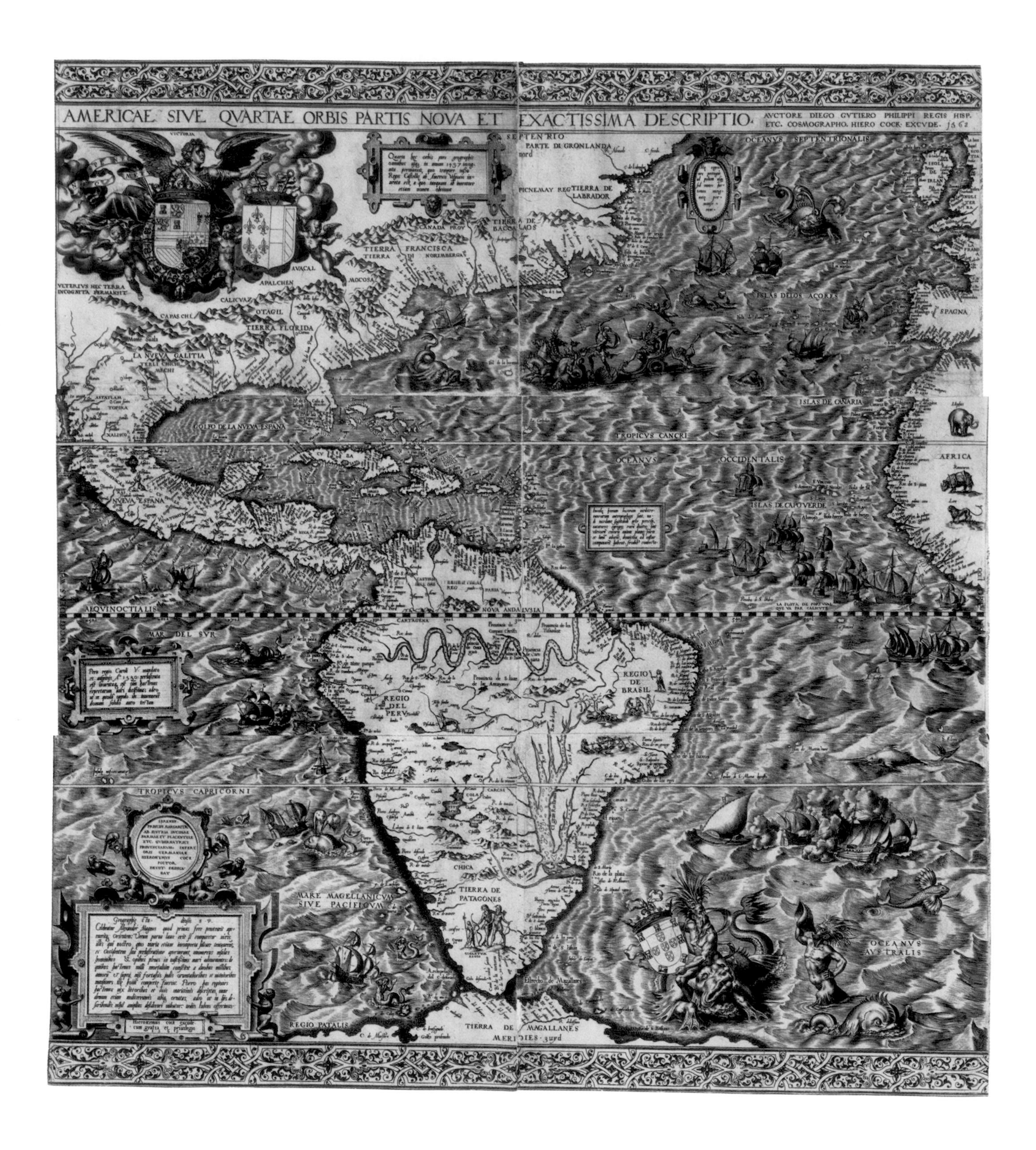

1562 Americae sive Quartae Orbis Partis Nova et Exactissima Descriptio (Map of the Americas) by Diego Gutiérrez
Copperplate Engraving, 36.6 x 33.8 in.
Library of Congress, Washington, D.C.

Official Spanish Map of the Americas

FIFTY-FIVE YEARS AFTER WALDSEEMÜLLER'S WORLD map was printed with its revolutionary view of the New World, a map of equal charm and originality appeared. Drawn by Diego Gutiérrez, it was engraved on six copperplates in the Antwerp workshop of Hieronymous Cock, a Flemish artist who popularized Dutch masters through his engravings. The largest, most detailed map of the New World published in the 16th century, it is the only Spanish wall map of the Americas printed before the late 18th century.

Both maps were derived from the master maps of discovery prepared and maintained by the hydrographic offices of Portugal and Spain. Known as the *carta padrão del el-Rei* (king's standard chart) in Portugal and *padrón real* (royal pattern) in Spain, these official charts depicted all coastlines explored and charted by dead reckoning and star sightings. Ships' masters and pilots were required to carry and update copies of master charts on their voyages. Their changes were registered with the *piloto-mayor* (chief navigator) upon their return, and incorporated into the master maps by skilled chartmakers. The official charts were carefully guarded and their reproduction a capital crime, but a few copies were somehow smuggled out of Lisbon and Seville. Italian diplomat Alberto Cantino in 1502 obtained an unauthorized copy of the Portuguese chart—now in the Biblioteca Estense in Modena, Italy. Waldseemüller had access to a Portuguese copy.

Similarly, Gutiérrez's map of the Americas was derived from the Spanish chart, but its publication may have been officially sanctioned, as suggested by the two coats of arms under the cartouche, or ornamental frame. One represents Phillip II, King of Spain (left), and the other his half-sister, Princess Margarite, the Spanish regent of the Netherlands, to whom the map is dedicated. Portugal's sphere of influence is suggested by its coat of arms, displayed in the lower right, east of the hypothetical papal line of demarcation that divided the world between Portugal and Spain.

Gutiérrez was a licensed chart and instrument maker for the Casa de la Contratación, Spain's hydrographic office, holding the position of cosmographer from 1534 until his death in 1554. His son, also named Diego, who succeeded him, may have ensured the publication of the map. Gutiérrez's map reflects the latest Spanish knowledge of the Americas. His portrayal of North America includes names of Indian tribes and topographic features divided by region—Greenland, Labrador, Baccalaos (named by Breton fisherman for the cod found off Newfoundland), Francisca, Norembega (Penobscot River drainage basin in Maine), Florida, and Nueva Galita. The name "C[ape] California" is the earliest map reference to California. A legend credits the discovery of America to Amerigo Vespucci in 1497, earlier than Vespucci claimed.

Only two complete copies of Gutiérrez's map survive—in the Library of Congress and British Library.

Chinese Scroll Map of Coastal Waters

PRIOR TO THE EUROPEAN AGE OF DISCOVERY, THE Chinese were considered the greatest mariners. Their ships far surpassed those plying the Mediterranean Sea and the Indian Ocean. Chinese sailors invented the rudder, watertight compartments, and the fore-and-aft rig and lug sails that allowed them to sail into the wind. Experienced coastal and ocean navigators in the time of Ptolemy, they were trading throughout Southeast Asia by the 13th century. Chinese naval power reached its pinnacle in the early Ming dynasty, when the Muslim Admiral Zheng He led seven maritime expeditions to Indonesia, the Arabian Peninsula, East Africa, and perhaps as far south as the Kerguelen Islands, near Antarctica. Some believe that his imperial fleet even reached America in 1421!

Chinese navigational charts undoubtedly mirrored nautical developments, but only a few surviving examples predate the introduction of European concepts of mathematical cartography into China by Jesuit mapmakers in the late 16th century. A coastal/ocean chart believed to have been associated with Zheng He's southern voyages has been preserved in published form. It was compiled in about 1422 by a member of Zheng He's staff, and later published in a treatise on military preparations. The original chart was in the form of a scroll map, about 8.1 by 220 inches. In design, it was similar to the scroll map displayed on the opposite page, but with the addition of Zheng He's sea route from Nanjing to East Africa, indicated by a dotted line.

More common are the hydrographic maps and charts associated with China's inland waterways, coastal areas, offshore islands, and river estuaries. The map reproduced here, only a portion of which is shown, extends from Hainan Island off the coast of South China (to the left) north to Shandong Peninsula. Signed by the commander of the Ling Shan military district, it was first drawn in about 1563 to aid officials in curbing piracy and smuggling. Recommendations for improving security are provided in legends. The 1705 copy shown here is formatted on folding boards, measuring 11.8 by 107.8 inches.

The map's geographic and topographic details are expressed in traditional Chinese style, combining images and text. Scale and orientation vary from one part to another, but north generally is on the left. Four symbols aid the reader: Circles enclose names of military districts; rectangles, names of departments; and squares, names of cities and prefectures; flagpoles indicate military garrisons. Mountains and islands are rendered as miniature landscape paintings.

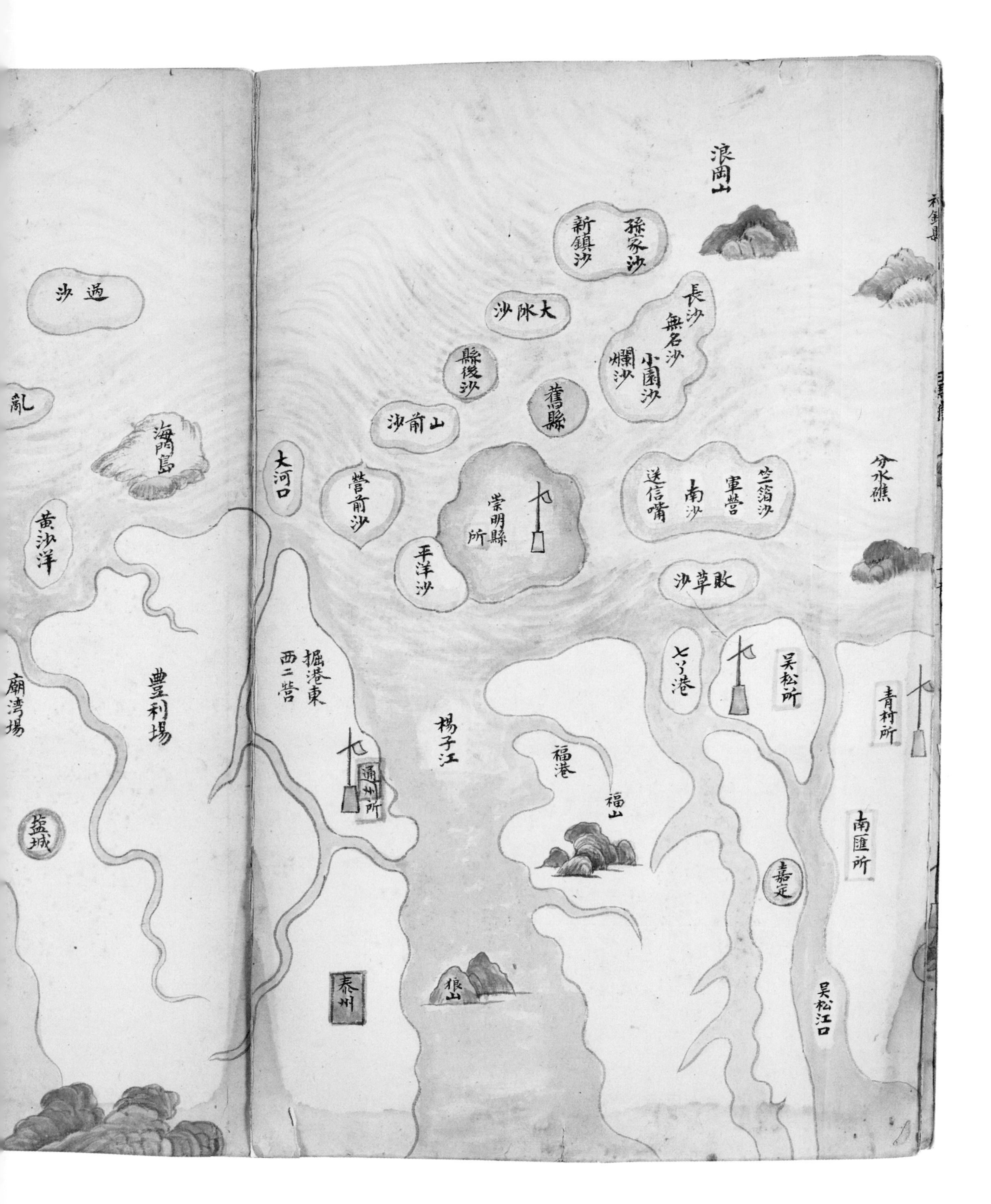

Circa 1563 Segment of Wanli Hai Fang Tu (Map of Maritime Defenses Along 3,500 Miles)
Probably Drawn by Zheng Ruozeng, Copied in 1705
Manuscript Scroll Map on Boards with Watercolor Wash, 11.8 x 107.8 in.
Library of Congress, Washington, D.C.

SECTION THREE

Maps for Royalty, Nobility, Clergy, and Merchant Princes

Map Shop in Paris, 1692
Frontispiece in Sanson's Atlas Nouveau

INTRODUCTION

SIX THEMES CHARACTERIZED WORLD MAPPING DURING THE 200-YEAR PERIOD FROM THE LATE 16TH TO THE late 18th century: a dramatic rise in the awareness and use of maps worldwide by court officials, soldiers, clergy, scientists, and an emerging middle class of merchants and tradesmen; the charting of the seas primarily by the Dutch and British rather than the Portuguese and Spanish; the exchange of geographic and cartographic information between Asia and Europe; the use of maps as tools of commerce and government; the role of science in European mapmaking; and the relationship between the visual arts and cartography.

Increased map awareness in Europe coincided with the shift of the center of mapmaking from Venice and Florence to Antwerp and Amsterdam. This began with the appearance of Gerardus Mercator's great world map of 1569, represented by Edward Wright's famous 1599 engraving, and Abraham Ortelius's atlas of 1570. The golden age of Flemish and Dutch cartography followed, culminating with Joan Blaeu's great 12-volume atlas in 1667. It was a period of cartographic innovation and achievement, conducted during a turbulent time of political unrest and religious fervor at the height of the Reformation. For the first time, professional mapmakers, engravers, and publishers prepared the majority of maps. Ortelius created the first printed atlas in 1570; Georg Braun and Frans Hogenberg produced the first atlas of city and town plans two years later; and, a decade later, Lucas Janszoon Waghenaer printed the first atlas of sea charts.

THIS GOLDEN AGE WAS SUPPORTED IN PART BY THE PRIVATELY OWNED DUTCH EAST AND WEST INDIA Companies, which came to control much of the world's trade by replacing the Portuguese as explorers and settlers in the Pacific and the Spanish in the Atlantic. Following the defeat of the Spanish, British privateers such as Francis Drake challenged both the Dutch and the Spanish. Drake's circumnavigation of the world is portrayed in Jodocus Hondius's 1595-96 world map. Dutch, Spanish, and British maritime explorers essentially completed the mapping of continental coastlines, with the exception of Australia. One of the more interesting features of this endeavor was the emergence of California as an island, shown by both Philip Eckebrecht in 1658 and Joan Blaeu. Inland exploratory mapping is represented by Guillaume Delisle's 1703 map of Canada and Ivan Kirilov's 1734 map of imperial Russia.

Asian mapmakers continued their ancient cartographic traditions, illustrated by a 1647 map of the Ming Empire, a 1700 Korean world map, and a 1736 Japanese scroll map, but for the first time the distinct European and Asian mapping cultures began to share geographical information. A world map published by a Japanese mapmaker in Nagasaki in 1645 revealed the world from a European perspective while Martino Martini's map of imperial China, published in Amsterdam in 1655, portrays the Middle Kingdom from a Chinese point of view.

Maps also began to be used as tools of empire, both in a commercial and a political sense. The Dutch East and West India Companies employed or used cartographers, including Blaeu and Martini, to promote their interests. John Mitchell's 1755 map of North America was commissioned by the British Board of Trade to document British territorial claims against France on the eve of the French and Indian Wars. Kirilov's map was prepared by direction of Tsar Peter the Great as part of his effort to modernize and control his vast domain.

The European scientific revolution that reshaped man's view of the universe and his place in it was reflected in new types of maps as well as new surveying and mapping instruments. Mercator's navigational chart, improved by mathematician Edward Wright, revolutionized maritime navigation. The invention of the telescope led to Eckebrecht's world map, commissioned by astronomer Johannes Kepler in 1658, and the first lunar charts, represented by Johann Gabriel Doppelmayr's map of 1708.

CONTRIBUTING TO THE EXPANDING INTEREST IN MAPS WAS THEIR CLOSE ASSOCIATION WITH THE VISUAL arts. Both the Dutch and the Japanese, for example, used maps as decorative wall hangings. In the West, images of people, places, and things continued to be expressed on maps by ornate drawings, but these illustrations were now moved from blank areas within the map to the map borders and cartouches. These were often further enhanced by exquisite copperplate engravings and dramatic colors. Typical is Ortelius's world map of 1570, embellished with interlaced strapwork. More elaborate is the baroque-style title and dedication cartouches found in Johann Baptist Homann's 1712 map of Salzburg. A new type of world map was designed specifically to provide more space for artistic embellishment. Rather than displaying the world as one image spread across a map, the world was divided into two circular hemispheres. Two exemplars of this device are world maps by Jodocus Hondius, dated 1596, and Joan Blaeu, from 1667.

Traditional Dutch landscape art is reflected in the topographical views of Cusco and Mexico City by Braun and Hogenberg, and the 1586 sea chart of the English coast by Waghenaer. Similarly, the Japanese style of art called *ukiyoe* is expressed in an anonymous scroll map of the land and sea route from Tokyo to Nagasaki, painted in 1736.

During this period, the ubiquitous mythical figures and cherubs of the early Renaissance begin to give way to drawings of real people in their native cultural attire (actual and imagined): South American Indians appear in Braun and Hogenberg; Hollanders in Pieter van den Keere's 1617 patriotic map of the Netherlands, which was drawn in the form of a lion, the symbol of Dutch resistance; Chinese mandarins in Martini; Jesuit missionaries and Canadian Indians, one boldly displaying a scalp, in Delisle; Russian merchants in Kirilov; and biblical figures in the 1695 Haggadah map by Avraham bar Yaacov. Most intriguing is the 1645 Japanese scroll map, which is accompanied by a woodblock print depicting 40 pairs of people from around the world.

Portraits of noted people adorn three maps, including Queen Elizabeth I (Hondius, 1595-96), the scientists Galileo Galilei and Tycho Brahe (Blaeu, 1667), and the Archbishop of Salzburg (Homann, 1712).

English Surveying Manual

AARON RATHBORNE'S *THE SURVEYOR IN FOURE BOOKES*, a surveyor's manual published in London in 1616, is but one of many such works that appeared in Europe in the 17th and 18th centuries. These describe and illustrate new instruments that were transforming surveying and mapmaking from a reliance on secondary sources to the preparation of original field surveys based on geodetic and astronomical observations. Two allegorical figures gracing Rathborne's frontispiece—Arithmetica and Geometria—symbolize the scientific foundations of these developments.

The telescope, critical to the precise measurement of distant objects, was invented in Holland about 1608 and first used for scientific purposes by Galileo Galilei a year later, when he sketched a lunar map. Its value as a surveying instrument increased with the development of the ocular lens by German astronomer Johannes Kepler, and the addition of crosshairs to this eyepiece by William Gascoigne in 1639. Thirty years later, French cartographer Jean Picard attached the telescope to a quadrant for field surveying.

Two instruments essential to field surveying are illustrated in this frontispiece: the plane table (bottom center), a portable drawing board mounted on a tripod with an alidade, or ruling bar with collapsible sights, for plotting topographical details; and a theodolite (top center), an instrument for measuring horizontal and vertical angles, created by English surveyor Leonard Digges in 1555. Other necessary tools included the spirit level, for determining elevation differences; the surveyor's chain, devised for accurate measurement of distances; and the gradient vernier scale for precision measurements.

Equally important was the development of geodetic triangulation for determining distances and directions. In triangulation, the surveyor—using a quadrant or a theodolite with telescope, a plane table, and the principles of trigonometry—develops an interconnecting network of unlimited triangles that can stretch across a region from a single baseline. Once a triangulation network is established, the location of any feature located within this framework can be precisely determined. First proposed by Dutch physician Gemma Frisius, the theory was tested in 1615 by Willebrord Snel, a mathematics professor at the University of Leiden, who used triangulation in an attempt to determine the true shape and circumference of Earth and the length of a degree of arc—the first steps in creating an accurate map of the world.

Triangulation was perfected in France during the reign of Louis XIV. Geographers and mapmakers carried out a series of triangulations in France, Peru, and Lapland from 1679 to 1743. The research proved that Earth is an oblate spheroid and that a degree of latitude increases as one moves away from the Equator. They also conducted a geodetic survey of France, a 65-year effort completed in 1744, which placed France at the forefront of scientific cartography.

1616 Title Page From Surveyor's Manual (The Surveyor in Foure Bookes) by Aaron Rathborne
Drawings of Theodolite and Plane Table
Library of Congress, Washington, D.C.

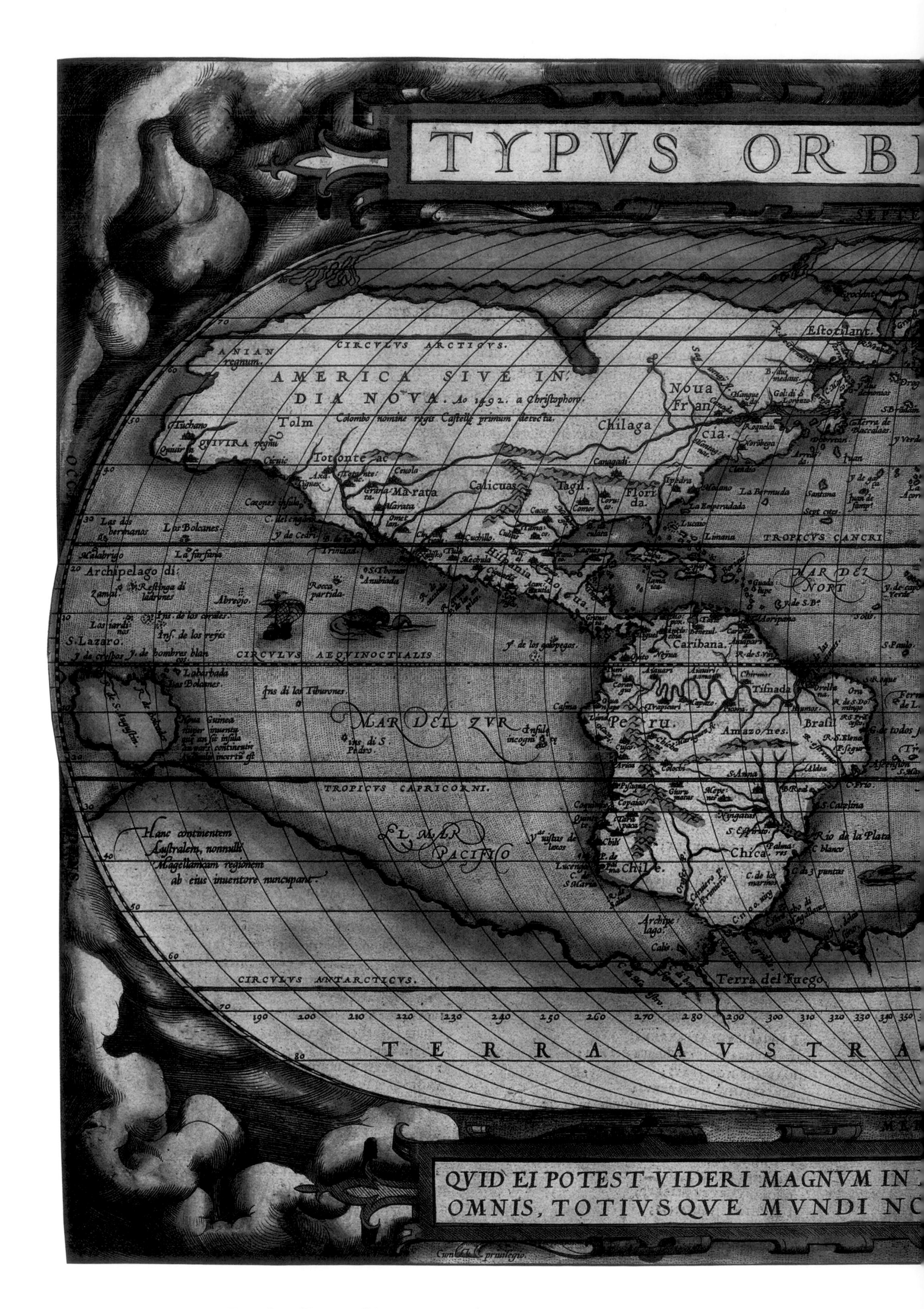

1570 Typus Orbis Terrarum (Map of the World) by Abraham Ortelius
Copperplate Engraving by Frans Hogenberg, Hand-Colored, 13.25 x 19.4 in.
Library of Congress, Washington, D.C.

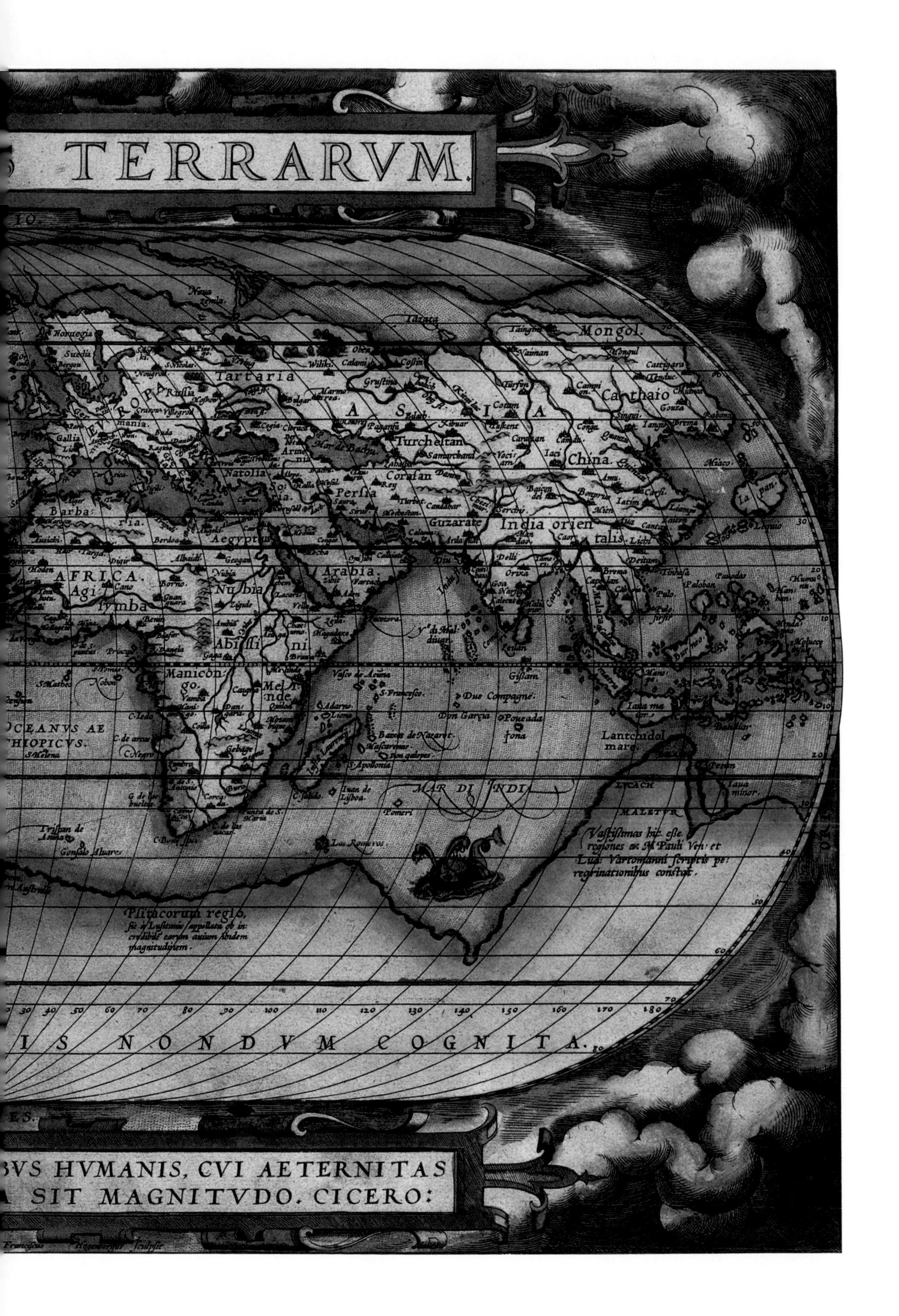
TERRARVM
EVROPA
ASIA
AFRICA.
Tartaria
Mongol.
Cathaio
China.
Turcheftan
Perfia
Arabia.
India orien talis.
Aegyptus
Nubia
Barbaria.
Manicongo.
OCEANVS AE THIOPICVS.
MAR DI INDIA
Lantchidol mare.
Vaftiflimas hic esse regiones ex M. Pauli Ven: et Lud: Vartomanni scriptis peregrinationibus constat.
Plitacorum regio, sic a Lusitanis appellata ob incredibilem earum avium ibidem magnitudinem.
IS NONDVM COGNITA.
BVS HVMANIS, CVI AETERNITAS
A SIT MAGNITVDO. CICERO:

First Modern Atlas

In 1570 Abraham Ortelius revolutionized the map trade with the publication of his *Theatrum Orbis Terrarum*, or *Theater of the World*, the first book of maps uniform in design and size. The most expensive book published up to its time, its appearance shifted the center of the European map trade from Rome and Venice to Antwerp. Its success laid the foundation for the golden age of Dutch and Flemish cartography.

Born in 1527, Ortelius, who Latinized his name from Ortels, began his career as a map illuminator and eventually assembled one of the best geographical libraries in Europe. He entered the map-printing business at age 37 with the publication of a wall map copied from an Italian map. From 1570, he devoted his efforts to the atlas. The first edition assembled 70 maps from the best available sources. Another innovation was a list of his source materials and the cartographers who produced them, providing modern scholars with the first map bibliography. The maps were redrawn by Ortelius in a uniform style and engraved by Frans Hogenberg. On the back of each map sheet, Ortelius included a description of the region, "since we thought it would be displeasing . . . to see the backside of the map sheets bare and empty," he wrote.

The idea for a book of maps in atlas format is credited to Aegidius Hooftman, an Antwerp merchant who wanted an easy reference book rather than unwieldy wall maps. Italian dealers had begun binding loose maps on orders from customers but had not redrawn them to a specific size.

The *Theatrum* was an immediate success. "You deserve great praise for having selected the best descriptions of each region . . . collected . . . into one manual, which can be bought at small cost, kept in a small space and even carried about wherever we please," wrote his friend Gerardus Mercator. Two editions quickly followed in 1571, with 31 editions issued over the next 41 years. Altogether, at least 7,300 atlases were printed, an enormous press run for the time. The original Latin edition was followed by publications in Dutch, German, French, Spanish, English, and Italian. To increase the atlas's appeal, Ortelius offered hand coloring for an additional charge, although wealthy collectors often hired their own illuminators.

Despite Mercator's reference to its modest price, a deluxe edition would cost $1,631 in today's currency. Two abridged versions of the *Theatrum* were introduced in 1577 for the less affluent. These "pocket atlases," containing reduced-scale maps, continued in print for nearly 150 years. The *Theatrum* opened with an illustrated title page and world map, followed by regional and country maps. The world map shown on pages 98-99, a one-sheet oval projection, is a reduced version of Mercator's great 21-sheet wall map of the world, published in 1569. It portrayed a northern sea passage between America, Asia, and the North Pole; an exaggerated North America; and a vast imaginary southern continent—Magellanica. Nevertheless, for the next 40 years, virtually every world map was based on either Mercator's original map or this reduction.

California as an Island

THE GOLDEN AGE OF DUTCH CARTOGRAPHY USHERED in by Ortelius and Mercator found its fullest expression during the 17th century with the production of beautiful multivolume world atlases in Amsterdam—spurred on by a rivalry between two mapmakers and neighbors: Willem Blaeu and Johannes Janssonius. They contended for the international atlas trade by adding maps and volumes to their world atlases. Joan Blaeu continued the rivalry after his father's death in 1638. Finally, when Janssonius printed an 11-volume work with more than 500 maps in German, Latin, and Dutch editions in 1658, Blaeu trumped him with a 600-map effort. Blaeu's 1662 Latin edition of his *Atlas Maior* was issued in 11 volumes, followed by Dutch, German, French, and Spanish editions in 9 to 12 volumes.

Willem Blaeu started out inauspiciously as a clerk in the herring trade, but an interest in mathematics led him to the Danish astronomer Tycho Brahe, with whom he also studied astronomy and land surveying. Returning to Amsterdam, he established a printing and publishing firm, specializing initially in maritime cartography. In 1633, Blaeu was appointed hydrographer of the Dutch East India Company, a position that provided him with the latest geographic information from the company's far-flung commercial interests. By the mid-17th century, the House of Blaeu was famous throughout Europe.

Joan Blaeu's *Atlas Maior* is generally considered the finest atlas ever published. "Every self-respecting Dutch merchant, every collector of books and maps, whether in the Netherlands or abroad, wished to possess this gigantic work," wrote Dutch historian Herman Verwey. The Blaeu workshop maintained the highest typographic standards, and its maps and atlases were hand-colored and prepared with bindings of vellum, morocco leather, or violet velvet.

Reflecting the artistic achievement of the House of Blaeu is the world map on pages 102-03 from the 1667 French edition. Surrounding the illuminated double-hemisphere map are majestic celestial figures and allegorical depictions of the four seasons. Galileo Galilei is portrayed in the upper left, Tycho Brahe in the upper right (shown without his nose of gold and silver alloy, which replaced what was lost in a duel).

Blaeu's world map is representative of the period. Japan is more realistically portrayed, and Australia's coastline has been extended. Most striking is the depiction of California as an island. This configuration began appearing on maps in 1622, although the earliest North America maps represented California as a peninsula. Its source is Father Antonio Ascension, who voyaged along the California coast in 1602. Ascension's map made its way to Amsterdam, where English mathematician Henry Briggs popularized it. His map of North America appeared in 1625. Janssonius revived interest in the island theory with his 1638 North America map. A reprentation of California as an island remained on maps until the mid-18th century when Ferdinand VII of Spain outlawed its delineation as an island by royal decree.

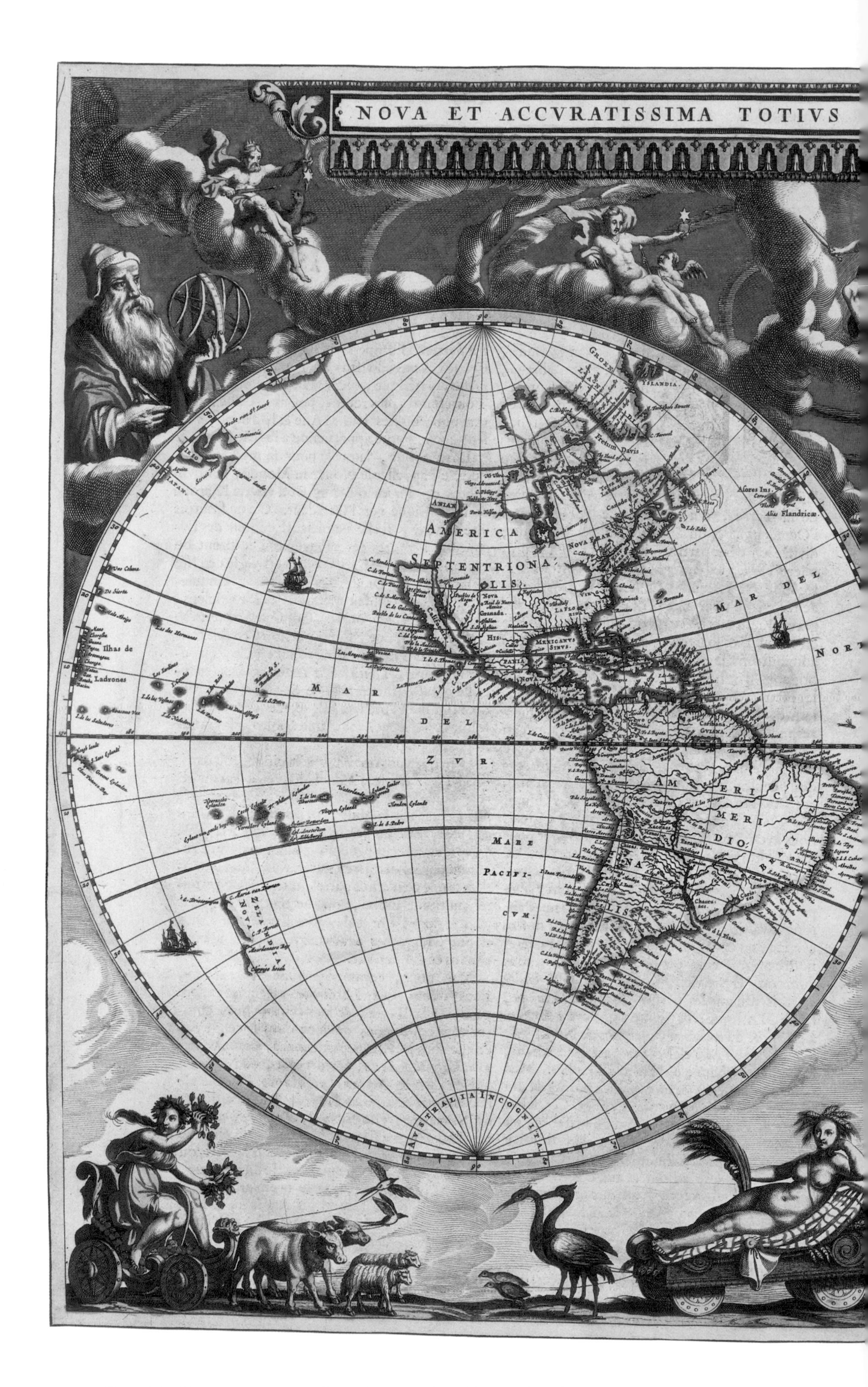

1667 World Map, Nova et Accuratissima Totius Terrarum Orbis Tabula in Joan Blaeu French Edition, Le Grand Atlas ou Cosmographie Blaviane

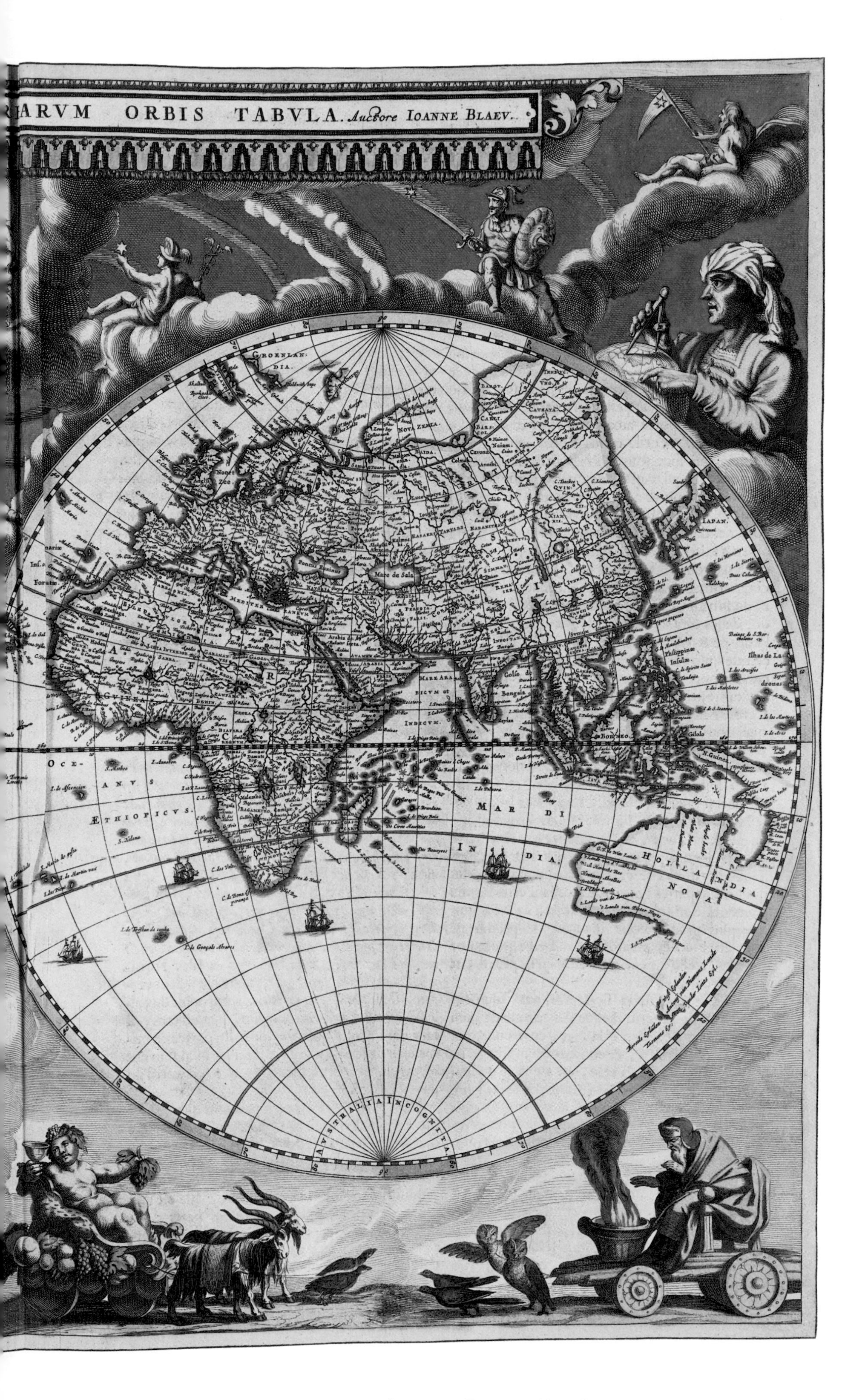

Copperplate Engraving, Hand-Colored. 15.7 x 21.6 in.
Library of Congress, Washington, D.C.

First Town Views

VIEWS OF MEXICO CITY AND CUSCO, PERU, THE greatest 16th-century cities of the New World, are but two of the 350 illustrations that grace the *Civitates Orbis Terrarum*, a six-volume atlas issued between 1572 and 1617. One of the great books of the later Renaissance, produced during a period of religious turmoil and political upheaval, the *Civitates* provided the first systematic, detailed images of the world's leading cities in atlas form. Inspired by Ortelius's world atlas, the *Civitates* was prepared for the curious merchant or student interested in other cities and towns but unable or fearful to travel to them.

German cleric Georg Braun, a canon of Cologne Cathedral, edited the volumes and wrote most of the text, but he notes that "the cunning hand of Frans Hogenberg" did the engraving. Driven from the Netherlands by religious persecution following his work for Ortelius, Hogenberg settled in Cologne, where the Protestant engraver collaborated with the Catholic priest and theologian to produce the *Civitates*—a remarkable partnership at the height of the religious wars. After Hogenberg's death in about 1590, his son Abraham succeeded him. Miniature-painter Georg Hoefnagel, and later his son Jacob, traveled throughout Europe, sketching town views for the engravers.

Editions were published in Latin for the educated reader and in German and French for the general public. Following Abraham Hogenberg's death, the copperplates were passed on through auctions and bequeaths to five map-publishing houses. Dutch publishers J. Covens and Cornelis Mortier were still producing "faded impressions" of these city views near the end of the 18th century.

Abraham Hogenberg engraved these views of Mexico City and Cusco in 1617, copying them from woodcuts that appeared in Antoine du Pinet's book of city maps published in Lyon, France, in 1564. The image of Mexico City can be traced to Hernán Cortez's map of Tenochtitlan, printed in 1524. Foreground figures portray typical city inhabitants. An Aztec ruler, for example, is depicted on the right, carried by slaves and escorted by soldiers. Another reason Braun introduced these figures was to keep his work "out of the hands of the infidel Turks, whose religion forbade the representation of the human figure," according to Raleigh A. Skelton, former head of the map library of the British Museum.

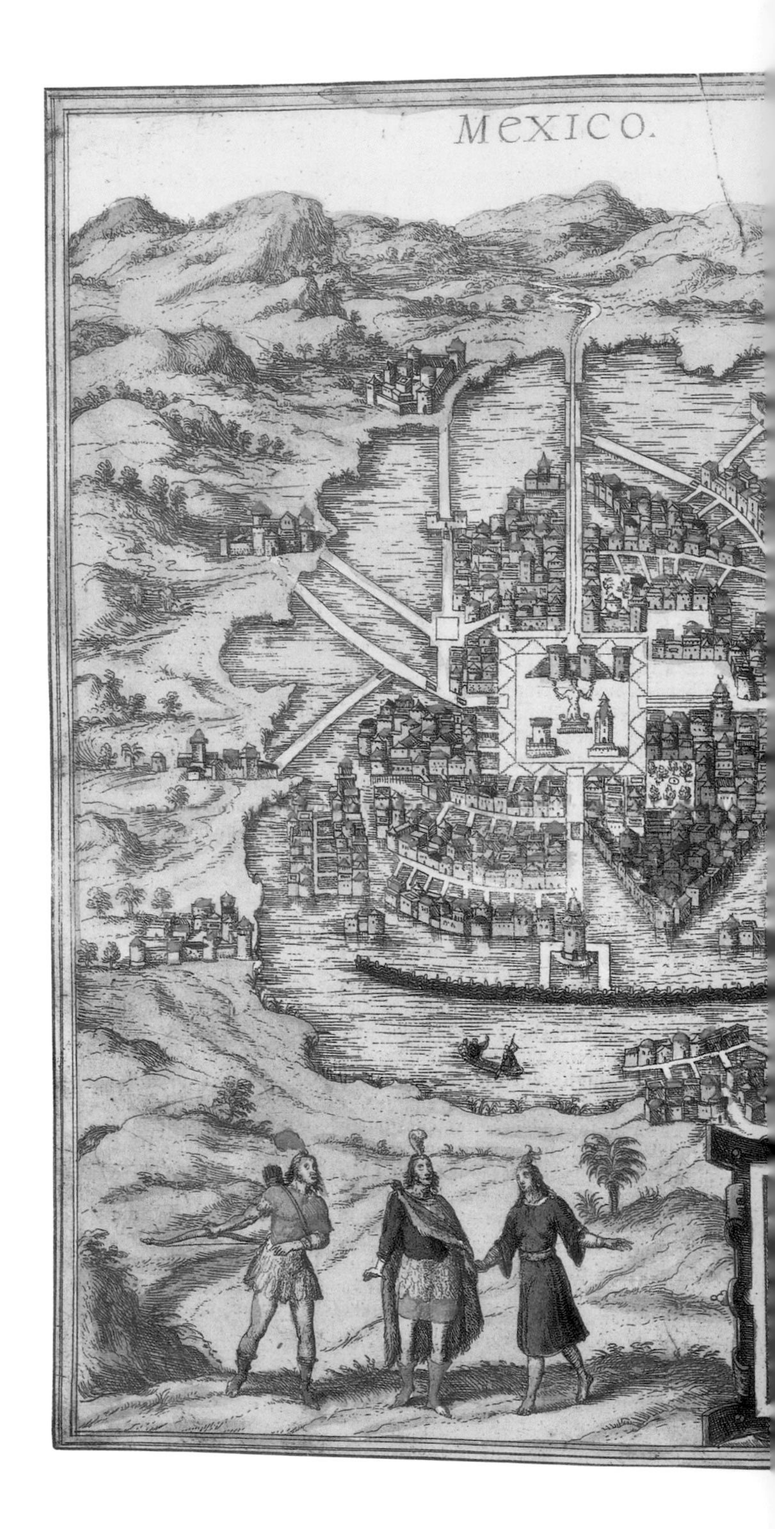

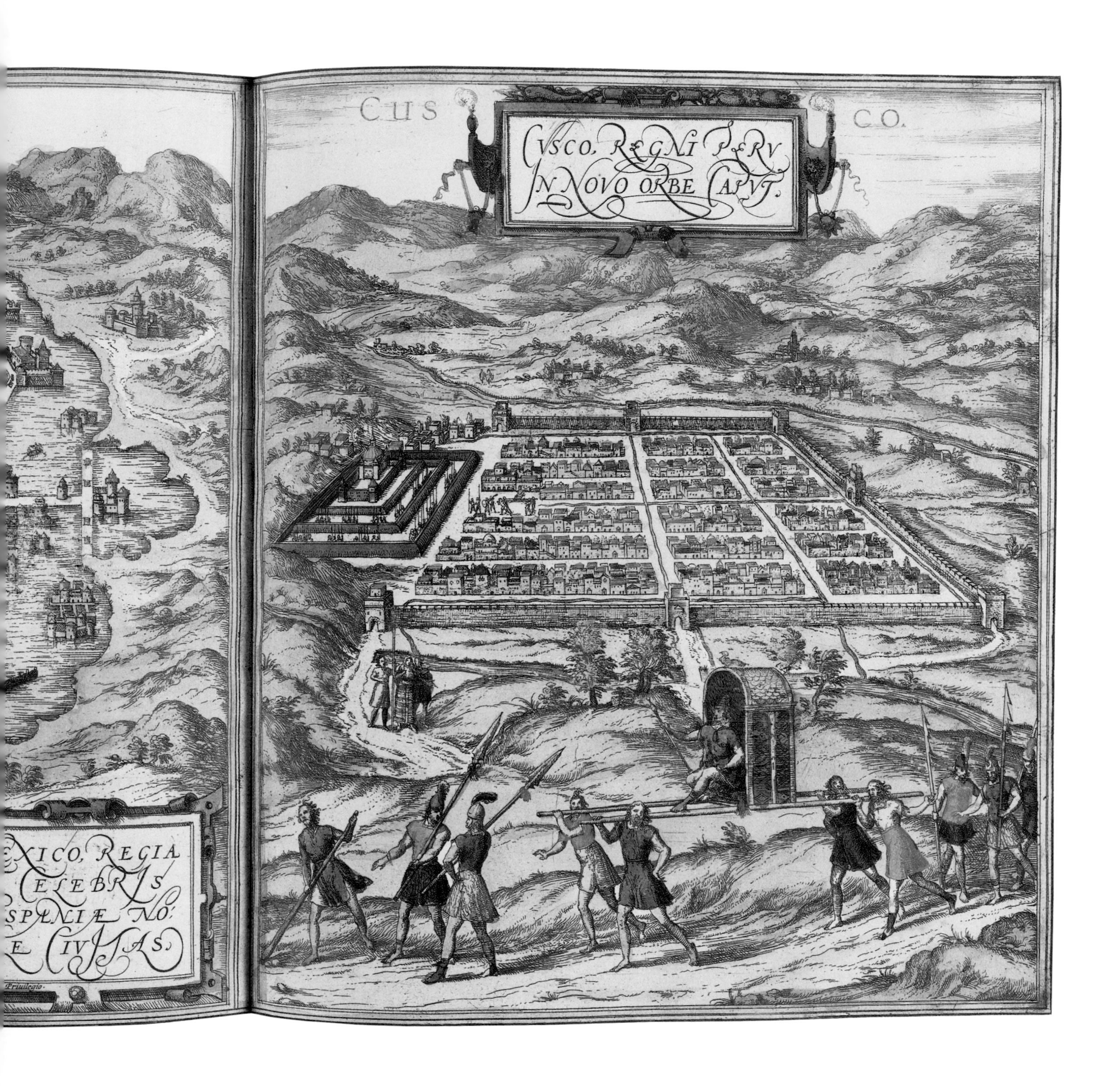

1572 Mexico, Regia et Celebris (Mexico, Regal and Famous)
and Cusco, Regni Peru in novo Orbe Caput (Cusco, Capital of the Kingdom of Peru in the New World)
in Georg Braun and Frans Hogenberg's Civitates Orbis Terrarum (Cities of the World)
Copperplate Engraving, Hand-Colored, by Abraham Hogenberg, 21 x 15 in.
Library of Congress, Washington, D.C.

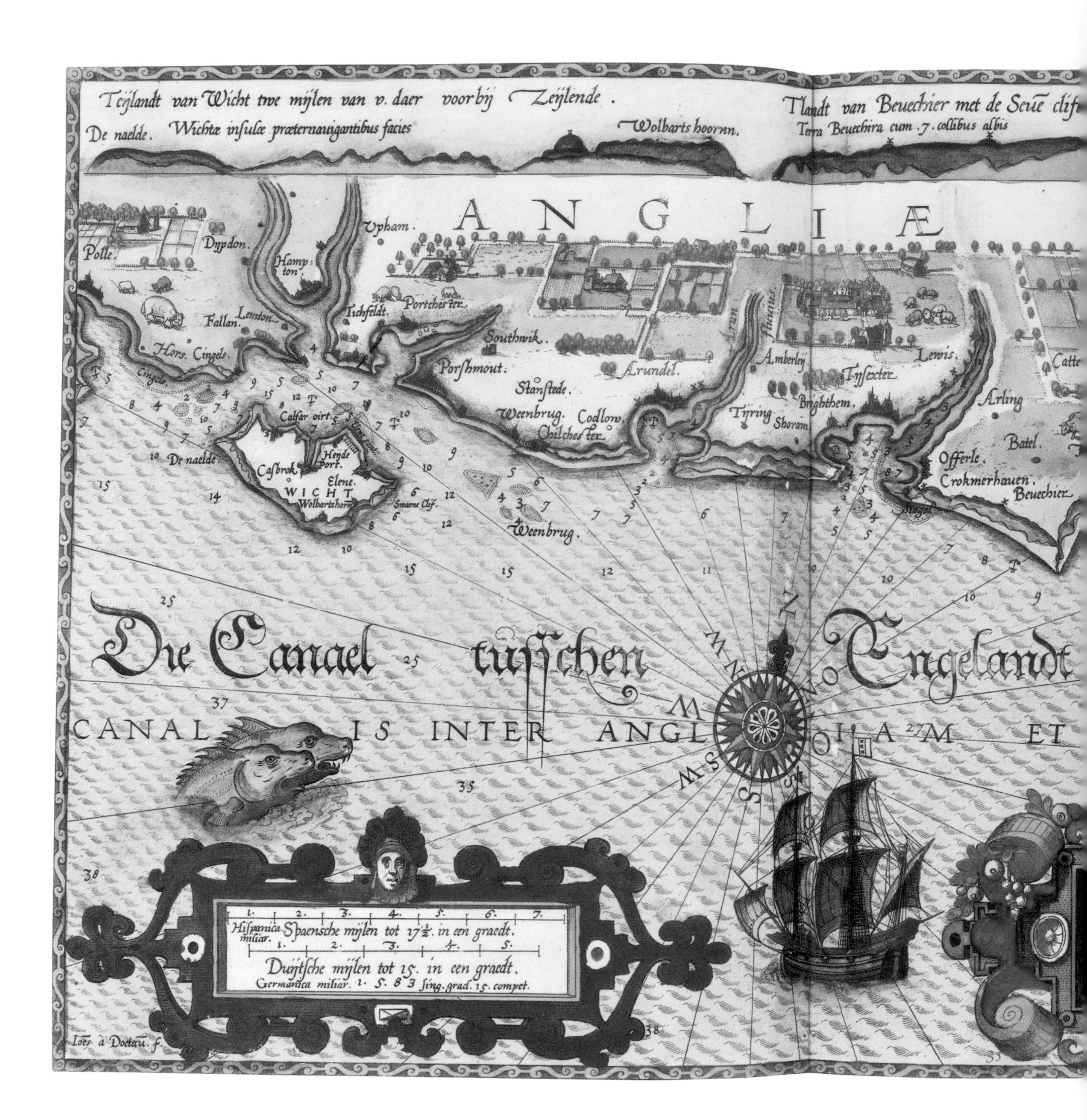

1586 Sea Chart From Speculum Nauticum (The Mariner's Mirror) by Lucas Janszoon Waghenaer
Copperplate Engraving, Hand-Colored, 16 x 24.5 in.
Library of Congress, Washington, D.C.

First Printed Sea Atlas

A NEW ERA IN SEA CHARTING BEGAN IN 1584 WITH the publication of Lucas Janszoon Waghenaer's *Spieghel der Zeevaerdt (Mirror of the Sea)*, followed in 1586 by a Latin edition, *Speculum nauticum (The Mariner's Mirror)*. During the 16th century, Dutch shipbuilders and ship owners emerged to dominate maritime trade within European waters, creating a market for accurate and up-to-date navigational aids. Popularly referred to as a "waggoner," the *Spieghel* was the first printed sea atlas for European coastal waters from Spain to Norway. It provided every navigational aid needed by a mariner, including a set of maritime charts on a common scale, a manual of practical navigation, and sailing directions.

Waghenaer was a Dutch hydrographer and chartmaker, born and raised in Enkhuizen, a bustling port city on the Zuider Zee. He prepared his charts from soundings and bearings collected during extended cruises as a navigator and pilot. Using his own savings, and perhaps additional funds from a wealthy merchant friend, Waghenaer selected master engraver Johan van Deutecum to etch the copperplates and Leiden printer Christopher Plantin to publish the atlas. Plantin had published the 1579 edition of Ortelius's *Theatrum* but then was forced to flee Antwerp for Leiden for religious reasons.

Dedicated to Prince Willem of Orange, leader of the Protestant revolt of the former Spanish provinces of the northern Netherlands, the *Spieghel* was published in two volumes, containing 45 charts altogether. Each chart covers two pages, with sailing directions and related information printed on its verso. Seamen valued these charts for their coastal details, since navigating inshore waters is the most treacherous part of sailing. Various navigational aids are displayed on this chart of the south coast of England, extending from the Isle of Wight eastward to Dover. These include, for example, soundings that mark depths at half-tides in fathoms; safe anchorages, indicated by anchor symbols; and profiles of coastal headlands and landmarks that could be seen from far out at sea. Because seamen are more interested in harbors than in the coastal stretches between them, harbors were exaggerated in size in relation to coastal information. The exquisite scale and text cartouches, uniquely designed for each chart, and the various embellishments—vessels, wind roses, sea monsters, animals, and buildings—display Deutecum's skill as an artist as well as his talents as an engraver.

Map of "The Queen's Pirate"

FRANCIS DRAKE'S VOYAGE AROUND THE WORLD from 1577 to 1580 was celebrated throughout Great Britain. Only the second circumnavigation of the Earth, this expedition, funded and supported by the queen of England, Elizabeth I, enhanced her reign with plundered silver bullion and gold. Upon his return, the buccaneer presented the queen with "a very large map" of his route, which she ordered to be kept from the Spanish "on pain of death." Drake's map, compiled with the assistance of his nephew John Drake, was later placed on display in the King's Gallery at the Palace of Whitehall, where it remained until destroyed by fire in 1698.

Mysteriously, two Dutch mapmakers obtained copies of Drake's suppressed manuscript and issued printed versions. The most famous is a double-hemisphere world map, drawn and engraved in the baroque style by Jodocus Hondius in about 1595-96. A native of Flanders, Hondius was an accomplished engraver as well as an artist and calligrapher. He spent eight years in London as a political refugee before returning home to the Netherlands in 1593.

In addition to showing Francis Drake's track (dotted line), Hondius displayed the route of Thomas Cavendish (light broken line), who followed in Drake's wake in 1586. Their voyages took them from England to South America, around Cape Horn north to California (which Drake claimed for the British crown), then across the Pacific to the Spice Islands, the Indian Ocean, around the Cape of Good Hope, and home. Drake's geographical contributions included the discovery that Tierra del Fuego, at the southern tip of South America, was an archipelago. He also found an open water passageway from the Atlantic Ocean to the Pacific Ocean south of the Strait of Magellan, reducing the size of the mythical Terra Australis in comparison with earlier maps.

Most intriguing are the ornate vignettes. A portrait of Queen Elizabeth I below the royal coat of arms, for example, guaranteed the map's authenticity. The *Golden Hind*, the only vessel to survive of the five that set out on Francis Drake's famous voyage from England, is prominently featured. Its depiction is based on a sketch of the ship by Hondius while it was moored on the Thames. In the upper left, Drake's ship is pictured in Portus Novae Albionis (Port New Albion), long thought to be San Francisco Bay, or Drake's Bay, but now believed to be the Mexican port of Acapulco.

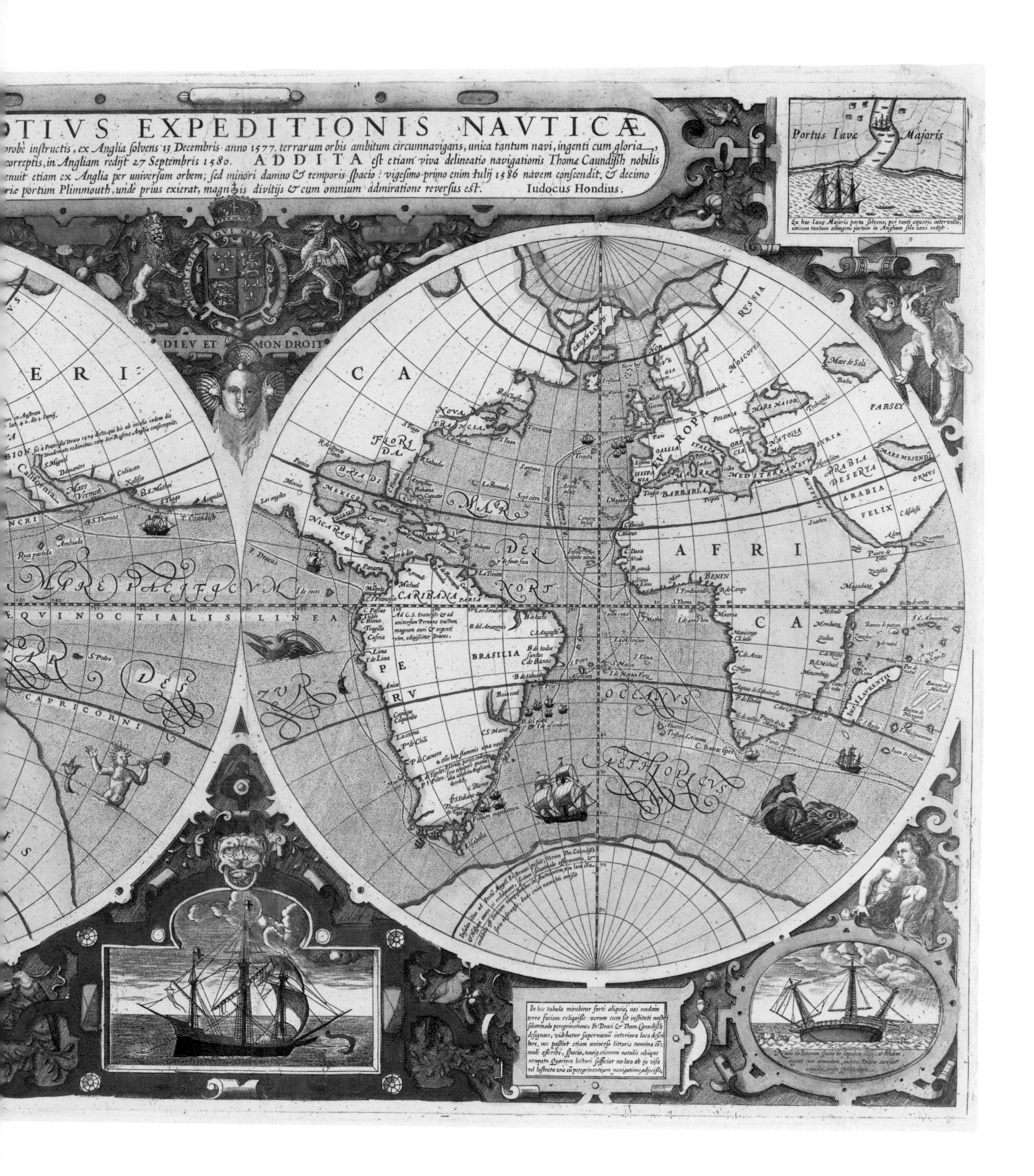

Circa 1595-96 Double-Hemisphere World Map Entitled Vera Totius Expeditionis Nauticae Descriptio by Jodocus Hondius
Copperplate Engraving, Hand-Colored, 12 x 18.5 in.
Library of Congress, Washington, D.C.

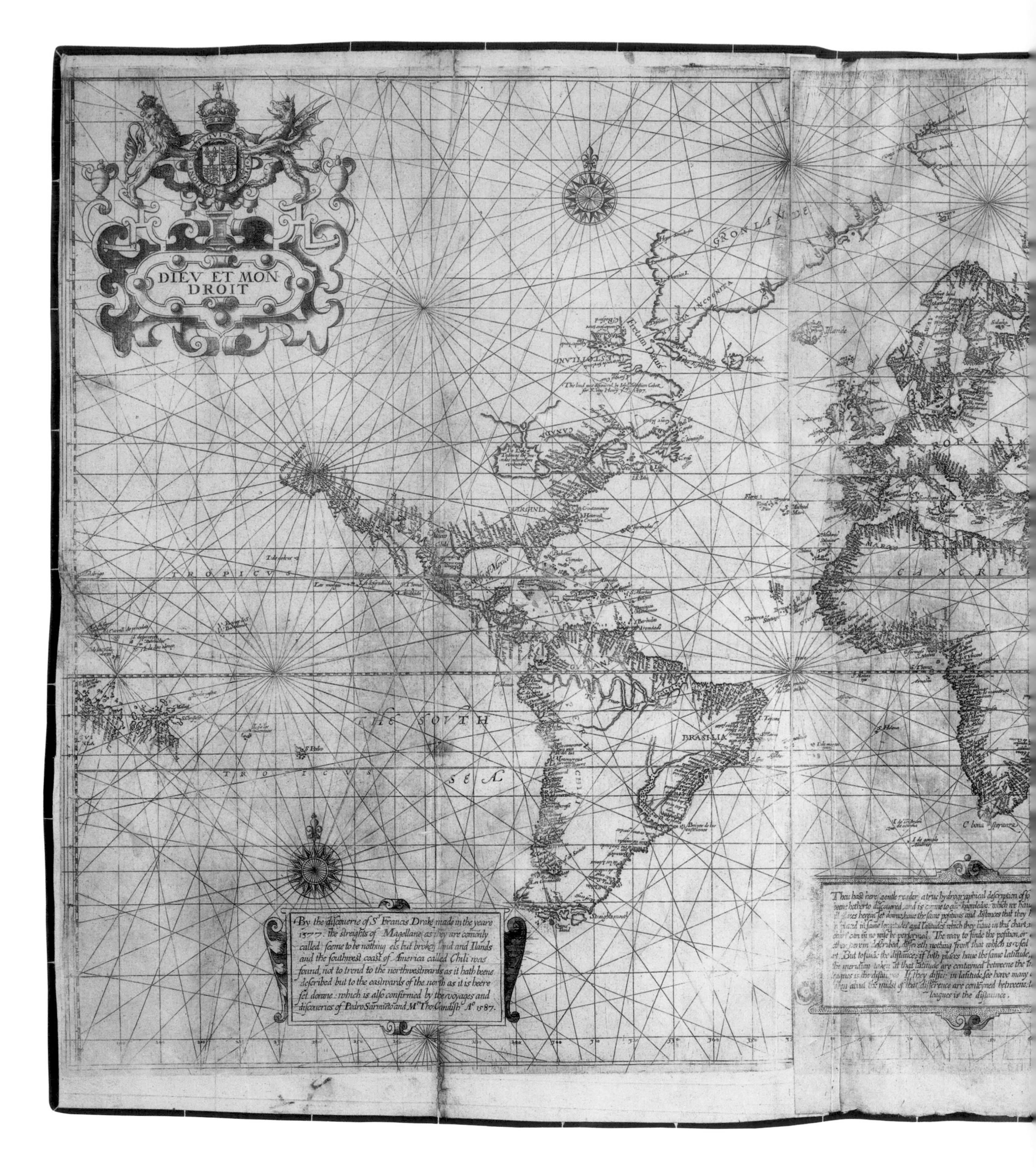

1599 World Chart on Mercator's Projection, Attributed to Edward Wright and Emery Molyneux
in The Principal Navigations by Richard Hakluyt
Copperplate Engraving, 25 x 16.5 in.
Newberrry Library, Chicago, Illinois

Shakespeare's "New Map"

"HE DOES SMILE HIS FACE INTO MORE LINES THAN is in the new map with the augmentation of the Indies. You have not seen such a thing as 'tis." With these lines from *Twelfth Night*, Shakespeare indirectly introduced the English public to Edward Wright and Emery Molyneux's "new map" of the world, based on Gerardus Mercator's famous projection.

For generations schoolchildren viewed world geography through Mercator's eyes. They learned that Greenland and Africa were similar in size, despite Africa's being 14 times larger. Mercator's projection, which grossly distorts geographical areas in the higher latitudes, was designed for navigation not for geographical instruction. Since all lines of constant compass bearings are shown as straight lines, sailors can easily plot their course from one point to another by a single compass setting.

Born in Flanders in 1512, Mercator was educated at the University of Louvain in philosophy. He later immersed himself in religion, mathematics, geography, engraving, and calligraphy. Arrested for heresy, he was fortunately released after seven months (four associates were executed). Mercator relocated to Duisburg, Germany, where he published his 18-sheet world map on the Mercator projection in 1569. Mercator's projection is reflected in the grid of straight lines of latitude and longitude crossing at right angles, with increasing spaces between latitude lines moving away from the Equator. He was the first to use the name "North America" on a map (1538), and he introduced the word "atlas" for his influential book of maps (1595).

Mercator's projection was improved by Edward Wright, who some credit as its true "inventor." Wright studied and lectured on mathematics at Cambridge University in England, later becoming an advisor on navigation for the British East India Company. His book *Certaine Errors in Navigation*, published in London in 1599, provided tables for plotting courses on Mercator charts.

Designed by Wright from a globe prepared by Emery Molyneux, this chart appeared in Richard Hakluyt's great work on English discovery and exploration, *The Principal Navigations*, published in London in three volumes (1598-1600). Confusing and somewhat obscuring the Mercator grid is a traditional network of intersecting wind roses and rhumb lines. Nevertheless, this "graphic epic," in the words of cartographic historian Kenneth Nebenzahl, is the most accurate and scientific map of the 16th century.

Emblems of Power

MAPS HAVE LONG BEEN USED TO ASSERT POLITICAL authority. One of the most intriguing examples in the history of cartography of the use of a map as a political act is the engraving by Pieter van den Keere of *Leo Belgicus*, a map of the Protestant provinces of the Low Countries, whose outline is highlighted to resemble a lion. During the Dutch revolt against Spain from 1566 to 1648, the lion became the official symbol of the Dutch Republic and later the Kingdom of the Netherlands. Austrian nobleman Baron Michael Aitzinger was the first to use this concept in a map, which appeared with his history of the Low Countries, published in Cologne, Germany, in 1583. Apparently Aitzinger chose the lion motif because most of the provincial shields or coats of arms of the provinces depicted a lion.

This decorative feature took on more obvious political overtones, becoming an "emblem of power," to use cartographical historian Brain Harley's apt phrase, when it appeared in van den Keere's *Germania Inferior*, the first national atlas of the Netherlands, whose publication in 1617 was a provocative act. A native of Ghent, van den Keere spent his formative years as a political and religious refugee in London, where he apprenticed as a map engraver under another Dutch émigré, his brother-in-law, the mapmaker Jodocus Hondius. Settling in Amsterdam in 1593, van den Keere became one of the most prolific Dutch map engravers of the early 17th century, a period coinciding with the Twelve Year Truce of 1609-21, when hostilities between Spain and the Dutch Republic temporarily ceased. At this time cartography flourished, as Amsterdam became Europe's leading commercial and cultural center.

The *Germania Inferior* contains 26 maps and plates of the 17 provinces that today include Belgium, the Netherlands, Luxembourg, and part of northern France. Pieter van den Berghe, another London refugee, wrote the text for both the Latin and French editions. Van den Keere's map expressed Dutch republican support not only through the lion motif but also by portraying provincial shields, many of which had been banned by Spanish authorities. Figures of men and women dressed in the simple everyday clothes of Dutch Calvinists and reformers further advanced nationalistic feelings.

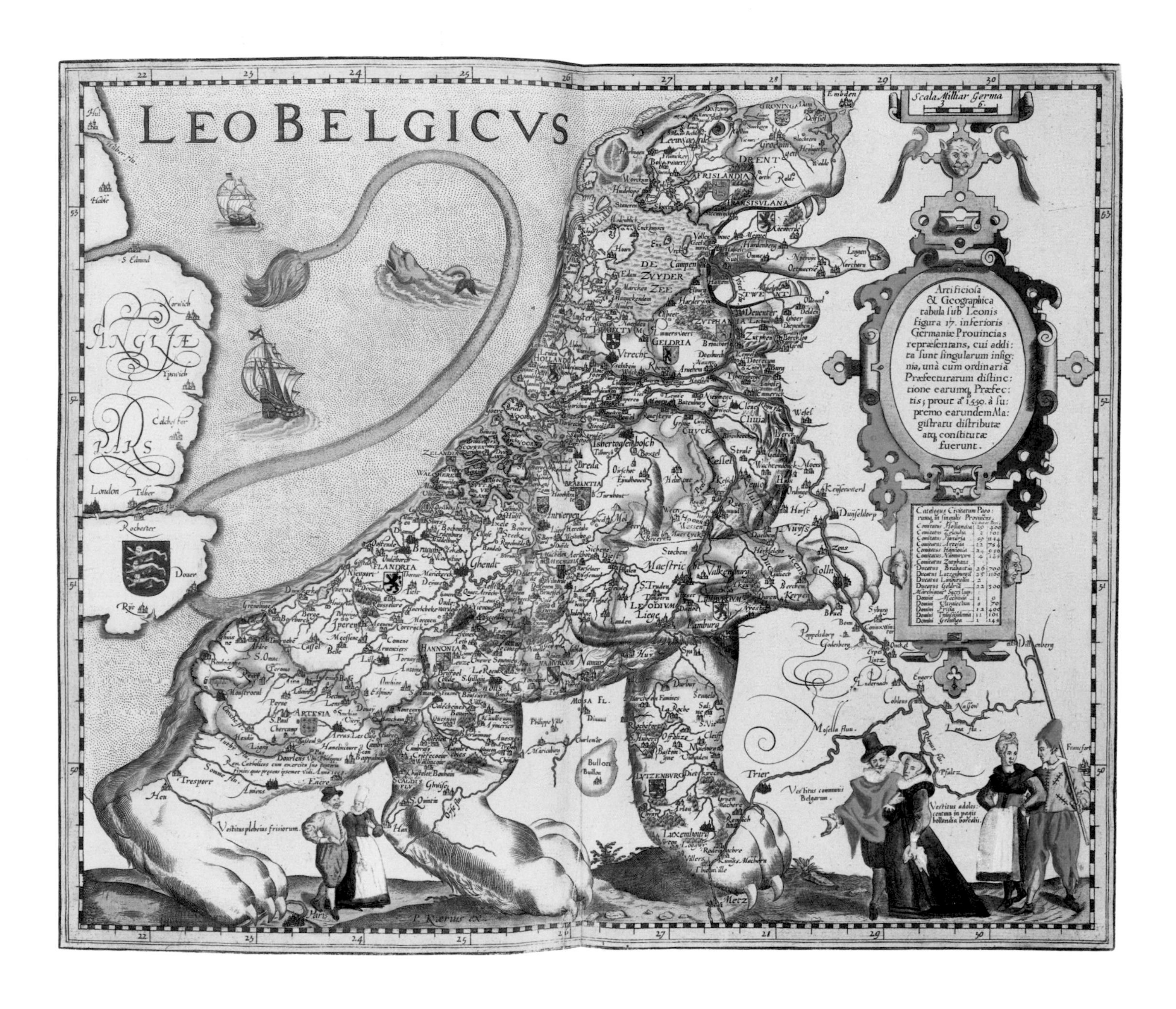

1617 Map of the Low Countries Entitled Leo Belgicus (The Belgian Lion) in Germania Inferior by Pieter van den Keere
Illuminated Copperplate Engraving, 14.5 x 17.75 in.
Library of Congress, Washington, D.C.

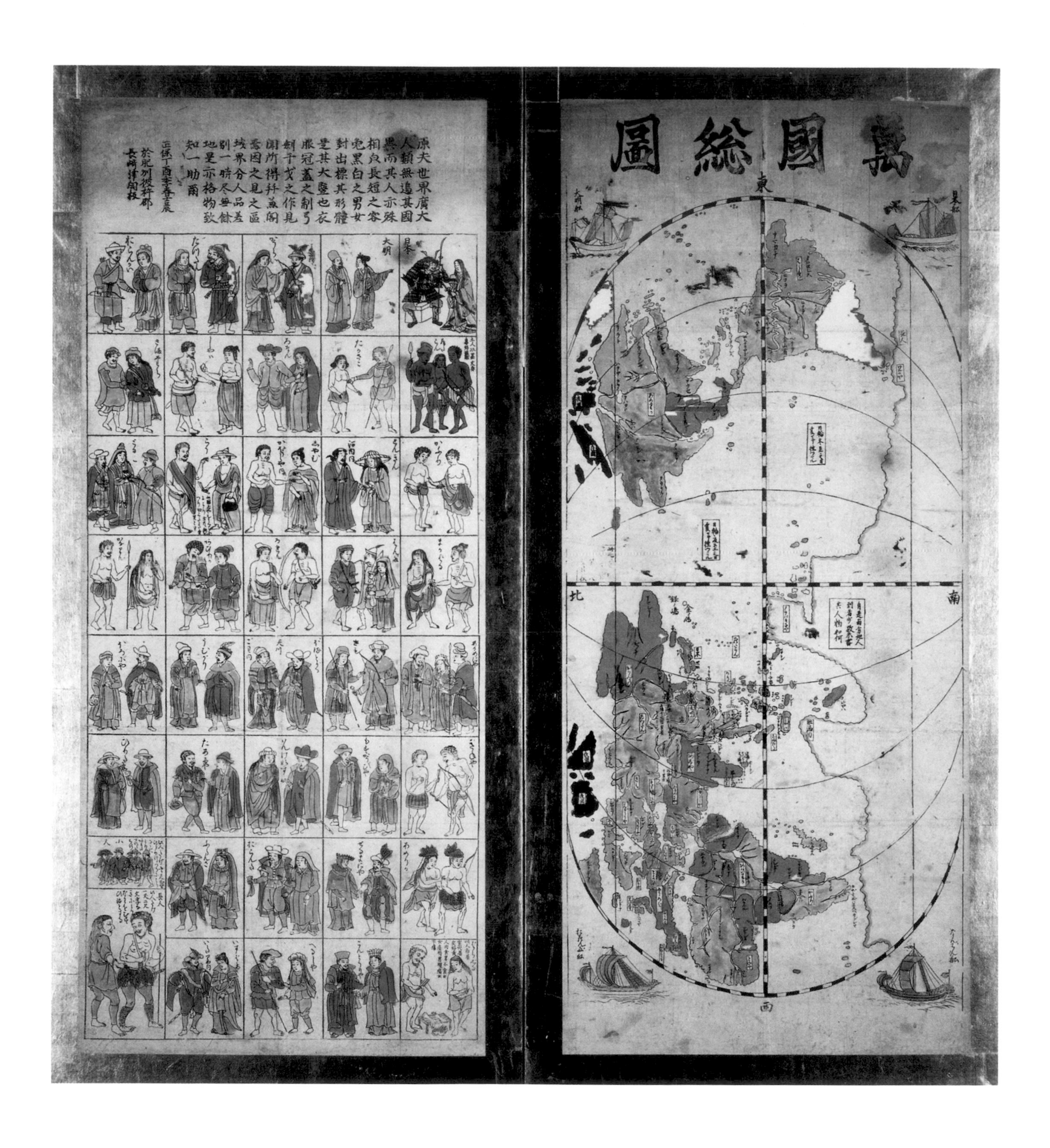

1645 Bankoku Sozu (Complete Map of the World) and Scroll of Peoples of the World, Anonymous
Scroll Woodcut, Hand-Colored, 53.8 x 23 in.
Kobe City Museum of Nanban Art, Kobe, Japan

"Southern Barbarians" World Map

THE JAPANESE WORLDVIEW EXPANDED DRAMAtically with the introduction of European world maps in the 16th century. Portuguese maritime traders first visited Japan in 1543, and it is likely that their navigators soon introduced Japanese sailors to portolan charts. The Japanese word for chart is *karuta*, derived from the Portuguese word for map, *carta*. By 1580 European terrestrial globes and world maps were readily available. Several years later, according to cartographic scholar Kazutaka Unno, "four young nobles representing the three Christian Kyushu feudal lords" were sent to Italy for further study. They returned to Nagasaki in 1590 with an astrolabe, a terrestrial globe, nautical charts, and a variety of maps and atlases, including copies of Abraham Ortelius's *Theatrum Orbis Terrarum* and the first three volumes of Georg Braun and Frans Hogenberg's *Civitates Orbis Terrarum*. Copies of Father Matteo Ricci's world map, compiled and printed in China, with place-names and legends in Chinese characters readily understood by Japanese, reached Japan shortly after its publication in 1602.

European maps were transformed into traditional Japanese map forms: large painted screens that served as room dividers, and scrolls designed to be hung and displayed in decorative alcoves in homes of the Japanese nobility. These were known as *nanban* maps, after the name for foreigners, *nanbanjin* (southern barbarians), or people who arrived in Japan from the south.

The *Bankoku Sozu (Complete Map of the World)* is a nanban world map in the form of a scroll. This anonymous map is the first European-style map published in Japan, and was printed from a woodblock in Nagasaki in 1645. Copied from Ricci's world map, the configuration of Japan is much improved, as it was based on Japanese sources. Place-names and legends were transliterated from Ricci's Chinese characters to the two standard types of Japanese lettering, *hiragana* and *katakana*. Some Portuguese place-names were added in Japanese by hand. The map is centered on Japan, a characteristic of nanban world maps. It is oriented to the east, with the Americas at the top. Japanese and European ships decorate its four corners. The *Bankoku Sozu* was paired with a second wood-block print illustrating 40 nationalities, each represented by a man and a woman in traditional attire.

For nearly a century cartographic works and surveying techniques were freely exchanged between nanbanjin and Japanese mapmakers, until Tokugawa Ieyasu issued the first edicts suppressing Christianity. A series of exclusion decrees soon followed. From 1641 until Commodore Perry's arrival in Tokyo Bay 200 years later, Japanese trade with nanbanjin was severely restricted. By 1668, European maps were banned from import, except for world maps. Despite Japan's isolation, world maps and ethnographic illustrations remained popular, reflecting continued Japanese fascination with other countries and peoples.

Traditional Chinese Maps

BEGINNING IN THE LATE 16TH CENTURY, FOREIGN influences penetrated China from the west and north. Jesuit missionaries established their first mission in Guangdong Province. They were followed by nomadic Manchu invaders who eventually conquered China in 1644, establishing the Qing dynasty, the last imperial dynasty of China, which survived until 1910. The Jesuits introduced European scientific developments, including astronomy, surveying, and mapping techniques, to Chinese intellectuals and government officials in an effort to reach the masses through their leaders. The Manchu conquerors also embraced Western cartography, recognizing its value for improving political control, tax collection, and military planning. Emperor Kangxi, the second emperor of the Qing dynasty, commissioned French Jesuit Joachim Bouvet, an astronomer and mathematician, to compile an imperial atlas based on scientific measurements. From 1708 to 1717, Father Bouvet directed a team of Jesuits and Chinese assistants that surveyed and mapped much of China. One of the few surviving copies of this imperial atlas is found in King George III's Topographical Collection in the British Library.

At the same time that Emperor Kangxi was espousing Western cartographic traditions, most Chinese scholars and officials continued to produce and use traditional Chinese maps, which had changed little in purpose and format since the earliest dynasties. This anonymous map of China, found in a manuscript atlas of 20 maps in the Library of Congress, is representative. Drawn in 1647 at the beginning of the Qing dynasty, it portrays China during the late Ming dynasty (1368-1644). The generalized nature of this map and the atlas suggests that they were prepared from earlier textual sources in keeping with *kaozheng*, a school of scholarship that arose in China during the Ming to confront the introduction of new forms of Confucian beliefs and Western scientific ideas. Its followers, focusing on recovering China's past to understand its present, relied upon information from earlier periods rather than contemporary field surveys.

Several distinctive features of traditional Chinese cartography are reflected in this map. Map scale and perspective vary. Mountains and the Great Wall are displayed pictorially in profile while rivers are shown as viewed from above. Undulating lines depict water features. Boundaries of the 15 provinces are generalized. Symbols for cities are geometrical and hierarchical—hexagons designate provincial capitals, circles indicate prefecture cities, rectangles mark county seats, and diamonds represent district cities. The Five Sacred Mountains south of the Great Wall were pilgrimage sites for generations of emperors and peasants. These sites form a system of cardinal points centered on Song Shan (Lofty Mountain), a 4,900-foot peak located in north-central China, which was home to Zhong Yue, the first Daoist temple.

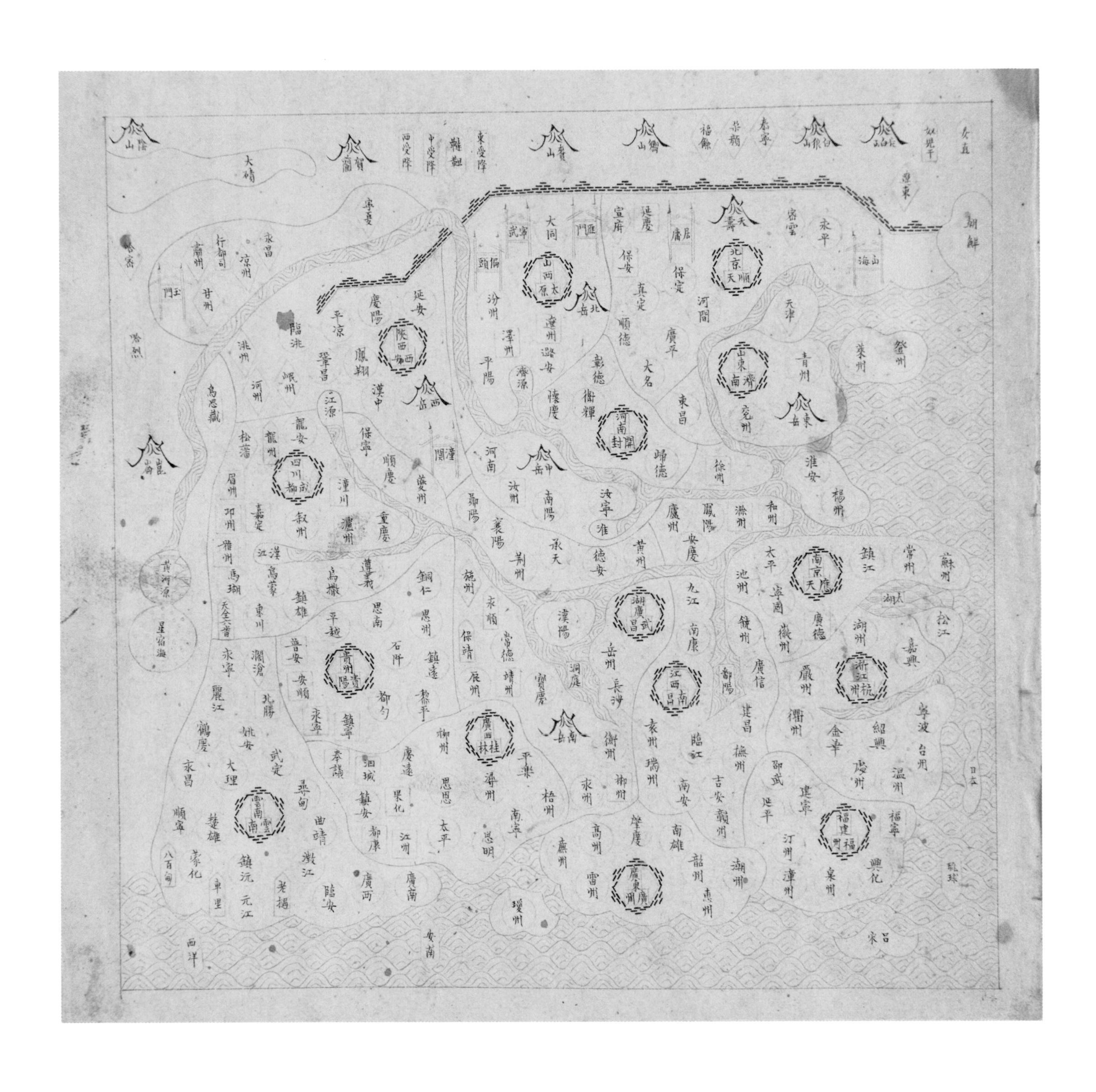

1647 Map of China in Xia Lan Zhi Zhang (Atlas of the Ming Empire)
Manuscript, 6.7 x 6.7 in.
Library of Congress, Washington, D.C.

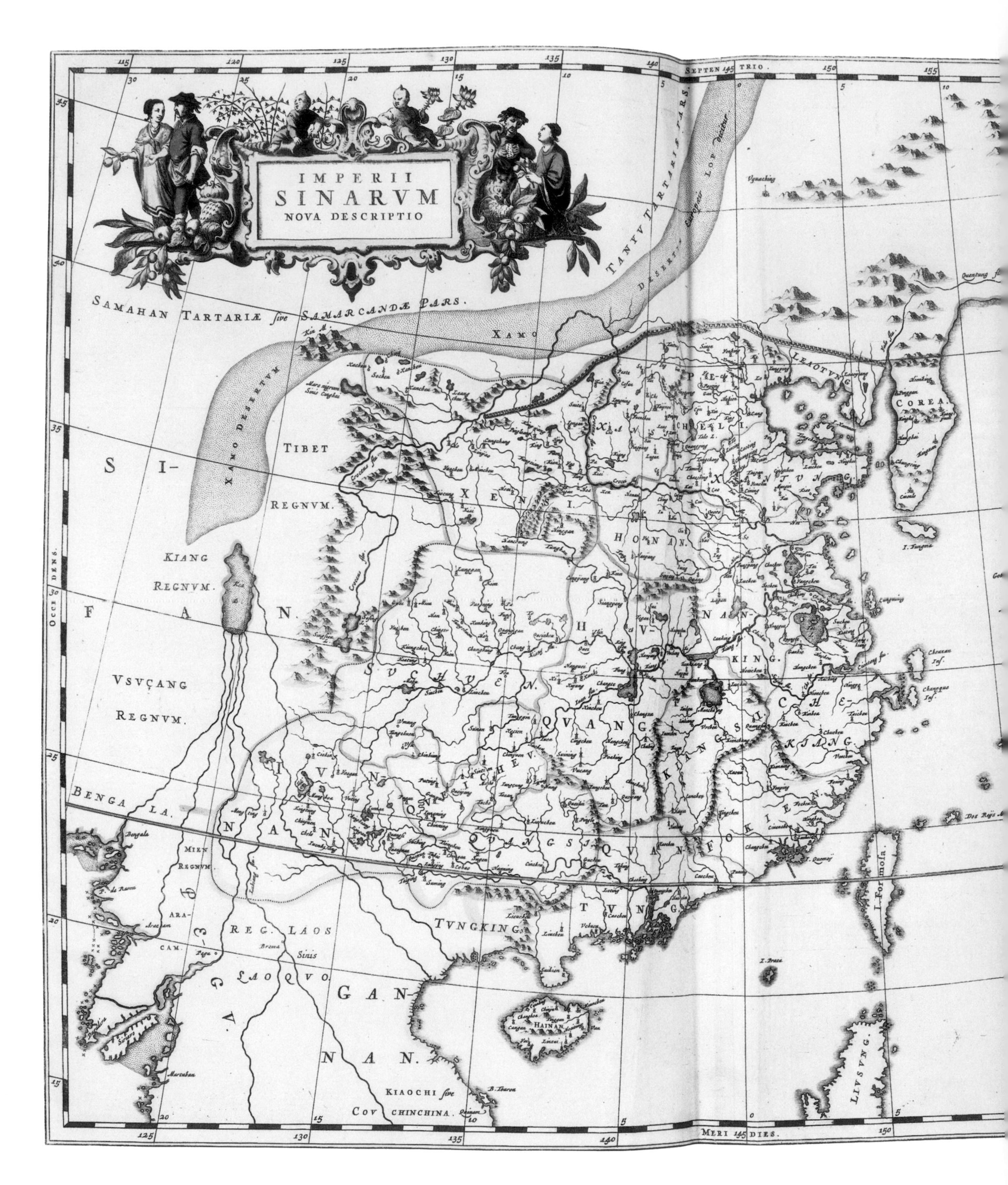

1655 Imperii Sinarum Nova Descriptio (New Description of Imperial China) by Martino Martini
Copperplate Engraving, Hand-Colored, 18 x 23.6 in.
Library of Congress, Washington, D.C.

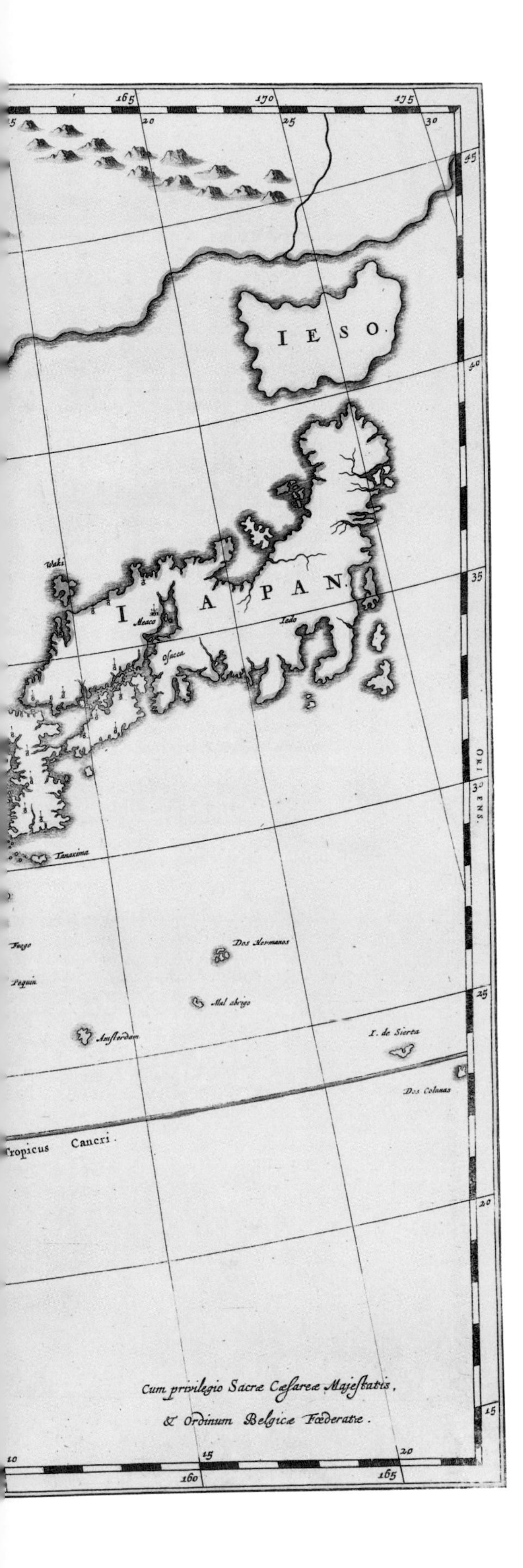

Empire of China

MISSIONARY CARTOGRAPHERS WERE MAJOR transmitters of mapping traditions. The Jesuits were the first to carry European cartographic and geographic concepts to China. Father Matteo Ricci led the way, establishing a mission in present-day Guangdong Province in 1583. During the next 20 years, the Italian prepared a number of terrestrial globes and large world maps. Deriving them initially from Abraham Ortelius's map of 1570, Ricci drew them at larger scales to allow for the addition of place-names and legends written in Chinese characters so that, he said, the maps would "speak Chinese." Others soon followed Ricci. Flemish Jesuit Ferdinand Verbiest, for example, a renowned astronomer, prepared celestial globes and a large double-hemisphere printed map for Qing emperor Kangxi. Through these maps and globes, Chinese scientists and officials were reintroduced to the concept of a spherical Earth and to the revelation that China was but a small part of a much larger world.

At the same time, missionaries returning to the West with indigenous Chinese maps provided more accurate portraits of the region's geography than contemporary European maps of the same land. Martino Martini's atlas of China provides an interesting case study. An Italian Jesuit scholar, Martini traveled widely in China for more than two decades, compiling manuscript maps of its 15 provinces in the process. Returning to Europe by ship in 1651, he was captured and briefly held by the Dutch East India Company in Batavia (Jakarta, Indonesia), the capital of the Dutch East Indies. Martini was not allowed to depart until his manuscript atlas was translated into Dutch. He then received free passage to Amsterdam, where the celebrated Dutch mapmaker Joan Blaeu, eager for new information about China, used the Dutch translation provided by Martini's captors in his atlas. Despite Blaeu's unseemly acquisition method, the printed atlas, *Novus Atlas Sinensis*, is considered a landmark work. Based on Chinese sources, it far surpassed any contemporary European atlas of China.

The atlas contains this map of China, 15 provincial maps, and a map of Japan. The snake-like feature stretching across the northwest frontier of China is the Gobi Desert (Xamo Desertum); it represents the first time a European map used a set of dots to indicate a desert, a convention borrowed from the Chinese.

Imperial Austrian Double-Eagle Map

THIS BIFURCATED DOUBLE-HEMISPHERE WORLD map, held in the grasp of the imperial Austrian double-headed eagle, was the first map to be prepared from longitude determined by differences in time through celestial observations. It was designed by German merchant Philip Eckebrecht for his friend, astronomer Johannes Kepler. Eckebrecht's map was made to be used with Kepler's *Tabulae Rudolphinae*, a set of tables that provided the positions of the planets and included a star catalog based on observations made by the Danish astronomer Tycho Brahe. The discovery by Kepler that the planets follow elliptical rather than circular orbits around the sun laid the foundation for modern astronomy. By comparing Kepler's tables with local time determined from observations of the moon and stars, the longitude of the point of observation could easily be found.

Eckebrecht's unique map projection of a central intact hemisphere separating two half-hemispheres was inspired by his desire to honor Brahe. The map is centered on the intersection of the prime meridian (marking 0° longitude) and the Equator, with the prime meridian running through Brahe's famous observatory at Uraniborg on Hveen Island, off the southwest coast of Sweden, where the astronomer spent 20 years recording the movements and positions of the planets and stars. A timescale along the Equator indicates the hours to be added or subtracted for determining longitude (one hour equals 15 degrees longitude).

Geographically, some of the latest discoveries were depicted, the information having been obtained from Hessel Gerritsz, the official cartographer of the Dutch East India Company. This map was once believed to portray the earliest Dutch discoveries along the northwest coast of Australia, located just above the scroll-shaped presentation legend in the right corner. Although engraver Johann Philip Walch dated his plate 1630, the same year that Kepler's *Tabulae Rudolphinae* was printed, the only surviving copies, according to Rodney Shirley in his comprehensive *The Mapping of the World*, are believed to have been printed after 1658, the year in which Leopold I, to whom the map is also dedicated, became the Holy Roman Emperor.

The striking engraving of the double-headed eagle, the emblem of the Holy Roman Empire, honors Emperor Rudolf II. The emperor was an astronomer who had appointed first Brahe and then Kepler to the post of imperial mathematician.

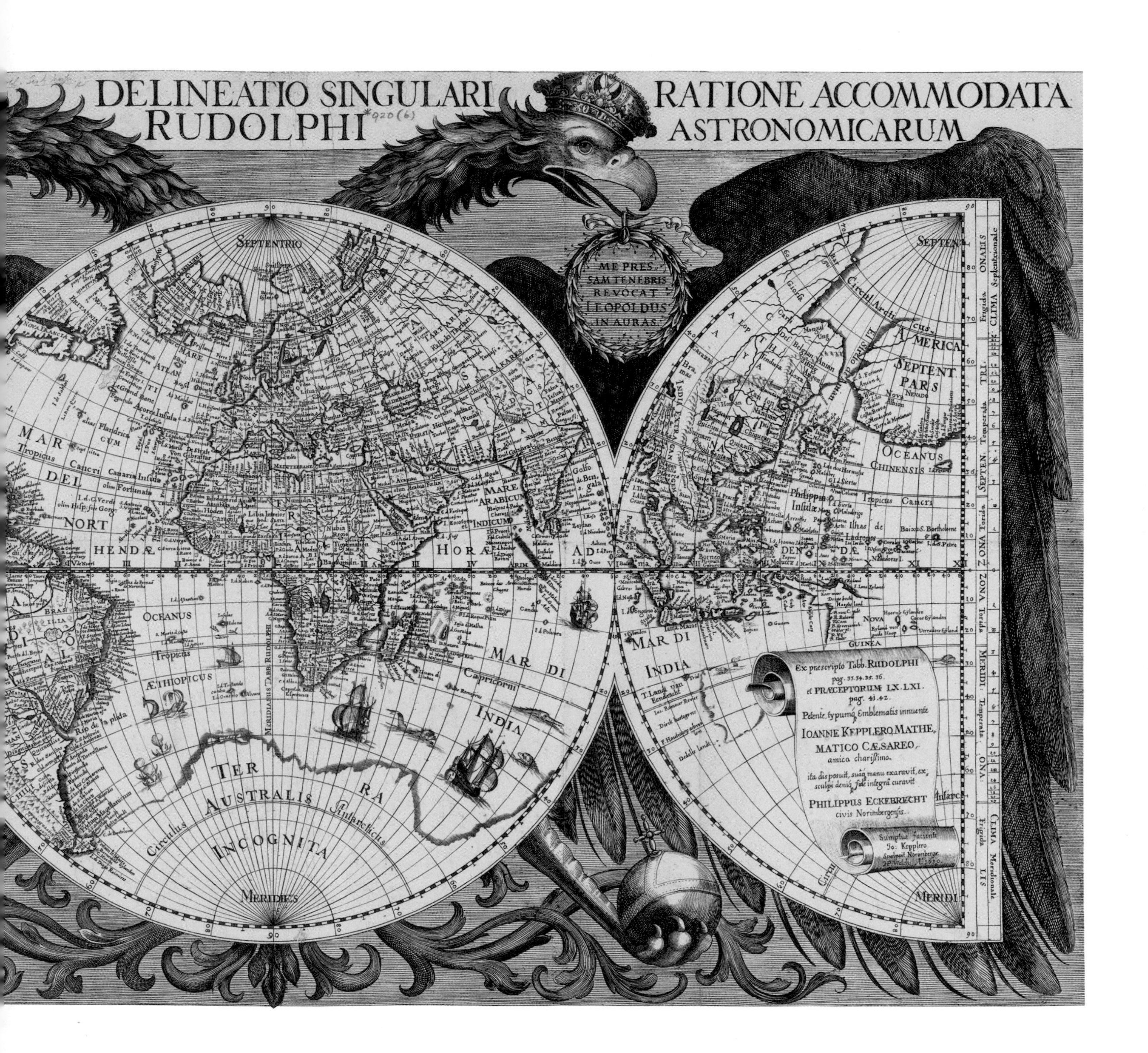

Circa 1658 World Map Entitled Nova Orbis Terrarum Delineatio Singulari Ratione Accommodata Meridiano Tabb. Rudolphi Astronomicarum
(New Image of the World Adapted to the Unique Meridian Calculations of the Rudolphine Astronomical Tables)
by Philip Eckebrecht
Copperplate Engraving by Johann Philip Walch, 27 x 15.2 in.
British Library, London, England

Concealed Messages

BIBLICAL SCENES AND HISTORICAL GEOGRAPHY ARE combined in one of the earliest Hebraic maps of the land of Israel. It was first prepared for the 1695 Amsterdam edition of the *Haggadah*, an illustrated book of the story of the exodus from Egypt, which is read at the seder, the ceremonial dinner that begins the Jewish Passover holiday.

The map was drawn by Avraham bar Yaacov, a Rhineland priest who converted to Judaism, relocated to Amsterdam, and then took up the trade of copperplate engraving. His objective, according to a note on the map, was "to make known to every reasonable person the journey of forty years in the desert, and the breadth and length of the Holy Land from the River of Egypt to the City of Damascus and from the Valley of Arnon to the Great Sea, and within it the territory of each and every tribe." Oriented to the southeast, this map portrays the mouth of the Nile in the lower right corner; the route through the desert; the 41 encampments, also listed in the legend; and the Land of Israel, divided into the territories of the 12 tribes. Its geographical features were derived from an earlier work, Christian von Adrichom's *Situs Terrae Promissionis*, which was engraved in 1592 by Georg Braun, but bar Yaacov added "graphic details that are characteristically Jewish and especially appropriate for Passover." These include, for example, Jonah being tossed overboard to the whale and then washing up on shore.

Several hidden messages are concealed within this map, according to geographer Harold Brodsky. The number four, for example, which has special meaning to the Passover seder ("the four questions, the four sons, and the four cups of wine—all related to the four expressions used in Exodus 6:6-7 to describe the deliverance of Israel") is revealed in the map: Solomon's four sailing ships towing rafts "of the trees of Lebanon" to build his temple; four cows symbolizing prosperity; and the four points of the compass. The semicircular loop of the Exodus route across the Red Sea suggests to Brodsky a Talmudic influence, while the figure of a seminude woman with a parasol sitting on an alligator, an ancient allegorical symbol for Egypt, reflects Christian art.

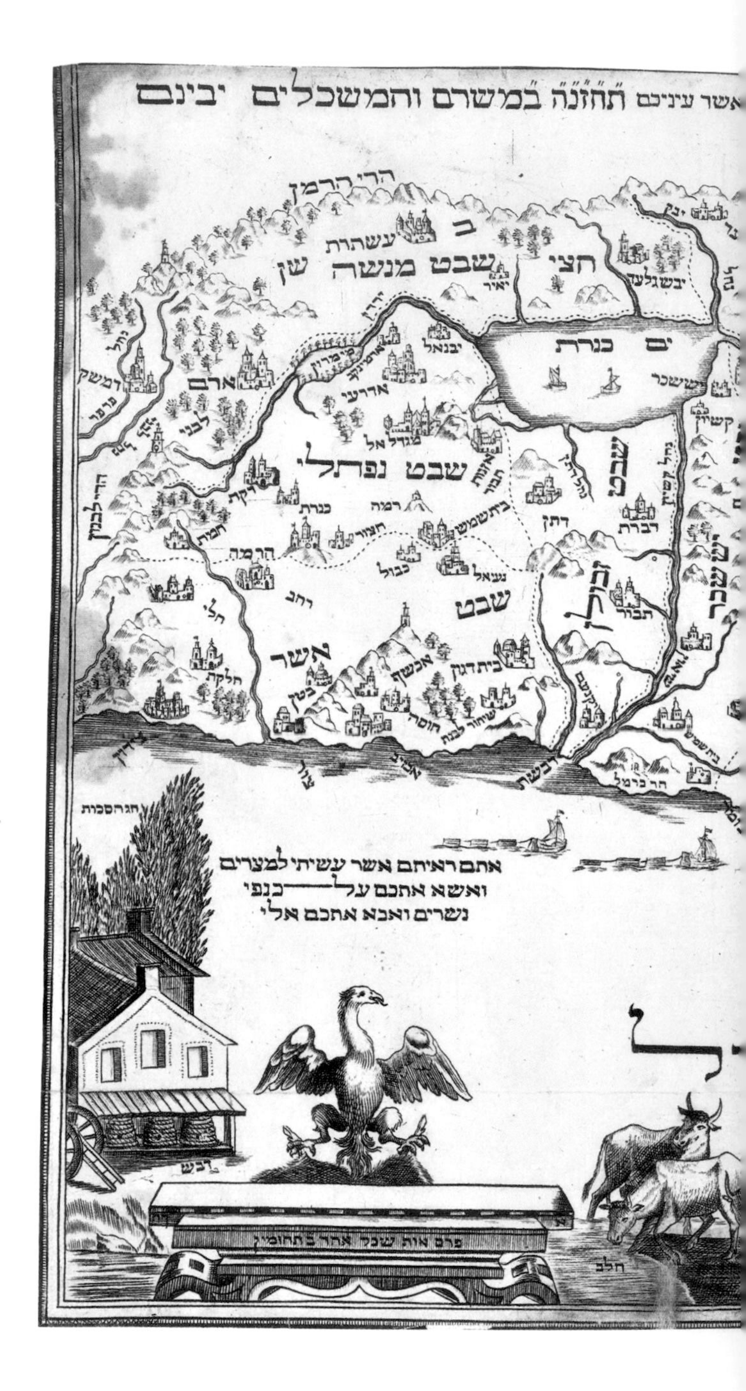

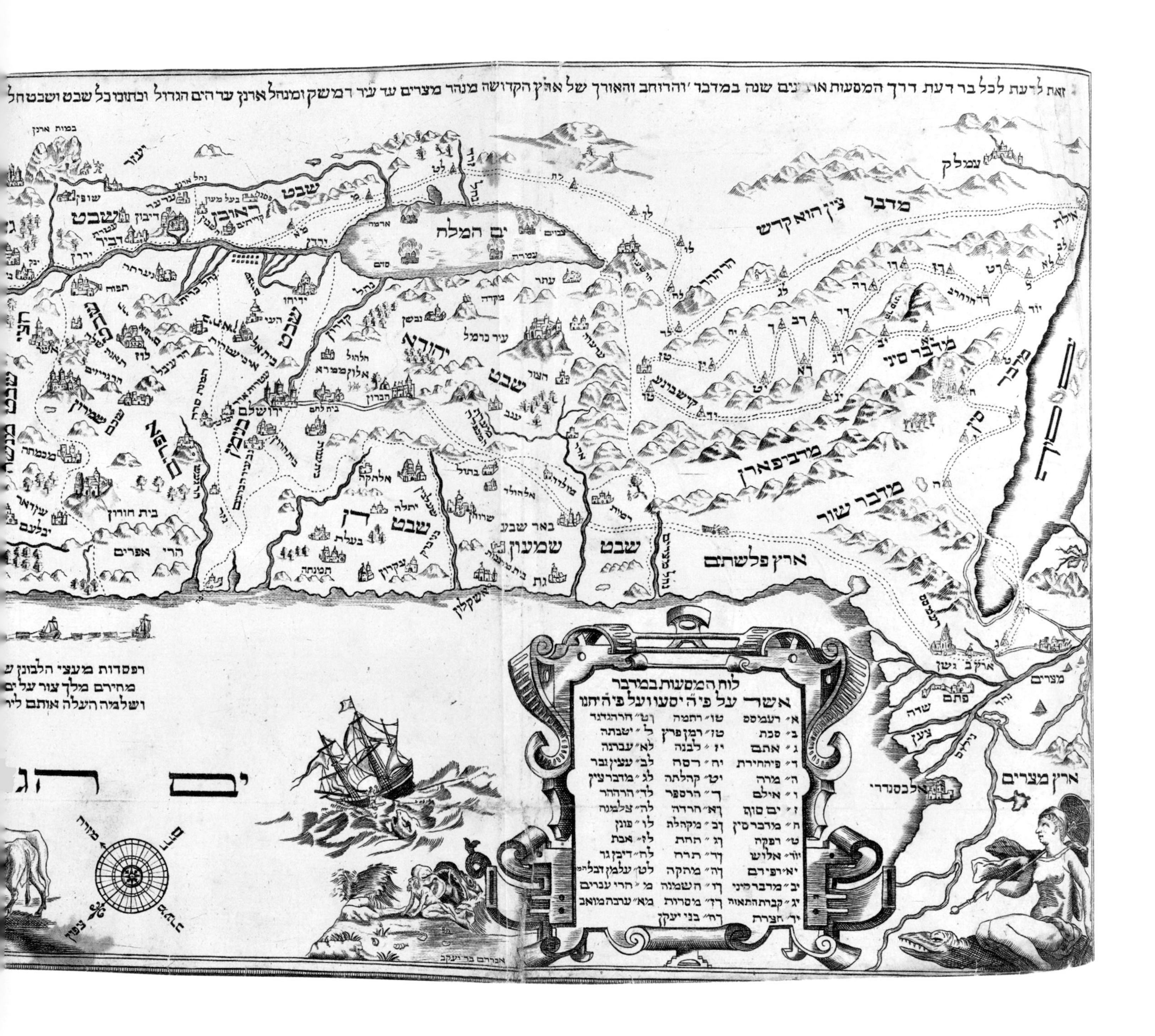

1695 Map of Eretz Hakodesh (the Holy Land) by Avraham bar Yaacov in Seder Haggadah Shel Pesah (Passover Haggadah)
Copperplate Engraving, 10 x 19.1 in.
Library of Congress, Washington, D.C.

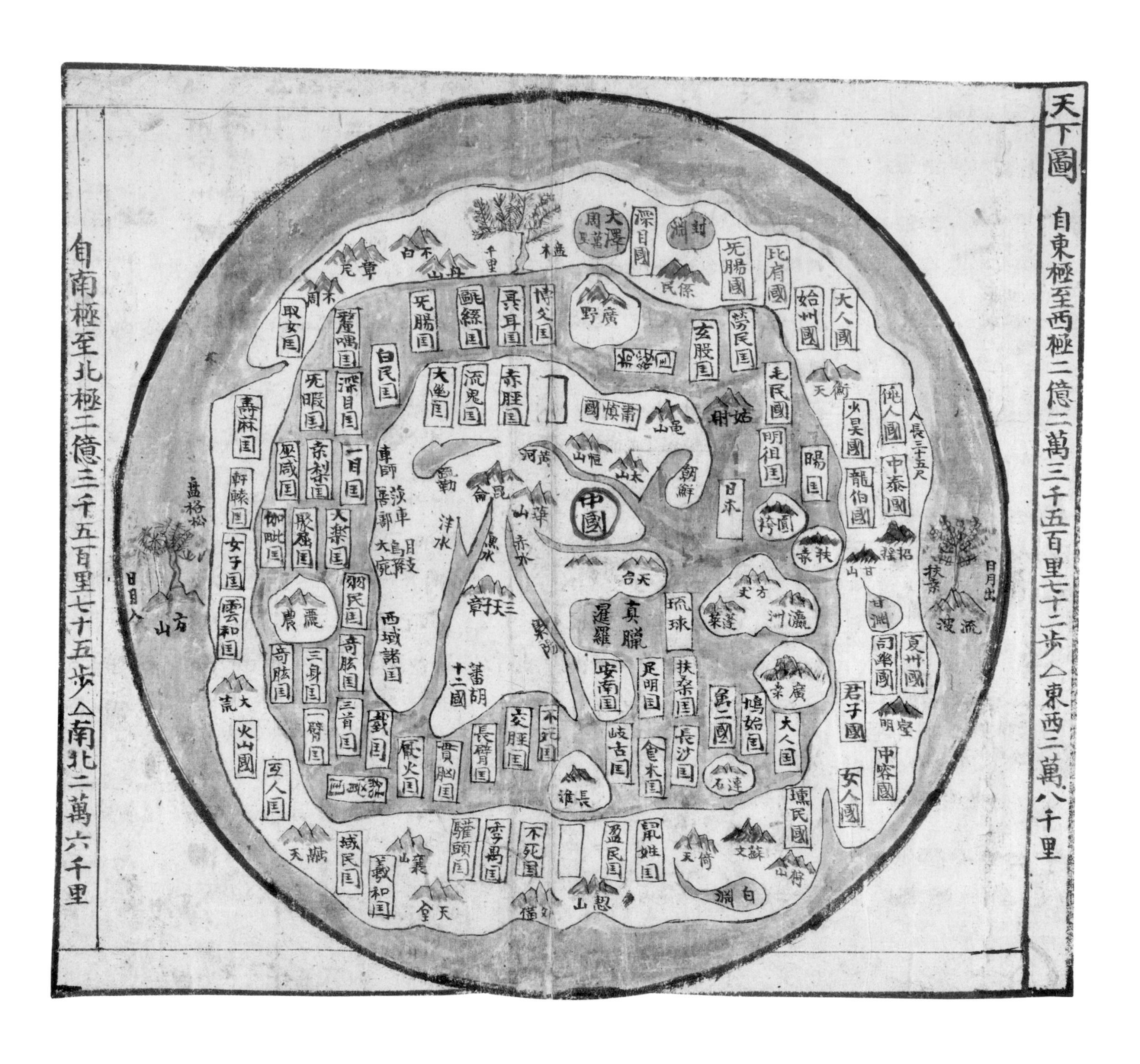

Circa 1700 Chonhado World Map (Map of All Under Heaven)
Woodcut, 12.5 x 15.5 in.
Library of Congress, Washington, D.C.

Wheel Map

DESPITE ITS NICKNAME—THE HERMIT KINGDOM—implying self-imposed introspection, Korea has a long history of interest in the wider world. That interest is reflected in a mapmaking tradition closely associated with Chinese cartography but transformed by Korean culture. The most common Korean map type from about 1700 to the end of the 19th century was the *chonhado*, popularly known as the "wheel map," a circular world map in both manuscript and woodcut formats. Uniquely Korean in concept and design, the chonhado (map of all under heaven) reflected a vision of the world that embraced Daoist and Confucian traditions as well as Chinese and Korean geographical concepts.

At the center of this chonhado an inner continent is depicted, which Korean historian Gari Ledyard believes evolved from the 1402 *Kangnido* world map described earlier. He suggests that this generalized continent represents, from top right to left, the Korean Peninsula; China (the name is highlighted with a red circle) and its great rivers, the Huang (Yellow) and Chang (Yangtze); the Tonkin Gulf; Vietnam; the "Twelve Barbarian Lands"; the Arabian Peninsula; and Africa. The focal point of the inner continent is Mount Kunlun in present-day Tibet; the peak is the center of the world in Daoist and Buddhist cosmology and Chinese classical literature. The lake located northwest of Mount Kunlun represents a combination of the Mediterranean and Black Seas, with Europe occupying the northwest section of the continent.

This inner continent is surrounded by an inner sea, which in turn is encircled by an outer land ring—the "wheel," which gives the map its name—and another ring of water. The inner sea depicts Japan to the east, as well as Cambodia and Thailand, which are represented as islands. Numerous place-names identify existing and fictitious countries, lakes, rivers, and mountains. Two trees, one on either side of the outer sea, represent a rising and a setting sun (East and West).

The chonhado dates from about the 16th century, but based on analysis by Nakamura Hiroshi, an authority on Korean maps, some features were derived from ancient Chinese geographical lore. What fascinates the historian of cartography is the continued popularity of this map type long after the introduction of more up-to-date world maps. Korean diplomats first acquired a Chinese world map in 1602, one prepared by the Jesuit cartographer Matteo Ricci. Additional purchases of Western-style maps followed, but they were not widely distributed. Not until 1834 did a Korean publisher print a Western-style map for general distribution. Ledyard attributes the continued popularity of the chonhado to the "security and familiarity" it gave Korean map readers. Steeped in Chinese classical literary traditions and Confucian principles, they took comfort in the world they found here: a flat Earth—with China, Korea, and Japan comprising much of the world's inner continent—and both classical and modern place-names of nearby countries.

River of the Dead

FRENCH EXPLORATION AND MAPPING OF THE NEW World during the second half of the 17th century coincided with the shift of the center of mapmaking from Holland to France. As the Dutch map trade was losing its vitality, French cartography reemerged with the support of royal patronage. Louis XIII hired Nicolas Sanson, the most influential 17th-century French cartographer, to teach him geography. Louis XIV, the Sun King, established the French Royal Academy of Sciences, which made notable advances in surveying, geodesy, and cartography. Under the academy's leadership, in 1669 Italian astronomer Giovanni Domenico Cassini and Jean Picard initiated a topographic survey of France—the world's first national mapping program—placing cartography on a scientific basis.

One of Cassini's students, Guillaume Delisle, working with his father, Claude, reformed French commercial cartography by introducing critical analysis of source material and by grounding their work within the astronomical and mathematical sciences. As members of the French Royal Academy of Sciences and geographers to the king, the Delisles had access to reports sent to Paris by missionaries, soldiers, and fur traders from western outposts in America.

The 1703 *Carte du Canada* represents a significant advance in the cartography of New France. The map's geographical grid of latitude and longitude was projected according to the latest measurements. Many documents were analyzed for accuracy. Journals and maps by explorers René La Salle and Father Louis Hennepin, and the royal hydrographer at Quebec, provided data about the upper Mississippi watershed and lower Missouri River. The Delisles' portrayal of the Great Lakes was the best to date, and the city of Detroit is depicted for the first time.

Most intriguing is the portrayal of the two rivers in the lower left quadrant, extending from the Mississippi northwestward to the Pacific, separated by a single range of mountains. Despite the Delisles' efforts, they fell victim to notions of Baron Louis de Lahontan, a French soldier of fortune. He claimed to have reached the "Shining Mountains" by a river he explored, which he named *Riviere longue*, or Long River, called by some Indians the *Riviere morte*, or Dead River. Although the Delisles noted the baron may have "invented these things," later readers, including Thomas Jefferson and Meriwether Lewis, perceived this water route as key to westward expansion. The *Carte du Canada* was distributed widely.

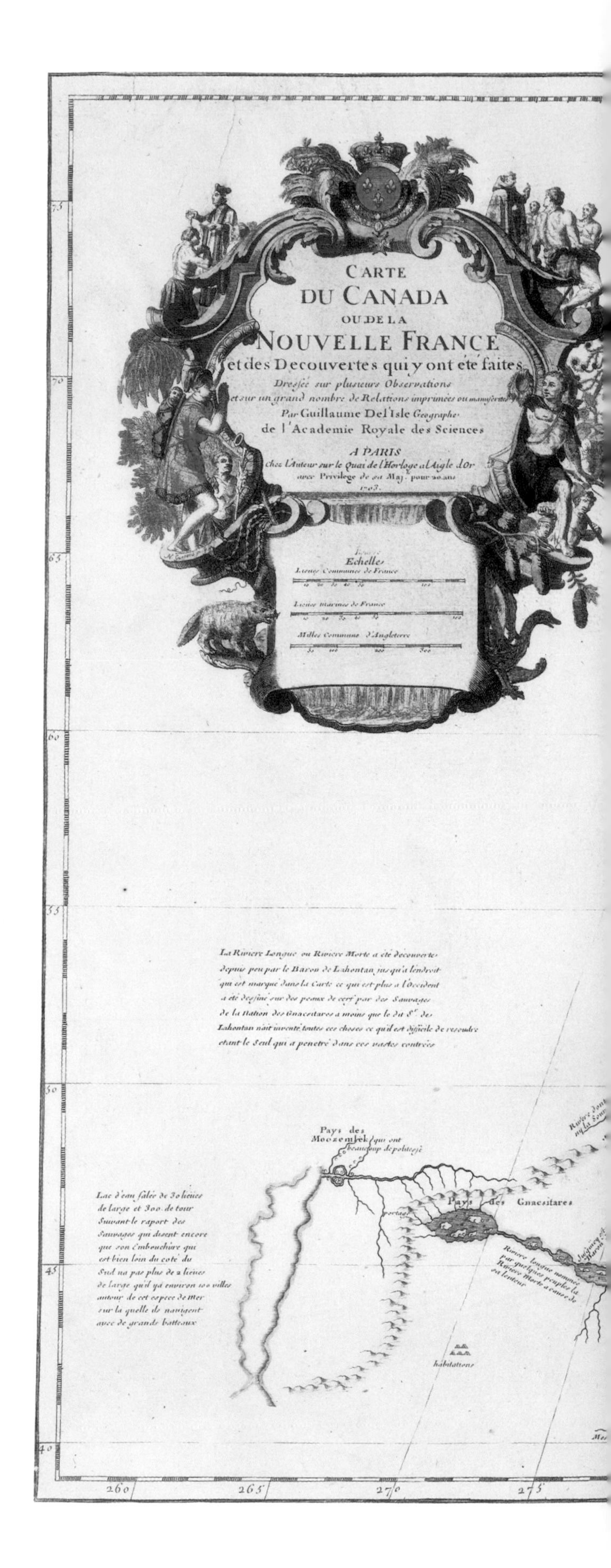

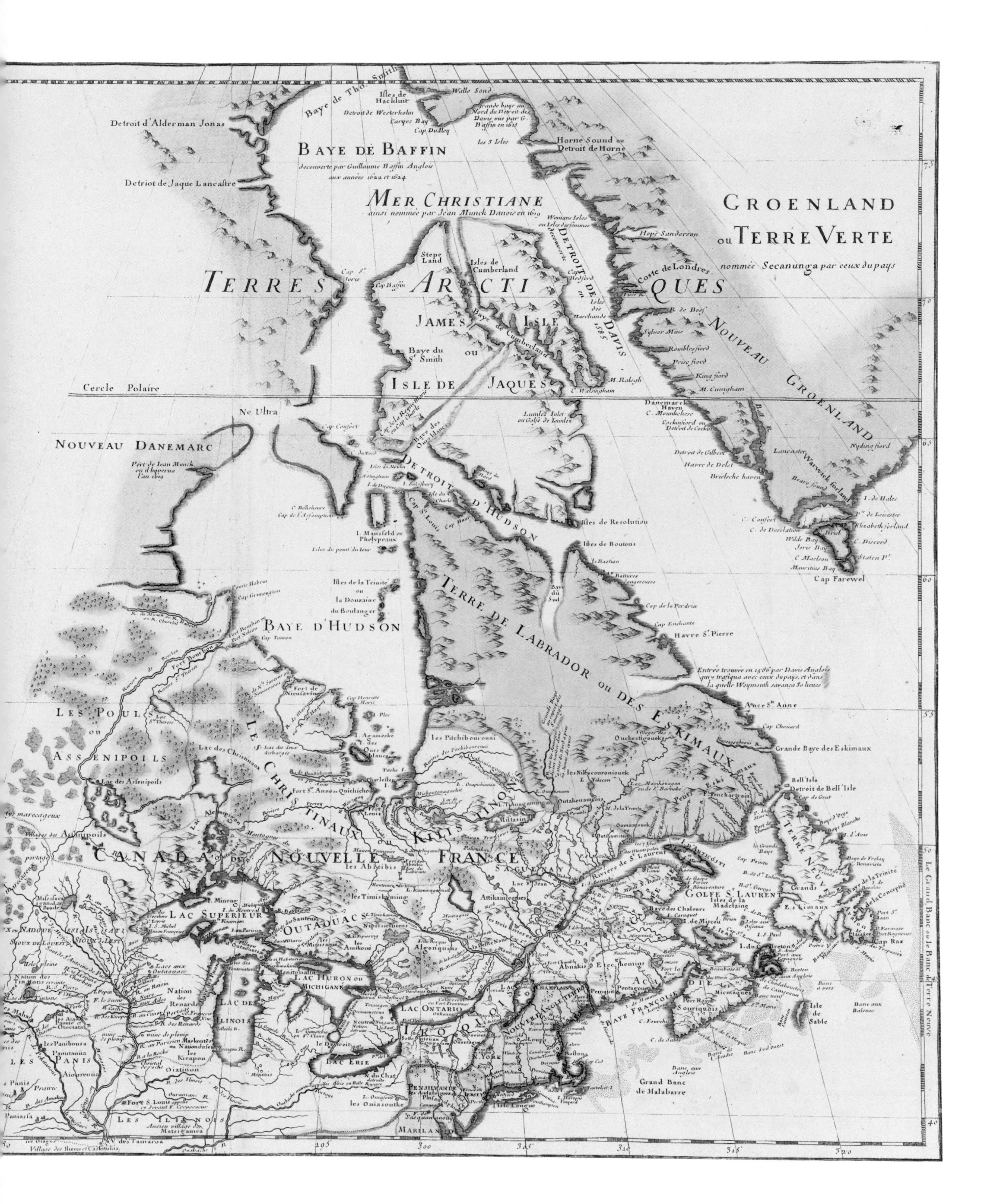

1703 Map of Canada (Carte du Canada ou de la Nouvelle France) by Guillaume Delisle
Copperplate Engraving Published by Jean Covens and Cornelis Mortier, 1729, 19.25 x 22.5 in.
Library of Congress, Washington, D.C.

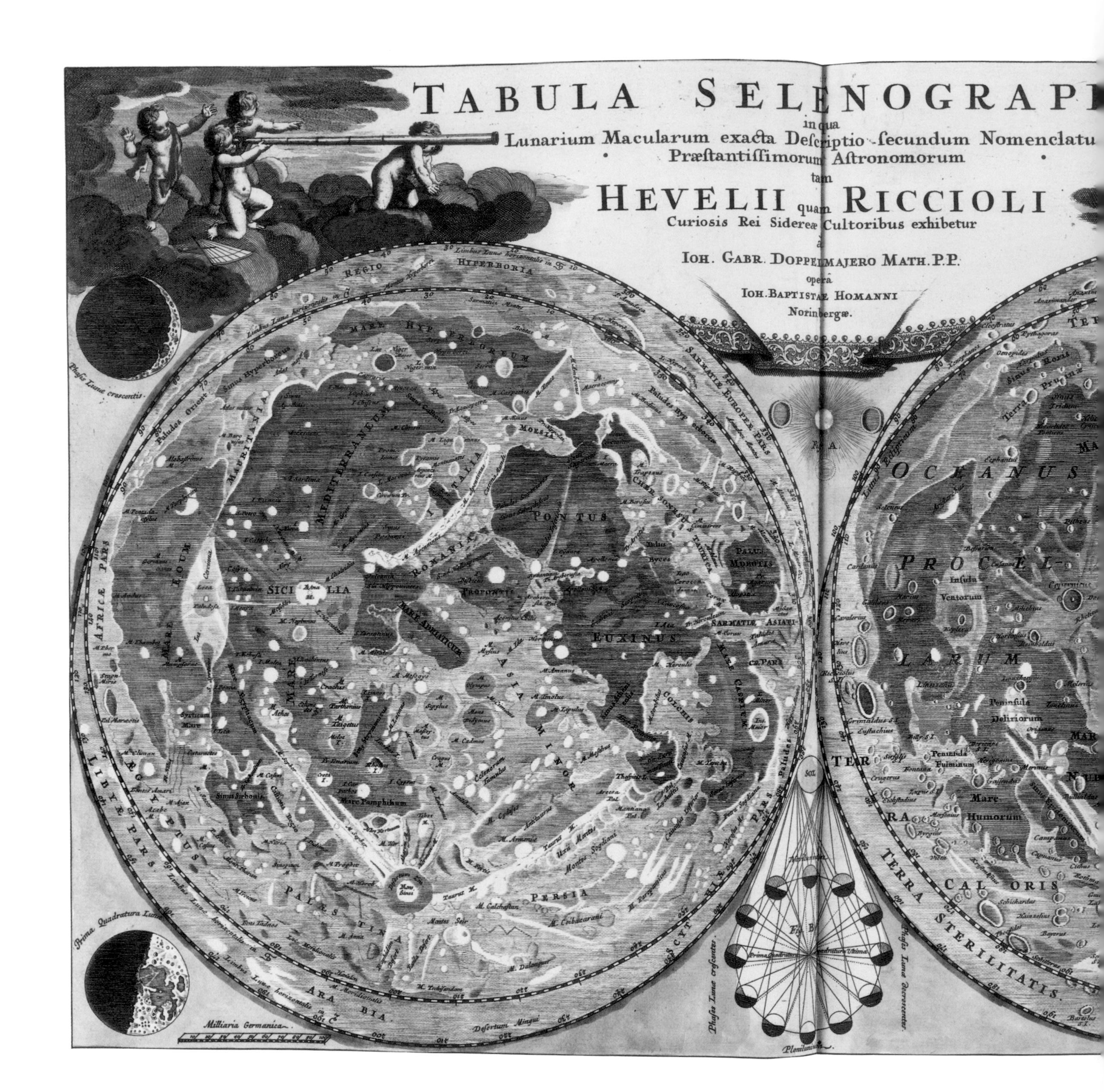

1708 Tabula Selenographica (Lunar Chart with Lunar Markings) by Johann Gabriel Doppelmayr
Copperplate Engraving by Johann Baptist Homann, 20 x 23 in.
Library of Congress, Washington, D.C.

Mapping the Moon by Telescope

MAPPING THE MOON AWAITED THE DEVELOPMENT of the telescope. Within a year of its invention in 1608, Galileo Galilei viewed the moon's "vast protuberances, deep chasms, and sinuosities." Using a telescope of his own design, the Italian scientist mapped "the ancient spots," which he speculated were vast seas. Despite rather poor results due to the limited power of his telescope, Galileo's dramatic woodcut of the four phases of the moon launched lunar mapping.

Johannes Hevelius advanced the discipline. Founder of the science of selenography, named after Selene, the Greek goddess of the moon, Hevelius was an accomplished astronomer, mathematician, artist, and copperplate engraver. He funded his research by managing the family brewery in Danzig. After six years of making lunar observations from an observatory he built behind his home, the Polish brewer personally compiled and engraved some 130 plates, including three maps, which he published in his 1647 atlas of the moon, *Selenographia, sive, Lunae descriptio*. Hevelius established the convention of portraying the entire visible lunar surface, which includes 59 percent of the moon because of the oscillation in its orbit as seen from Earth. Less successful was his choice of place-names, selected from classical names adapted from places around the Mediterranean Sea. Hevelius's pioneering work remains as one of the most comprehensive lunar atlases published. Pope Innocent X so admired the *Selenographia*, it has been said that "he wished it hadn't been written by a heretic."

Pope Innocent X got his wish when two Italian Jesuits collaborated on a lunar map in 1651. Cartographer Francisco Grimaldi and astronomer Giovanni Battista Riccioli introduced new nomenclature that remains in effect to this day. The dark areas on the moon, thought to be seas, were named after weather conditions, such as the Sea of Tranquility. The names of other features commemorated scientists and scholars.

Cartographers unable to choose between these two systems of moon nomenclature portrayed the two lunar maps side by side. In this map compiled by Nürnberg astronomer Johann Gabriel Doppelmayr and engraved by Johann Baptist Homann, Hevelius's lunar map is depicted on the left, Riccioli's on the right. By the time Doppelmayr's map appeared, the surface of the moon was better mapped than many parts of Earth.

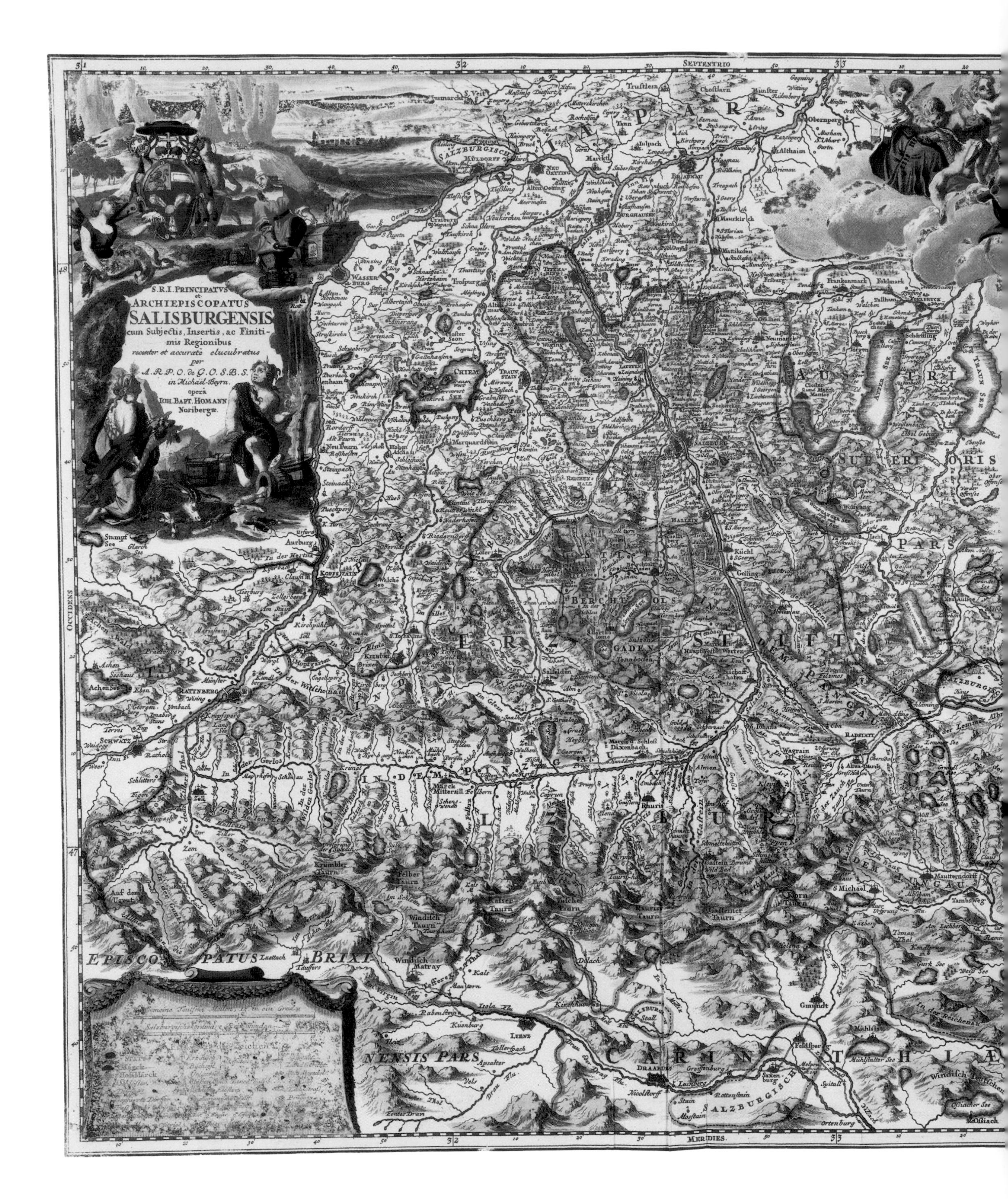

1712 Archiepiscopatus Salisburgensis (Archbishopric of Salzburg) by Johann Baptist Homann
Copperplate Engraving, Hand-Colored, 18.9 x 22.8 in.
Library of Congress, Washington, D.C.

Angels and the Archbishop

JOHANN BAPTIST HOMANN LED THE REVIVAL OF German cartography at the beginning of the 18th century. Trained for the priesthood, he "escaped" in 1687 from a Dominican friary in Würzburg, after his conversion to the Evangelical faith, according to a contemporary account. To support himself, he became a map colorist and an engraver. He was banished to Leipzig several years later for returning to "the papists," and there he engraved 34 maps for Christopher Cellarius's celestial atlas. Reinstated as an Evangelical, he returned to Nürnberg and established a map-publishing firm in 1702 that thrived for a century. Homann was named geographer to the Holy Roman Emperor and honored with membership in the Prussian Royal Academy of Science.

To capture the German map market, which had been dominated by the Dutch and French throughout the 17th century, Homann specialized in cheaper maps of quality that could be sold individually or combined in atlas format. Separate title pages and indexes were also printed. Customers created their own atlases by selecting maps they wished to include, with the result that each Homann atlas is unique.

Most of Homann's maps were copied directly from Dutch and French maps, generally without attribution, but a few were original, including this beautiful 1712 map of the Archbishopric of Salzburg, an ecclesiastical state of the Holy Roman Empire. It was based on a small manuscript map drawn by Benedictine monk Odilo von Guetrather, author of a course on the teaching of geography and the use of maps.

Homann emphasized political geography and place-names, subjects popular in Germany. Regions were colored according to political, ecclesiastical, and legal characteristics, but Homann differentiated his maps from their foreign counterparts by coloring an entire area, using shades to show subdivisions. Mountains were portrayed in profile or perspective, as viewed from the south, often with some artistic merit.

The engraved baroque-style title legend and vignette, boldly hand-colored for impact, was designed to amuse, instruct, and perhaps appeal to spiritual sentiments. In the upper right, Count Franz Anton Harrach, Archbishop of Salzburg (1709-29), is portrayed. He is surrounded by protective church officials, angels, and figures representing the dioceses of Salzburg, identified by their coats of arms. The title cartouche illustrates attractions for which the region is known, including mineral baths, silver, gold, and a salt mine. The shield above the title displays Salzburg's coat of arms above the Harrach family emblem.

Picture of the Floating World

ITINERARY, OR ROUTE, MAPS ARE AS OLD AS WORLD maps, and are found in every major mapping culture. In the Western tradition, road itineraries such as the 3rd-century road map of the Roman world or the 12th-century map of Britain by Matthew Paris were enhanced with small pictures of prominent way stations or pilgrimage sites. The European cartographer used art to enhance the map and direct the reader. Pictorial images were valued but not primary to the map's structure. In the Japanese tradition, the artistic and emotional elements of map design often surpassed the practical. Japanese map users, nurtured on the *ukiyoe* (pictures of the floating world) style of art, preferred maps that evoked the landscape. This tradition continued well into the 19th century.

The first Japanese itinerary maps were prepared during the Edo period (1603-1868), when feudal lords were required to spend every second year in Edo (present-day Tokyo). In the course of the lords' traveling with large retinues of retainers and servants, towns and cities developed along their routes. Enterprising mapmakers soon devised route maps. They were prepared in manuscript and woodcut form, and ranged from pocket-size to long scrolls.

The information for these maps was obtained from a variety of sources. Among the most famous routes was the Tokaido, the road extending from Edo to Kyoto, Japan's ancient capital, and then on to Nagasaki by a land and sea route. It was officially surveyed under the direction of the Tokugawa shogunate in the latter half of the 17th century. In addition to the route, represented by a single or double line, these maps display pictures of travelers, inns, historic sites, and landscape views drawn by ukiyoe artists such as Hishikawa Moronobu, Ando Hiroshige, and Katsushika Hokusai. These scroll maps with their vivid details, such as this image of the Mount Fuji region—a segment of a much larger map—are voyages in themselves, historian Unno Kazutaka explains: "Without moving from his chair, one can gaze at misty mountains in the distance and then pass on to inspect the remains of famous sites along the road; he can visit soaring castles and come right down to the travelers on their way and the boatmen rowing across the water."

1736 Segment of Japanese Itinerary Map Showing Mount Fuji Region
Rice Paper Scroll Map, Hand-Painted, 13.75 x 369.5 in.
Library of Congress, Washington, D.C.

1734 Imperii Russici Tabula Generalis (General Map of the Russian Empire) by Ivan Kirilov
Copperplate Engraving, 36 x 23 in.
Houghton Library, Harvard University, Cambridge, Massachusetts

Peter the Great's Map of Russia

Inspired during a visit with the French cartographic innovator Guillaume Delisle, Tsar Peter the Great set about to map the vast Russian Empire. He established mathematical and navigational schools in Moscow and St. Petersburg and engaged foreign teachers to prepare "the sons of princes, nobles, and their serfs" as navigators, surveyors, geodesists, and mapmakers. Six years of theoretical studies in French mathematical geography and English surveying techniques were followed by six to nine years of practical experience in England, Holland, or Italy. Peter himself administered their final examination.

The framework for a national map was established in 1720 when Peter inaugurated a national instrument survey under the direction of the Russian Senate, where the national cartographic service was located. Peter himself probably wrote the basic instructions for conducting the cartographic survey, which was placed under the direction of Ivan Kirilovich Kirilov. A Senate secretary, Kirilov was educated at the Naval Academy in the Kremlin from age 13 to 18, then trained in both London and Amsterdam. From 1720 to 1733, teams of geodesists, surveyors, and topographers fanned out over Russia, developing a skeletal network of traverse surveys along roads and streams. Latitude was determined by quadrant, longitude by dead reckoning, and distances were measured by chains.

Kirilov originally planned to publish the results of these surveys in three atlas volumes, each with 120 regional maps, but his efforts were opposed by the Russian Academy of Sciences, which supported the work of Guillaume Delisle's brother Joseph, whom Peter brought to Russia in 1725 to conduct a geodetic survey of the country. Only 37 maps were engraved, funded personally by Kirilov. The most important was this general map of the Russian Empire, compiled in 1731-32 by six Senate geodesists under Kirilov's direction. It covers the area from Kiev east to the Pacific coast, and from the Arctic coast south to Turkey, Persia, Mongolia, and China. Lacking the accuracy of later maps, it nevertheless was the first to portray the results of Vitus Bering's expedition in 1728, which confirmed the separation of Asia and America. Kirilov's map, dedicated to Peter's niece, Empress Anna Ivanovna, who reigned from 1730 to 1740, was suppressed by the Academy of Sciences, and only a few copies survive.

Propaganda and Political Map of North America

JOHN MITCHELL'S MAP OF NORTH AMERICA generally is considered the most significant map in the history of the United States. Embracing nearly half of the North American continent, its coverage extends from Hudson Bay south to central Florida and from the Atlantic coast to the present Missouri-Kansas border. Mitchell was one of the British colonies' leading scientists, a recognized authority on botany, zoology, and climatology. He practiced medicine in his native Virginia for 12 years before relocating to London. There he pursued a variety of scientific interests, including the preparation of a map of North America to illustrate a planned book on the natural history of the continent. Over time the focus of the map changed from science to politics, as Mitchell grew concerned that the vast unexplored region west of the Appalachians might fall into French hands.

Mitchell's manuscript brought him to the attention of the Board of Trade, the governing body of the British colonies, who shared his concerns. The board commissioned Mitchell to revise his map and gave him access to its vast archives of maps and reports, where he found and copied a manuscript map of the Ohio Valley prepared by Maj. George Washington. The board also directed colonial governors to furnish Mitchell with new maps and descriptions of their provinces. Joshua Fry and Peter Jefferson, father of Thomas Jefferson, drew the map of Virginia. Latitudes and longitudes of the Atlantic coast were obtained from ships' logs from the British Admiralty.

The map was engraved on eight plates by Thomas Kitchin, one of London's finest engravers, and printed on February 13, 1755. Mitchell's map provided the first detailed topographic image of the future United States in a relatively accurate manner. Because his purpose was promotion of British territorial claims, Mitchell portrayed provincial boundaries stretching from sea to sea. Highlighting in solid colors strengthened the impression of ownership. He buttressed British claims with notes documenting English frontier settlements, such as "English Ft Establish'd in 1748" or "Log's T[own] built & Settled by the English." French claims were reduced by restrictive boundaries (shown in blue) or challenged.

Issued at the onset of the French and Indian War, the map found a ready audience. A second edition followed in 1757. The map was reprinted at least 21 times in four languages and reproduced many more times without credit. Negotiators used the map to establish U.S. boundaries for the Treaty of Paris (1783) that concluded the Revolutionary War.

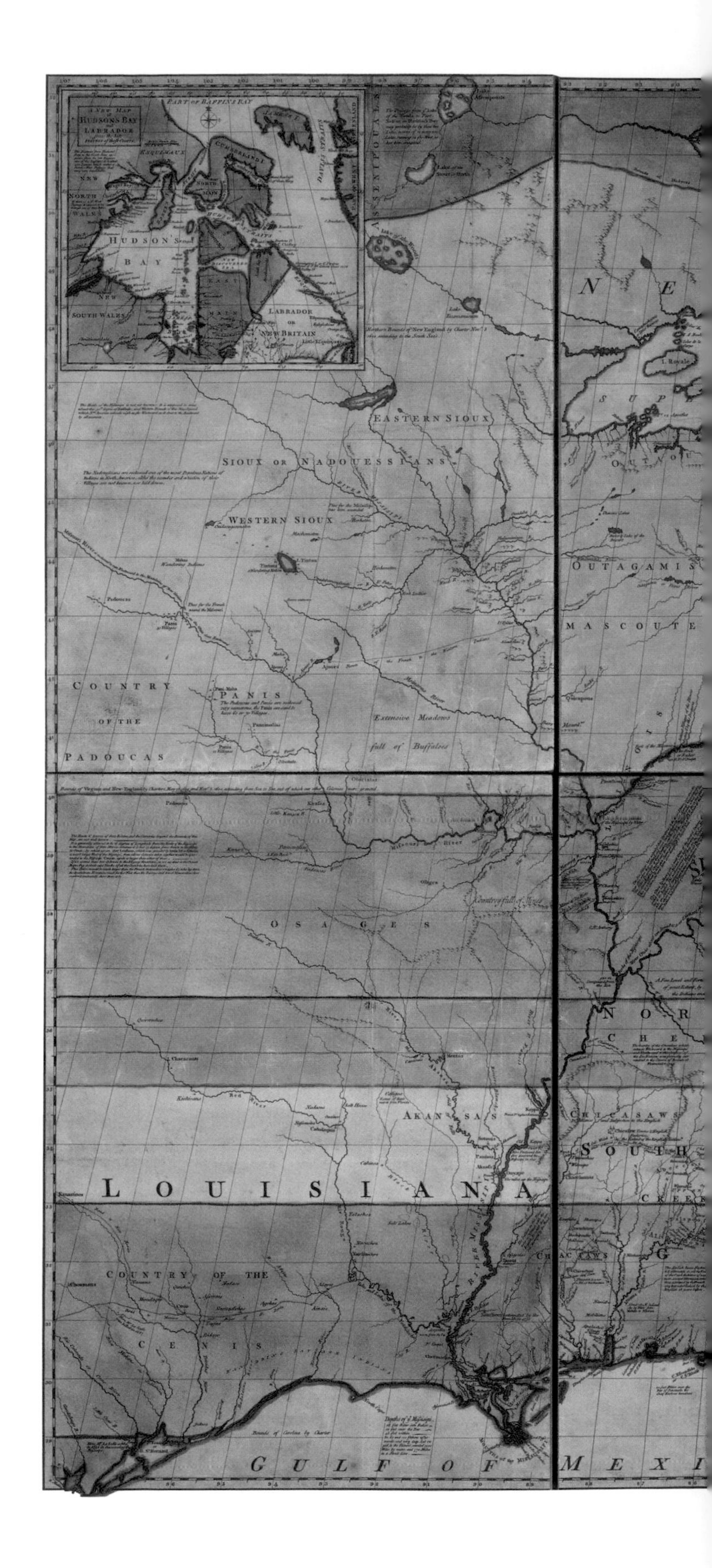

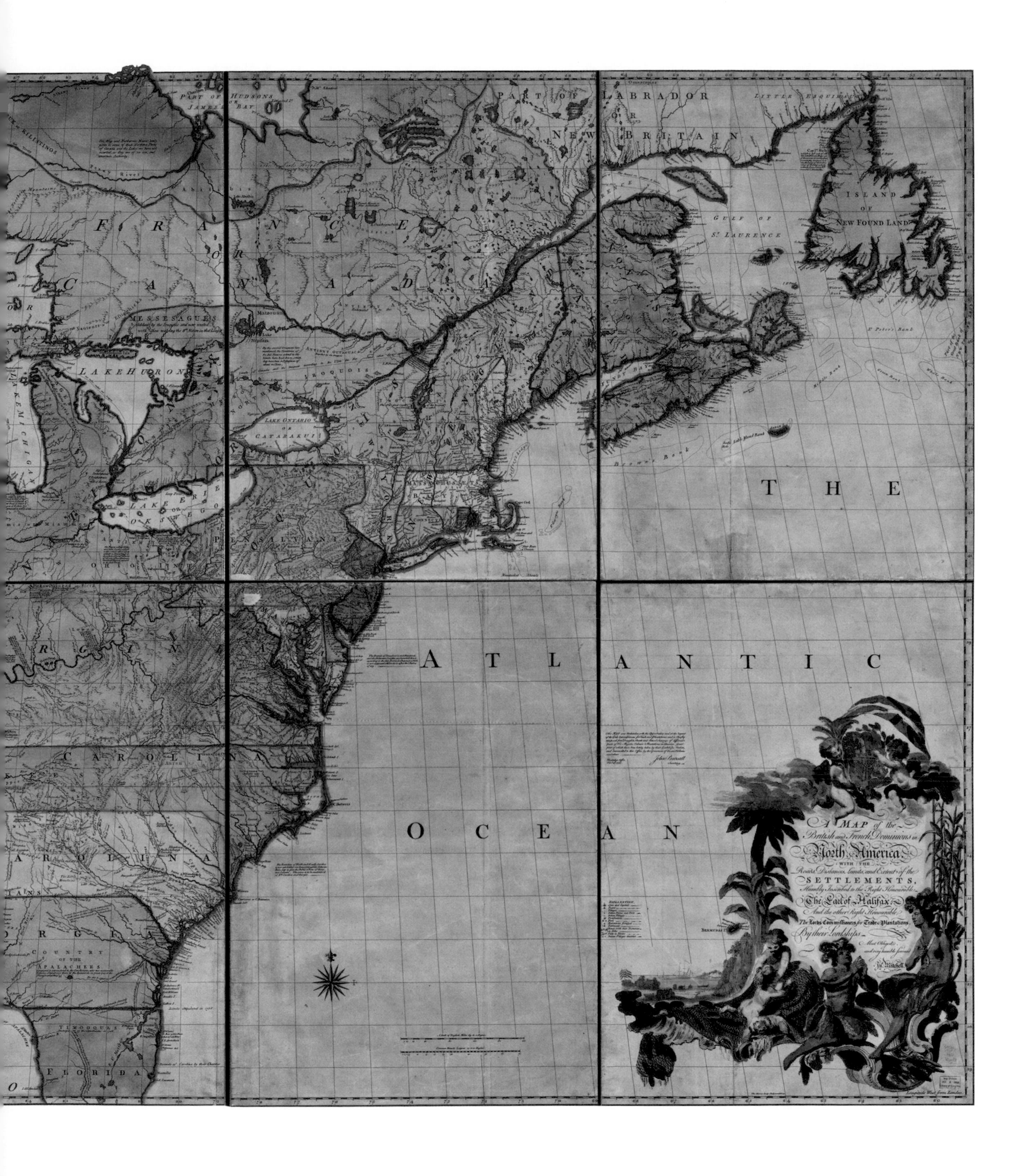

1755 Map of the British and French Dominions in North America with the Roads, Distances, Limits, and Extent of the Settlements
by John Mitchell
Copperplate Engraving, Hand-Colored, 152 x 75 in.
Library of Congress, Washington, D.C.

g
f

SECTION FOUR

National Surveys and Thematic Cartography

Panoramic View of the Rockies, 1877 by William Henry Holmes

Introduction

From the latter half of the 18th century through much of the 19th, the Western mathematical style of mapmaking spread throughout the world, inspiring the standardization of techniques, symbols, projections, and scales. By the 1880s, maps worldwide began to look alike. However, traditional cartography continued alongside this Western style in some cultures, as exemplified by an anonymous 1850 Chinese scroll map of Hainan Island. In other societies, Western maps were accepted for reproduction, but indigenous place-names and legends were substituted for Western ones. An 1803 map of Africa, issued in Istanbul by the Engineering College Press, illustrates this.

New developments in printing processes and technologies, the initiation of national surveying and mapping programs, and the introduction of thematic maps contributed to the diffusion of Western mapping traditions. Concurrently, the vast interior regions of North and South America, Africa, Australia, and the Pacific Basin were being explored and mapped for the first time.

New printing methods were introduced to reduce costs, improve production times, and properly display new kinds of data. Copperplate engravings could not be matched in quality, as illustrated by Cassini's 1789 map of the Paris region or the British Survey of India's 1833 map of the Nilgiri Mountains, but the process was labor-intensive and required skilled craftsmen, and press runs were limited. Steel engraving, which produced longer press runs, became popular for the production of atlases. It is illustrated by John Tallis's 1851 map of Egypt. Another effort to reduce engraving costs was wax engraving, represented by Gaylord Watson's 1875 U.S. centennial map. It became the dominant method of commercial map reproduction in the United States during the last quarter of the 19th century.

The introduction of the lithographic process in 1798 by Alois Senefelder, a Munich printer, transformed the map-publishing industry by simplifying the work of the mapmaker, making production less expensive, and providing technical options for conveying a wider range of data, including symbols. Representing the lithographer's art is Gouverneur K. Warren's 1857 military map of the American West and Charles Minard's 1862 world map, both hand-colored. Color printing or chromolithography, first developed in France and Germany in the late 1820s, is illustrated by Matthew Maury's 1847 wind and current chart of the Atlantic and Scribner's 1883 statistical atlas of the United States.

Photography, invented by Louis Daguerre in 1839, ultimately had a major impact on cartography but was adopted only slowly. One of the first to take advantage of this new technology was a Civil War officer, Capt. William C. Margedant, who began producing "sun print" maps in 1863 for the Union Army.

National topographic surveying and mapping programs were first conducted in France under the brilliant direction of four generations of the remarkable Cassini family. Similar programs soon were undertaken in

Denmark, the Netherlands, Norway, across the Channel in England, and in its colonial empire. By the end of the 19th century, most countries had begun producing official topographic maps.

These programs changed the way maps were made. The single mapmaker of earlier periods was replaced by groups of specialists—field men, surveyors, astronomers, mathematicians, hydrographers, draftsmen, terrain artists, letterers, photographers, engravers, and lithographers. Their product often consisted of several hundred map sheets, representing hundreds or thousands of miles. Designed primarily for administrative and military purposes, their work also provided basic data for commercial mapmakers.

Official hydrographic surveying was initiated with the establishment of the British Admiralty in 1795. Maritime explorer James Cook, who charted the Pacific Basin, laid the foundation with charts such as his 1769 nautical chart of the Society Isles. James Alden's 1854 chart of California was produced for the U.S. Coast Survey. The United States was unique among major maritime nations in having two government agencies responsible for nautical charting: the Coast Survey, limited to coastal waters, and the Navy's Hydrographic Office, for waters beyond the coasts.

THEMATIC MAPS FOCUS ON THE DISTRIBUTION OF A PARTICULAR PHENOMENON. THEIR DEVELOPMENT coincided with the emergence of scientific disciplines such as meteorology, geology, and, later, the social sciences. Often designed by scientists rather than cartographers, these maps introduced innovative symbols and designs to display their data. On Benjamin Franklin's 1769 map of the Gulf Stream, for example, arrows were used to show the direction of flow. William Smith, on his 1815 geology map of England and Wales, employed a novel use of color. Other creative thematic maps include Heinrich Berghaus's 1845 map of world plant geography; Maury's 1847 chart of ocean winds and currents; French statistician Charles Minard's dynamic flow map in 1862, designed to illustrate European emigration; and American geographer Henry Gannett's 1880 map of presidential election results.

The focus of exploratory mapping and charting during this period was on the Pacific Basin, first charted extensively by Cook, and the interiors of the Americas, Australia, and Africa. In North America, the West was initially surveyed and mapped by mountain men and Army topographers, often with the help of Indian maps such as the one prepared in 1801 by Blackfeet chief Ac ko mok ki. The first comprehensive map of a section of the trans-Mississippi West, printed in 1814, came from William Clark's manuscript of the Lewis and Clark expedition. Forty years of military mapping culminated with Warren's 1857 War Department map. An 1818 plan of the U.S. Military Academy at West Point by Cadet George Washington Whistler (father of artist James McNeill Whistler) depicts the first school in the United States to conduct formal classes in cartography.

1840 Field Surveying Party on the Northeastern Boundary of the United States by Philip Harry
Original Watercolor
U.S. National Archives and Records Administration, Washington, D.C.

The Surveying Party

TWO INSTRUMENTS USEFUL FOR INLAND SURVEYING and mapping during the 19th century are illustrated in this watercolor of a field surveying party, painted during the resurvey of the international boundary between Maine and Canada in 1840 and 1841. It was drawn at the site by one of the surveyors, Philip Harry. An accomplished artist as well as a mapmaker, the English immigrant later served as chief cartographer for the U.S. Army's Topographical Bureau in Washington, D.C.

Harry pictures a topographer using a reflecting circle with an artificial horizon to determine longitude by measuring the angular separation, or distance, between the moon and some other celestial body such as the sun, a star, or a planet. These lunar distances were recorded either in the early morning or late at night. It was a tedious, time-consuming procedure that required a succession of measurements, spaced a few minutes apart. The entire process could take several hours. Often trees had to be cut down to provide sight lines, which required a team of axmen.

The reflecting circle, later known as the repeating circle, consisted of a full circle with one or two telescopes. The German astronomer Johann Tobias Mayer invented it in the mid-18th century for both navigation and surveying purposes. French naval engineer Jean-Charles de Borda made further improvements in 1787. The instrument was first used extensively for the triangulation of the meridian arc in France.

For surveying in mountainous regions or heavily wooded areas, the reflecting circle was used with an artificial horizon, illustrated in the lower right-hand corner. The artificial horizon consisted of a triangular-shaped box with two mirrors and a small container that held a reflecting medium—normally mercury, but water also could be used. The flat, still surface of the reflecting medium served as the horizon when the true horizon was not visible. The surveyor measured the apparent angle between the sighted object and its reflection, then divided by two. The resulting calculation was highly accurate.

Because the artificial horizon was set on the ground, great care had to be taken to protect it from the vibrations of men and animals. An Army topographical engineer reported on one occasion, "When I placed my horizon on the ground, I found that the galloping of a horse five hundred yards off affected the mercury, and prevented a perfectly reflected image of the stars."

Benjamin Franklin's Gulf Stream

In late 1768, the London-based deputy post-master for the American colonies was asked to respond to a complaint by the Customs Board at Boston questioning why British mail packets sent to New York took two weeks longer to cross the Atlantic than more heavily laden merchant ships returning to Rhode Island. British Maj. Gen. Thomas Gage raised a similar concern, believing that the delay of his mail was due to a resentful packet captain. Fortuitously, this deputy postmaster, the person in charge of ensuring the timely delivery of the transatlantic colonial mail for the British Postal Service, was none other than Benjamin Franklin, one of the most scientifically curious and inventive men of his time.

Intrigued, Franklin consulted with his cousin Timothy Folger, a Nantucket whaler who often sailed to London. Folger laid the blame on the Gulf Stream, that strong river of seawater that flows at the rate of four miles per hour from the Gulf of Mexico to Europe through the Straits of Florida and along the eastern coastline of the United States and Newfoundland. Folger himself had witnessed British packet ships sailing to New York on the Gulf Stream "being carried back by the current more than they are forwarded by the wind." As Franklin later put it, Folger told him "the difference was owing to this, that the Rhode-Island captains were acquainted with the gulph stream, which those of the English packets were not." Franklin, always the practical man, searched for an answer. He found one in maps. "I then observed that it was a pity no notice was taken of this current upon the charts," he noted in a letter to a friend. At Franklin's request, Folger sketched out the limits, direction, and rate of flow of the Gulf Stream on a map, which the deputy postmaster then submitted for printing.

Folger's sketch, along with his sailing instructions, was added to an existing chart of the Atlantic Ocean by reengraving the original copperplate, a common practice during the copperplate era. Although the Franklin-Folger chart generally was ignored by the English packet captains, who, Franklin's cousin observed somewhat sarcastically, "were too wise to be counselled by simple American fishermen," a smaller French version of the chart was issued after the Revolutionary War to aid French shippers. And in 1786, at the age of 79, Franklin had a new version of the chart engraved for the scientific community. It accompanied his article on the Gulf Stream, which appeared in the *Transactions of the American Philosophical Society*.

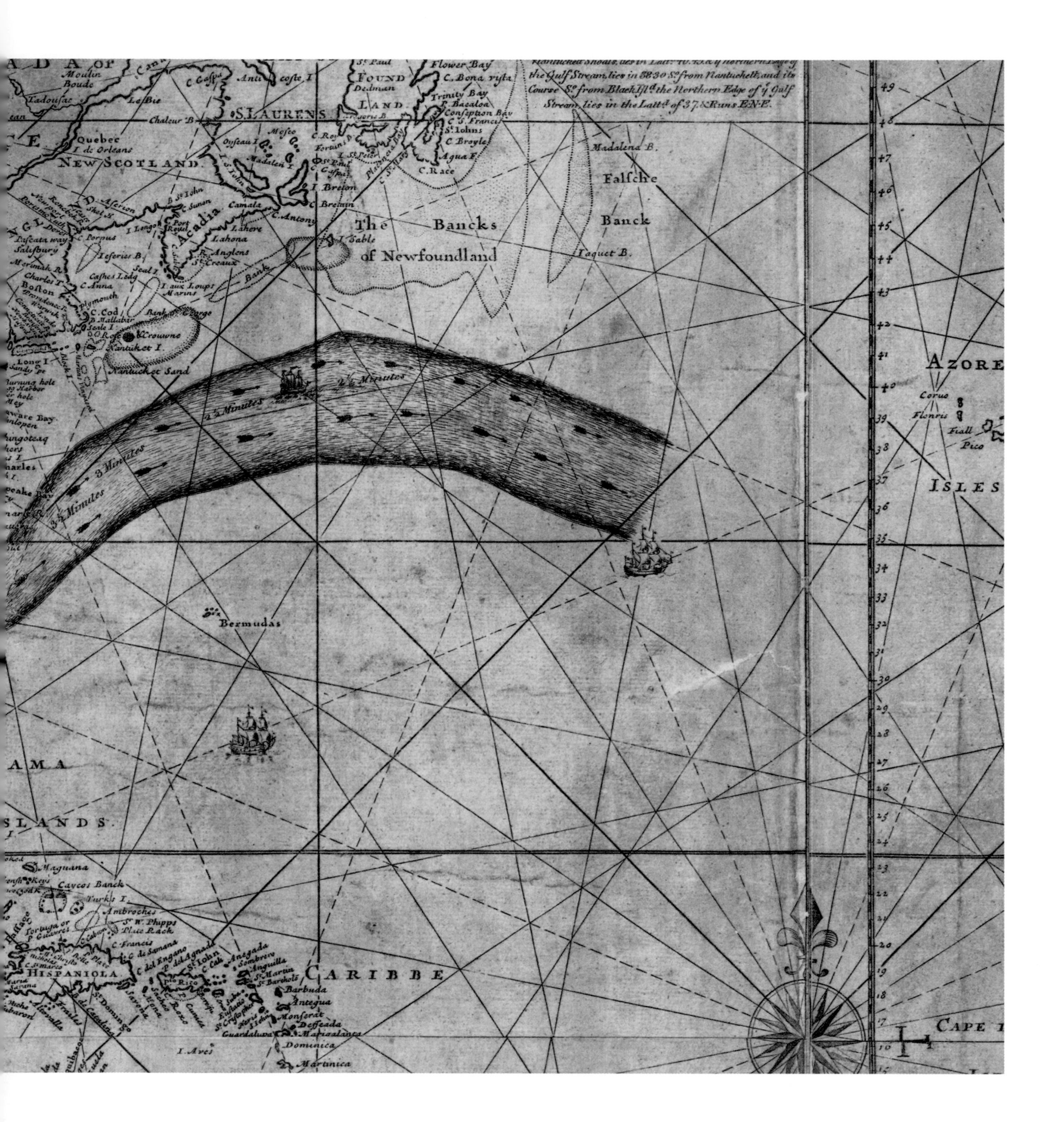

1769 Detail of Gulf Stream Map by Timothy Folger and Benjamin Franklin
Copperplate Engraving by John Mount and Thomas Page, 38 x 34.25 in.
Library of Congress, Washington, D.C.

Cook's Society Isles

THE PACIFIC OCEAN BECAME THE PRIMARY FOCUS of European geographical discovery and exploratory charting during the last half of the 18th century. In little more than 50 years, this immense region, covering more than one-third of Earth's surface, was essentially mapped by French, English, and Spanish maritime expeditions supported by government funds and scientific groups such as the Royal Society in England and the Museum of Natural History in Paris. The images of mythical monsters and sailing ships that inhabited earlier maps of the Pacific were swept away along with the fictitious geographical concepts of the great southern continent of Terra Australis and the Northwest Passage that many believed connected the Pacific with Hudson Bay. These were replaced by detailed charts of actual continental coastlines and extended island groups.

During three extended expeditions, James Cook discovered and charted more of the Pacific Basin than any explorer before or since. From 1768 to 1779, the British naval officer crossed the South Pacific three times, twice venturing into the Antarctic Circle. He cruised the North Pacific, entered the Arctic Sea through the Bering Strait in search of the Northwest Passage, and found his way to Hawaii, where he was slain with four of his men. The Cook expeditions produced the first accurate nautical charts of the Pacific Ocean, based on thousands of astronomical sightings. Marine chronometers, provided by the Board of Longitude, were used extensively for the first time to determine longitude. The charts were produced primarily by Cook and William Bligh, later fated to command the *Bounty.*

Typical is this chart of the Society Isles, located a few miles northwest of Tahiti. Guided by Tupaia, a Tahitian chief, Cook "discovered" and charted these islands in July 1769. Cook learned the basic elements of sea surveying from master navigators and through self-study. He perfected his skills during a five-year survey of the coasts of Newfoundland, establishing basic techniques that were to be followed by British naval surveyors for many years. In the vast Pacific, Cook had to adapt, substituting "running surveys" for the traditional detailed coastal survey. Most running surveying was done aboard ship. Latitude and longitude were observed at intervals; compass bearings, recorded during each change of course, were taken from fixed features on the island; the ship's speed was constantly recorded; and coastal details were sketched from the deck.

CAPTAIN JAMES COOK BY NATHANIEL DANCE
OIL PAINTING, 51 X 40.5 IN.
NATIONAL MARITIME MUSEUM, LONDON, ENGLAND

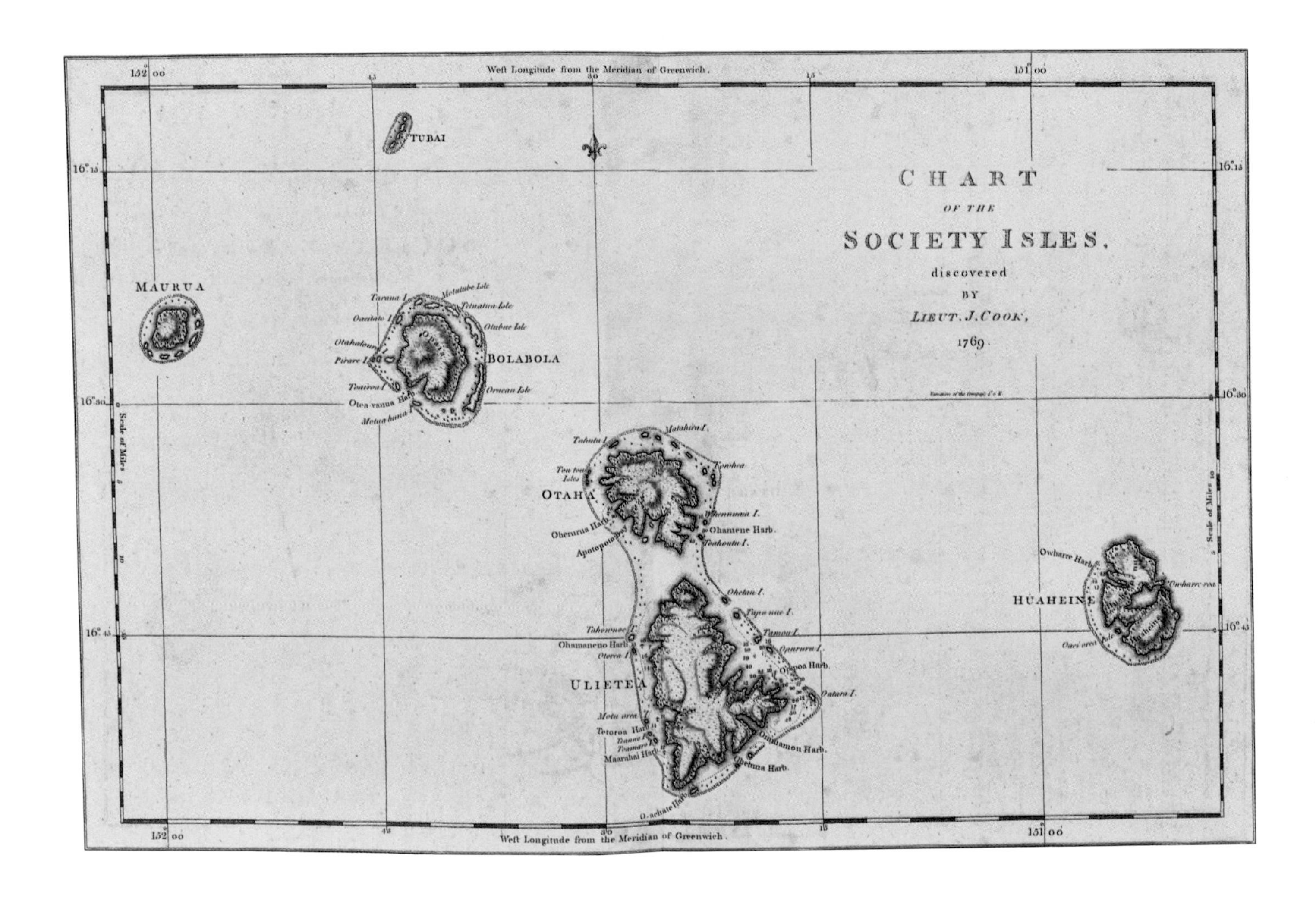

1769 Chart of The Society Isles Discovered by Lieut. J. Cook
Copperplate Engraving by J. Cheevers, 12.75 x 15 in.
Library of Congress, Washington, D.C.

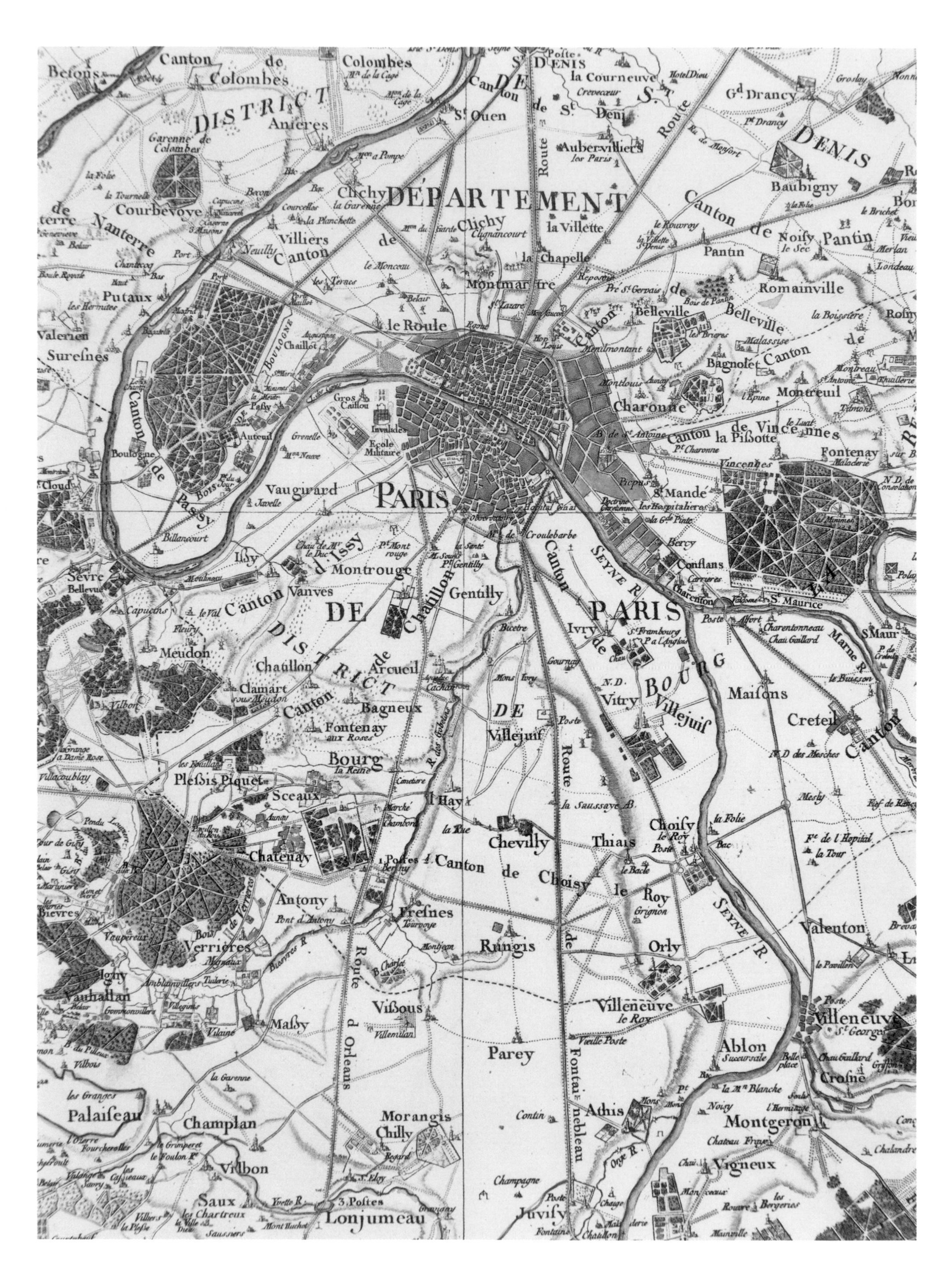

1789 Detail of Paris, Carte géométrique de la France (Map of France)
by César François Cassini de Thury and Jacques-Dominique Cassini
Copperplate Engraving, Hand-Colored, 24.5 x 37 in.
Library of Congress, Washington, D.C.

Carte de Cassini

THE FIRST NATIONAL MAP OF A COUNTRY constructed on modern principles of surveying and astronomy was the multi-sheet map series *Carte géométrique de la France*, or more commonly, *Carte de Cassini*. Prepared during the second half of the 18th century, this "monument to the taste and intelligence of the French Enlightenment," in the words of historian Josef Konvitz, was the legacy of four generations of a remarkable family of mapmakers—the Cassinis. The patriarch was Italian astronomer Giovanni Domenico Cassini, lured to Paris by Louis XIV's finance minister, Jean-Baptist Colbert, to work in the National Observatory and with the Royal Academy. In 1683, Jean-Dominique, as he became known, was assigned the task of completing the triangulation of France begun by Jean Picard four years earlier. Interrupted by the War of Spanish Succession, this work was finally completed in 1744 by his son Jacques and grandson César François Cassini de Thury. In all, some 800 primary triangles and 19 base lines had been measured, calculated, and printed on 18 map sheets, giving land surveyors, road and canal engineers, topographers, and cartographers the first nationwide geodetically controlled mapping network.

Three years later, Cassini de Thury was asked to prepare a more detailed map of France by Louis XV, who was inspired, some believe, while conducting a review of his victorious troops. This project was to take nearly 70 years. More than 80 persons worked on the project, supervised by Cassini de Thury and later his son, Jacques-Dominique. Surveyors and cartographers were held to the highest standards, enforced by stiff fines for errors. Initially funded by the French government, Cassini de Thury later formed a company to raise money by selling shares to nobles, officials, and leading members of society, including Louis's mistress, Madame de Pompadour. In 1793, the revolutionary government expropriated the map, dissolved the company, and beheaded its former president and treasurer. Shareholders were paid when Napoleon interceded on their behalf, but Jacques-Dominique lost control of the project that his family had nurtured for more than a century.

The map of France consisted of 180 map sheets printed at a scale of 1:86,400, somewhat more than one inch to the mile. Most of the sheets measured 41 by 29 inches and represented 50 by 31 miles. If placed side by side, they would measure approximately 36 by 36 feet. Normal print production was 2,500 per map sheet, but popular individual sheets were engraved in higher numbers. The Paris and environs copperplate, for example, was nearly worn out by the time the last map in the series was printed.

During the long development of the national map of France, the Cassinis professionalized the field of cartography, placed mapmaking on a firm scientific foundation, and established mapmaking standards that other countries quickly adopted.

The Indian Map That Guided Lewis and Clark

NATIVE AMERICAN KNOWLEDGE OF GEOGRAPHY played a very significant role in the exploration of America. Indians shared concepts of their cultural, physical, and sacred landscapes, using a variety of cartographic devices. They drew maps on animal skins, molded three-dimensional models with dirt and stone, and communicated spatial concepts with sign language. Sometimes these transitory maps were copied or collected and the information added to printed maps. One such case involves a map prepared in 1802 by a Blackfeet chief named Ac ko mok ki. Peter Fidler, a Hudson's Bay Company trader and surveyor working in present-day Alberta, Canada, requested information from Ac ko mok ki and several other Indians. The company often used maps to develop trading strategies with tribes and determine the most efficient transportation networks.

Ac ko mok ki used a combination of symbols to portray his country, which encompassed the front ranges of the Rocky Mountains and the headwaters of the Missouri River and south fork of the South Saskatchewan. Single lines represent mountains and rivers. Pictorial symbols mark topographic features that served as landmarks for hunting and war parties and as sites for sacred rituals. Landscape features are portrayed in terms of their human and animal attributes. In a letter to the Hudson's Bay office in London, Fidler noted that there "are remarkable & high places at the Mountains that the Indians fancy has [sic] the same appearance as the names given." Drawings of inverted hearts, for example, depict Heart Butte, Montana, and Heart Mountain, Idaho; and a curved tooth symbolizes Beartooth Mountain, on the northern border of today's Yellowstone National Park.

Map scale is measured in terms of time rather than miles. On an accompanying map by Ac ko mok ki, the distance from King Mountain, on the Canadian border, to Owl Head, in present-day central Wyoming is given as "33 Days walk—for young men."

Fidler forwarded a copy of Ac ko mok ki's map to the London office. "This Indian map," he wrote, "conveys much information where European documents fail." London map publisher Aaron Arrowsmith used Ac ko mok ki's sketch to revise his large wall map of North America. In 1803, President Thomas Jefferson purchased a copy of Arrowsmith's map. A copy was then drawn for Meriwether Lewis. Lewis and Clark carried both maps with them on their expedition to the Pacific coast. They entered Blackfeet country two years later, guided in part by Ac ko mok ki's geographical information.

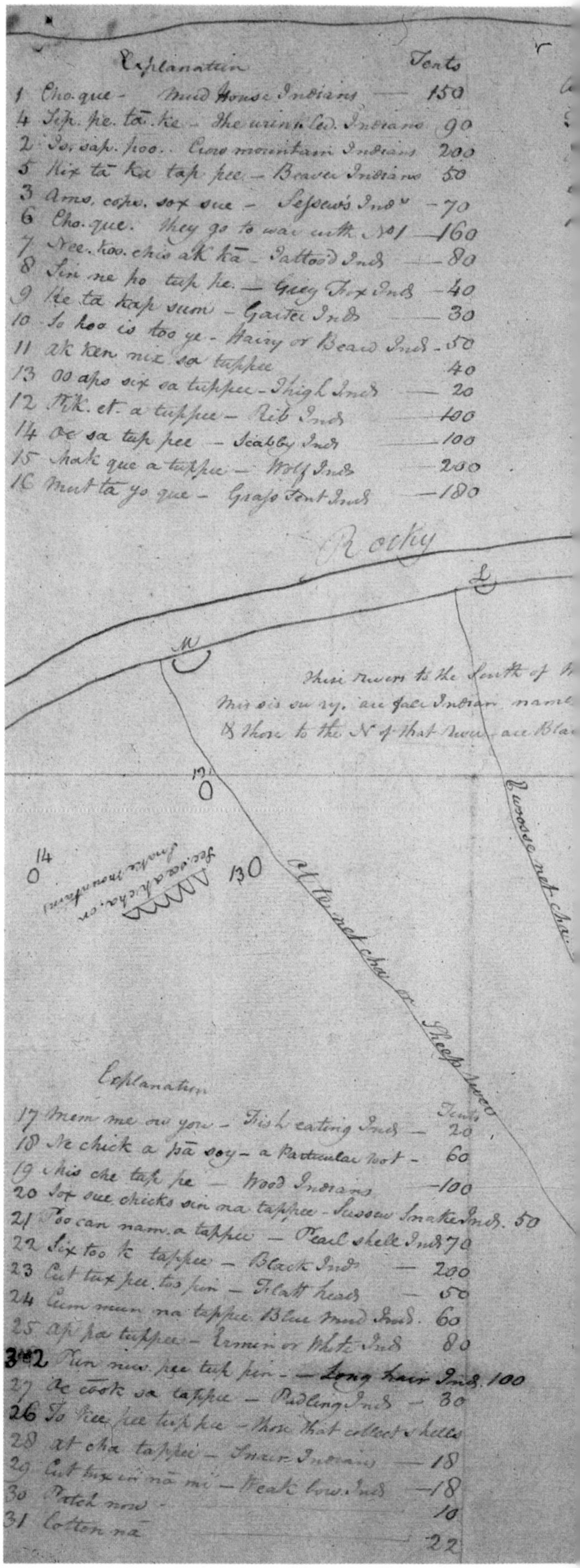

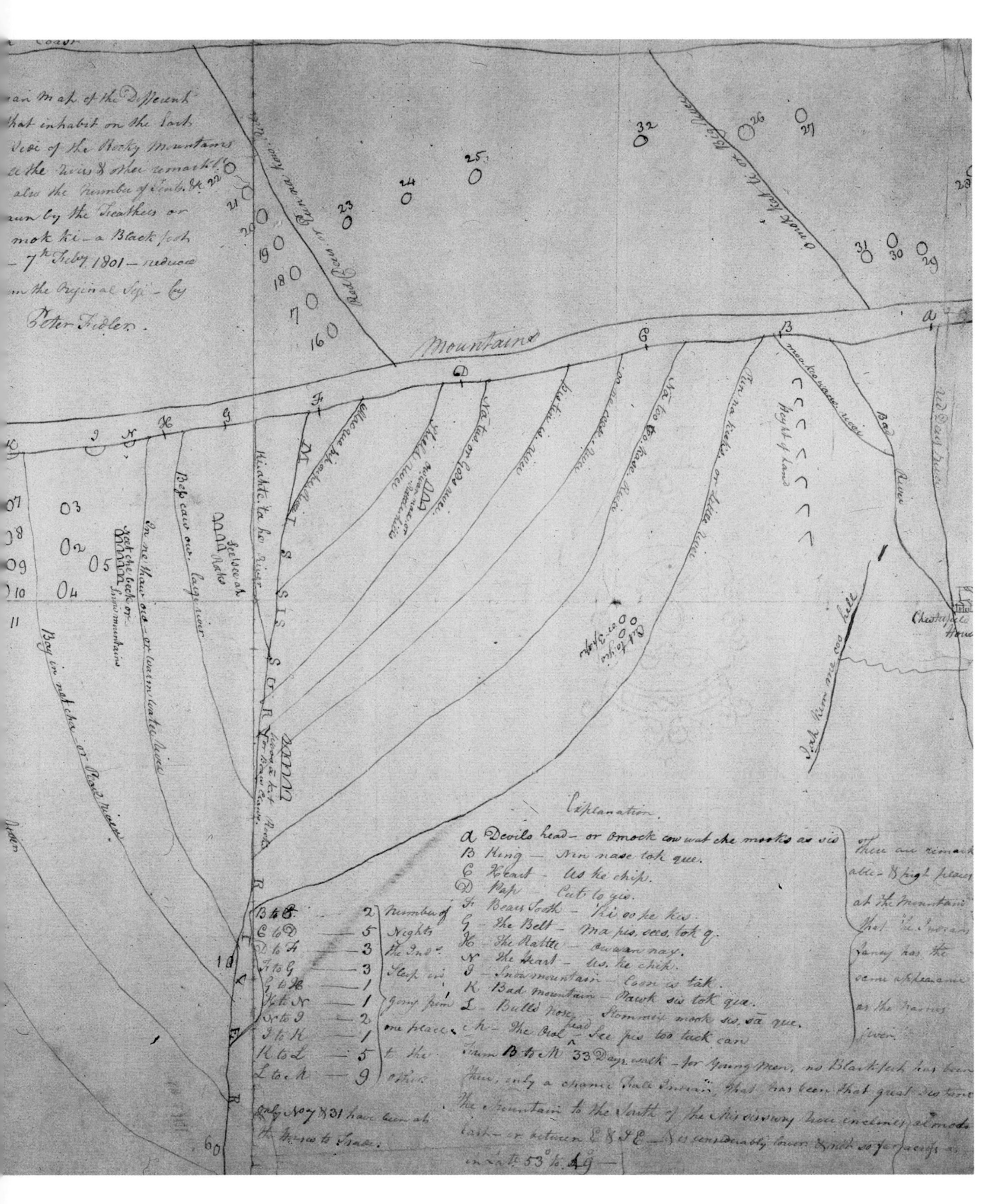

1802 Map of the Rocky Mountains by Ac ko mok ki from Peter Fidler's Journal of Exploration and Survey, 1789-1804
Manuscript, Pencil and Chalk on Paper, 9.5 x 14.75 in.
Hudson's Bay Company Archives, Provincial Archives of Manitoba, Canada

1803 Iklim Afrika (Map of Africa) by Anonymous, after L.S. de la Rochette's Map of Africa
Copperplate Engraving, Hand-Colored, 19 x 21.75 in.
Library of Congress, Washington, D.C.

Iklim Afrika

THE MIDDLE EAST WAS SLOW TO INTRODUCE MAP printing, lagging behind the West by some 200 years. Graphic designer and historian Sait Maden believes this delay was due to the complexity of the Arabic alphabet "rather than religious constraints as some still claim." Printing Arabic script required casting 500 letter cases, compared with 100 for Latin text. Although Christian and Jewish printers had published works in Arabic since the early 1500s, it was not until 1727 that an Islamic printing house was established in Istanbul. Despite opposition from the clergy and professional calligraphers, Sultan Ahmed III granted permission to Ibrahim Muteferrika to print nonreligious works. A Hungarian convert to Islam, Muteferrika published a world atlas in 1732 entitled *Cihan-numa,* or *Mirror of the World.* Geographer Katip Celebi, a widely traveled Ottoman army officer, contributed to it. The atlas was completed by Abu Bakr ibn Bahram ad-Dimashqi after Celebi's death in 1657. Both authors used European sources, including Ortelius, Mercator, and Hondius. Celebi, Abu Bakr, and probably Muteferrika collected and edited the maps of the Middle East.

Muteferrika's atlas was out-of-date when published; still, it was the standard geographic work in the Ottoman Empire during the 18th century. More timely was the *Cedid atlas tercumesi (The New Atlas),* printed by the Ottoman Military Engineering School Press in 1803. Influenced by westernizing reforms under Sultan Selim III, the school provided the latest geographical information for military students, officers, and government officials. The *Cedid atlas* was copied directly from William Faden's *General Atlas,* which was acquired in London by Mahmud Raif, when he was secretary to the first Ottoman-Turkish ambassador to Great Britain, in the 1790s.

The choice of Faden's *General Atlas* was not surprising. Faden was England's leading commercial mapmaker. His work was highly regarded throughout the world. Apprenticed as an engraver at age 14, he opened his own shop four years later, importing and reselling maps from Europe. Later, he commissioned and engraved his own maps. Appointed geographer to the king in 1783, he is most remembered for his two works relating to the American Revolution—*North American Atlas* (1777) and *Atlas of Battles of the American Revolution* (circa 1793)—and for his large-scale canal and railroad maps of England that document the British industrial revolution.

Mahmud Raif and the Engineering School engravers copied Faden's map of Africa, compiled by L. S. de la Rochette, and replaced the English text with Turkish written in Arabic script. It is a striking example of European and Muslim knowledge of the Dark Continent on the eve of the search for the headwaters of the Nile, Niger, Congo, and Zambezi. Note caravan routes across the Western Sahara centered on Timbuktu, a focal point of Muslim expansion.

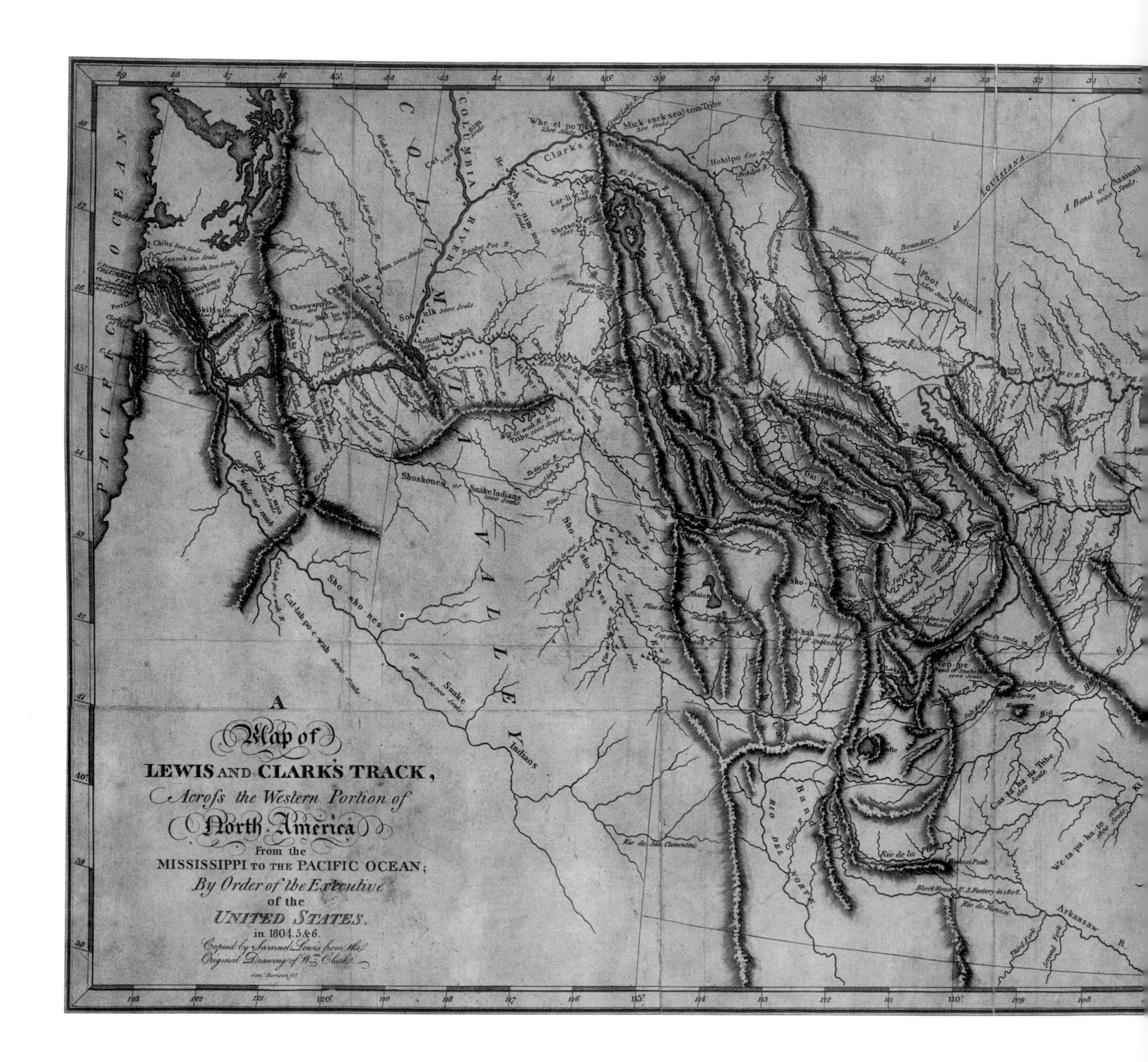

1814 A Map of Lewis and Clark's Track in 1804-06 by William Clark, Reduced by Samuel Lewis
Copperplate Engraving by Samuel Harrison, 11.8 x 27.6 in.
Library of Congress, Washington, D.C.

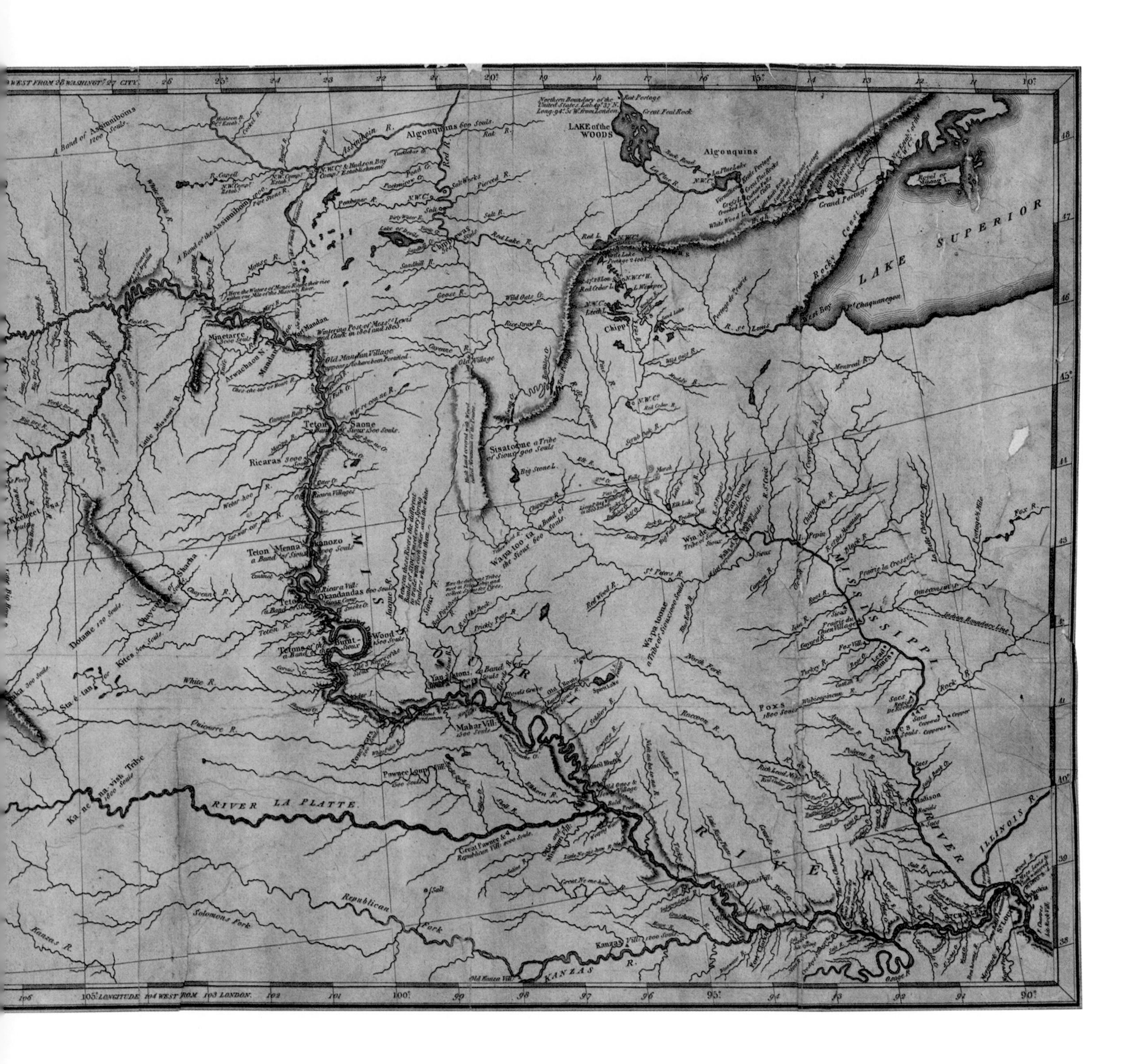

LAKE SUPERIOR
LAKE of the WOODS
Algonquins 600 Souls
Algonquins
Grand Portage
Rocky Coast
Pt. Chaquanegon
A Band of Assiniboins 1200 Souls.
Assiniboin R.
Red R.
Red Lake R.
Wintering Post of Messrs. Lewis and Clark in 1804 and 1805.
Old Mandan Village
Minetarre 3000 Souls
Teton
Saone
Ricaras 3000 Souls
Sisatoone a Tribe of Sioux 900 Souls
Big Stone L.
St. Peters R.
MISSISSIPPI
Foxs 1800 Souls
Madison
ILLINOIS R.
Mahar Vill. 1300 Souls
Pawnee Loup Vill.
Council Bluffs
RIVER LA PLATTE
Great Pawnee & a Republican Vill. 2000 Souls
Republican Fork
Solomons Fork
Kanzas R.
Kanzas Vill. 1300 Souls
Old Kanzas Vill.
KANZAS
White R.
Rock R.
Fox R.
105 LONGITUDE 104 WEST FROM 103 LONDON

Lewis and Clark Map the West

WHEN THOMAS JEFFERSON AUTHORIZED THE Louisiana Purchase in 1803, little was known about the new western territory, a region nearly one-fourth the size of Europe. No maps accompanied the treaty of cession. While its eastern and southern boundaries were understood to be the Mississippi River and the Gulf of Mexico, the treaty language was purposely vague with respect to its northern and western limits. Jefferson soon sent a team to explore the region. Led by Army Captains Meriwether Lewis and William Clark, a cadre of 33 men, one woman, and one child explored more than 8,000 miles inhabited by some 40 Indian tribes, most of whom had never seen Anglo-Americans. From May 1804 to September 1806, the Corps of Discovery conducted diplomatic relations, collected hundreds of specimens of unrecorded flora and fauna, and made the first maps of the trans-Mississippi West. Further, they lost only one man and had only one hostile engagement with the native inhabitants. A remarkable achievement of leadership and discipline, the expedition set the standard for future U.S. military explorations.

A primary objective of the expedition was mapping. The son of noted colonial surveyor and mapmaker Peter Jefferson—author of a map of Virginia—Thomas Jefferson wanted Lewis and Clark to prepare an outline map of the route, with details to be added later. Production of the final printed map took 11 years. Jefferson taught Lewis the basic principles of determining latitude by observing the altitude of the sun or a star with an octant and hired leading surveyors and astronomers to tutor him further. Clark was the cartographer. Apparently self-taught, he drew all but three of the 140 surviving maps made during the expedition. His daily route sketches were based on compass bearings and estimated distances. In winter quarters, Clark combined his route maps with Indian maps to create composite regional maps.

After their return, Clark prepared a large manuscript map of the West. Now at Yale University, it measures almost three by five feet and records both data from the journey and additional geographical information gathered from subsequent interviews with trappers, traders, and Indians. One of these was mountain man John Colter, a member of the Corps of Discovery. He was the first Anglo-American to see Jackson and Yellowstone Lakes, which are recorded on Clark's map along with his route. Clark's manuscript map was reduced and recompiled by Philadelphia cartographer Samuel Lewis in 1813 and engraved by Samuel Harrison a year later.

The legacy of the Lewis and Clark map is monumental. It was the first map to portray with some correctness the Missouri and Columbia River systems and their tributaries, the first to illustrate the multiple ranges and valleys of the northern Rockies, and the first to show the Teton-Yellowstone region. It laid the foundation for future mapping of the West.

Celebrating the American Centennial

EIGHTEEN SEVENTY-SIX WAS AN AUSPICIOUS YEAR for Americans, marking the one hundredth anniversary of the Declaration of Independence. Gaylord Watson, a New York publisher, issued the railroad wall map shown on pages 158-59 using a centennial theme. Railroad maps expressed at a glance the dominant transportation system of the 19th century as it spread across the country, linking geographical sections and revolutionizing travel. Such maps were critical in all phases of railroading, from planning to marketing.

The most popular railroad maps were general maps of the continental United States, sectional maps, and individual state maps. Ornate and detailed, they normally provided additional information on canals and roads. Railroad lines also issued promotional maps.

Several leading map publishers owed their early success to this genre. The Rand McNally Company, in Chicago, the central hub of North American railroads, first specialized in railroad maps, using wax engraving, or cerography. This relief printing process, invented independently by Sydney Edwards Morse in the U.S. and Edward Palmer in England, was popular with commercial map publishers because it did not require great skill to prepare the printing plates. Wax was used as a mold to cast the plates by electrotyping. Lettering set in type was stamped into the wax while tinting was done with the aid of ruling machines. Backed with wood, the plates were used with letterpress printing machines, which was more versatile than other printing methods. Wax engraving flourished in the U.S. until about 1940, when it was replaced by photomechanical methods. Its impact on the style of commercial cartography was significant. Its unique look—distinct from the contemporary European model—pervaded U.S. maps well into the 20th century. Erwin Raisz, an eminent 20th-century cartographer, described this style as "over-lettered and mechanical-looking."

Watson's *Centennial American Republic and Railroad Map* was produced by wax engraving and copyrighted by the Library of Congress just five years after those laws first went into effect. Measuring three by four feet, it was designed to hang in schools, homes, and offices. The borders are enriched with statistical tables and patriotic images and messages: the Main Building of the 1876 Philadelphia Centennial Exposition, George Washington, the Capitol, and Independence Hall above the text of the Declaration of Independence. Most revealing of its time is the title cartouche. An expression of Manifest Destiny and industrialization, it depicts the marvels of the age: telegraph lines and a railroad steam engine belching smoke as symbols of progress. These are flanked on the right by a thriving port and a prosperous family farm, and on the left by an untamed wilderness of open prairie and dark mountains occupied only by Indians on horseback in pursuit of buffalo.

1875 Centennial American Republic and Railroad Map of the United States and of the Dominion of Canada by Gaylord Watson
Wax-Engraved, Hand-Colored, 37 x 51 in.
Library of Congress, Washington, D.C.

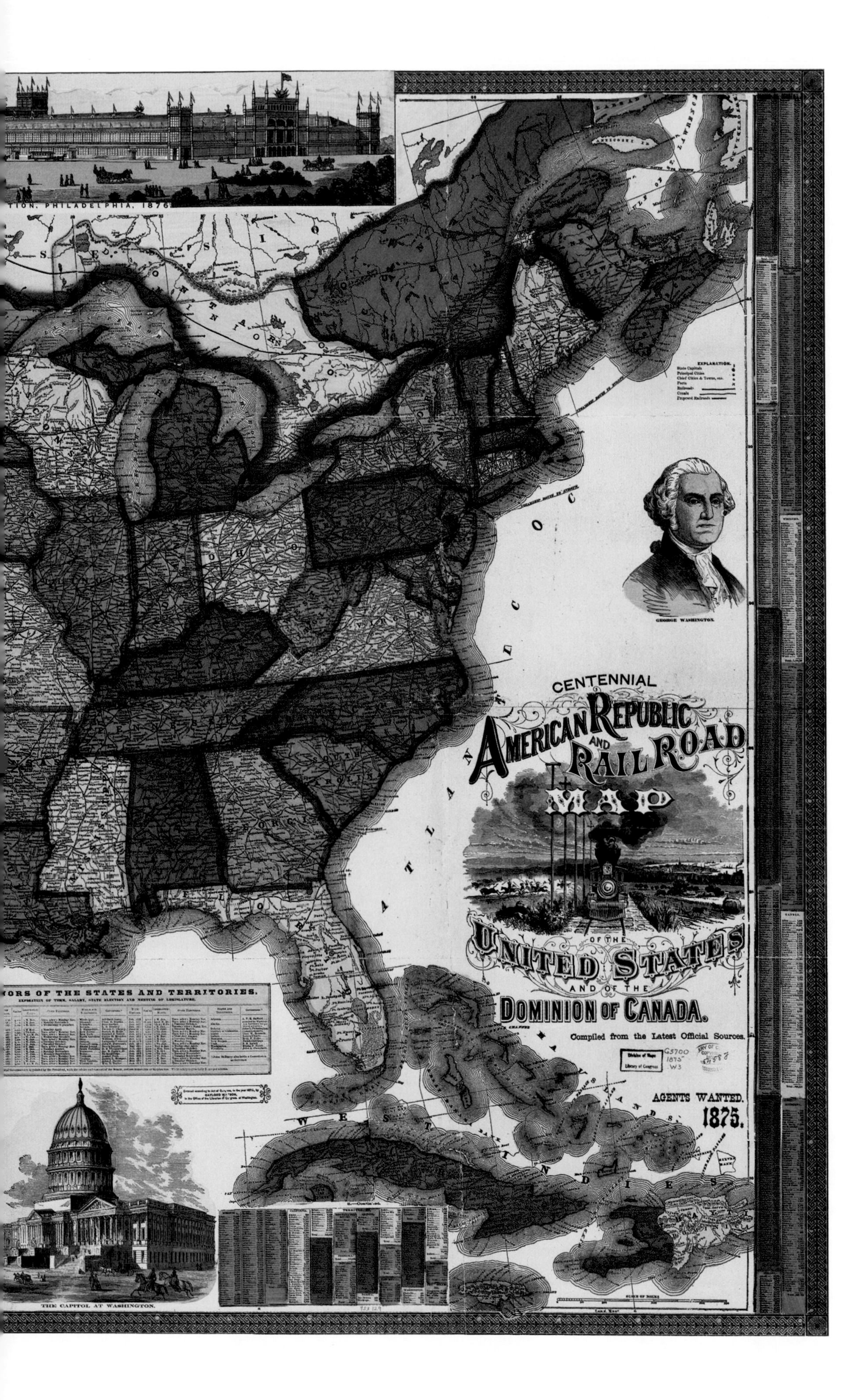
CENTENNIAL
AMERICAN REPUBLIC
AND
RAIL ROAD
MAP
OF THE
UNITED STATES
AND OF THE
DOMINION OF CANADA.
Compiled from the Latest Official Sources.
AGENTS WANTED.
1875.
GEORGE WASHINGTON.
THE CAPITOL AT WASHINGTON.
TION, PHILADELPHIA, 1876
NORS OF THE STATES AND TERRITORIES.
ATLANTIC OCEAN
WEST INDIES
EXPLANATION.

1815 Detail of A Delineation of the Strata of England & Wales, with Part of Scotland by William Smith
Hand-Colored, Copperplate Engraving by John Cary, 76 x 109.6 in.
Library of Congress, Washington, D.C.

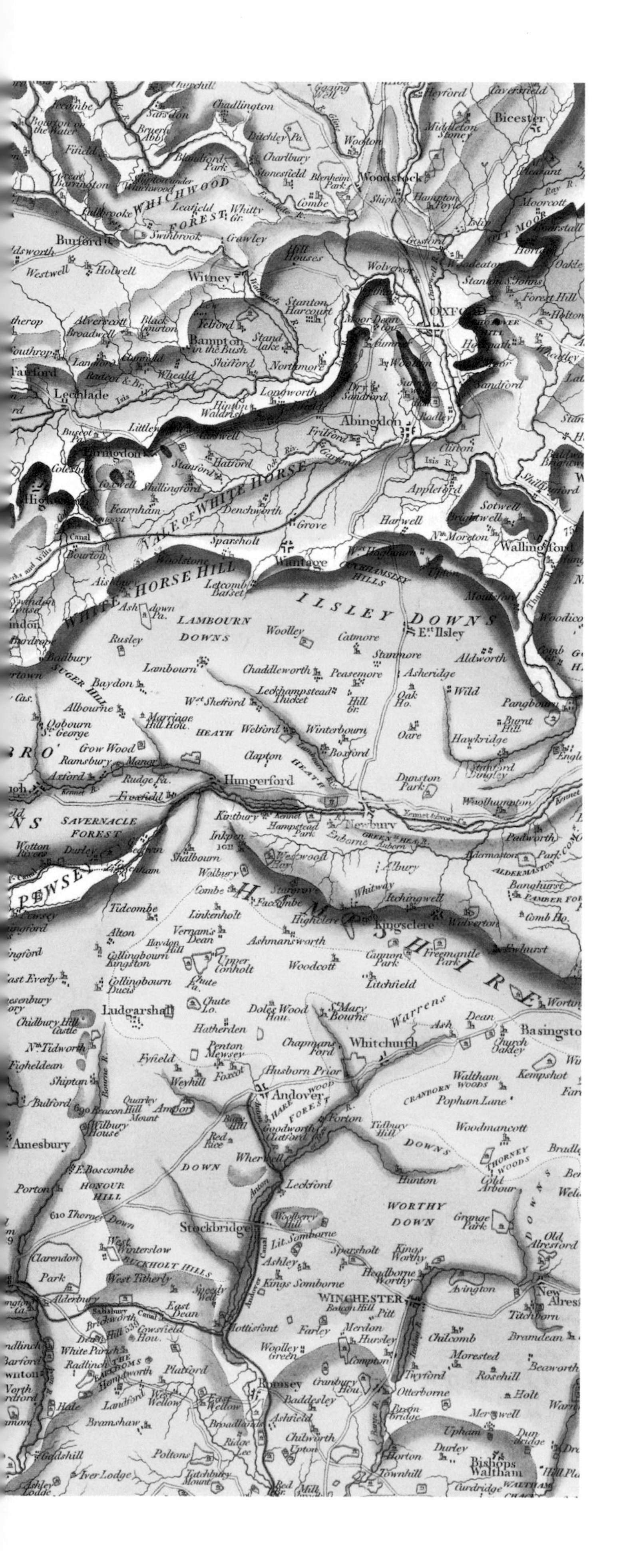

Mapping Britain Underground

WILLIAM SMITH'S PASSION WAS ROCKS AND THE fossils imbedded in them. For 20 years he crisscrossed England and Wales, following road-building crews, tracing streambeds, and examining coal mines. Smith sought breaks in the earth that exposed the underlying rocks. His tools were a pickax, pencil, and paper—and a creative mind. By recording and comparing the fossil content of one layer of rock with another, he pieced together the first geological map of an entire country. This massive work carries the title *A Delineation of the Strata of England and Wales, with Part of Scotland;* it measures six by nine feet when the 15 sheets that compose the map are laid side by side.

Smith's fascination with fossils began as a young boy on his uncle's farm, where fossil sea urchins served as measuring weights on butter scales. Later, the self-taught land and mineral surveyor discovered that characteristic fossils differentiated layers of rock. Inspired by a map of soil and vegetation types, Smith set out in 1798 to map the rock strata of England based on fossils. Sir Joseph Banks, president of the Royal Society of London, supported him. John Cary, one of England's most distinguished map publishers, gave additional funds in 1812 and furnished Smith with a topographic base map of the country, a necessary first step in preparing a geological map.

Smith devised an ingenious hand-coloring system, using tonal variations to portray rock formations on the basis of their physical characteristics and fossil content. Darker hues depict rock strata at Earth's surface, while paler tints of the same color show a dip underground. Each of the 15 sheet maps required seven to eight days to color. A memoir with the map had further information on its contents. Four hundred copies were printed in 1815 and 1816, with Smith constantly updating as new information was obtained. Although a great personal accomplishment, the map failed to sell. To raise funds, Smith sold his fossil collection to the British Museum; still, he went briefly to debtor's prison. He resumed mapmaking in 1819 with a reduced copy of his grand map and a series of geological maps of single English counties.

While Simon Winchester's best-selling book on this subject is called *The Map That Changed the World,* science historian Cecil J. Schneer notes that Smith's map was neither the first geological map of the 19th century nor the first to portray rock layers based on ordering fossil content. German and French geologists preceded him. "What Smith had accomplished," Schneer concludes, "was to open up a science that was saturated with theory but starved of data."

Whistler's Father

JAMES MCNEILL WHISTLER'S OIL PAINTING "Arrangement in Grey and Black: Portrait of the Painter's Mother" ("Whistler's Mother") is known worldwide. But few know that this etcher and painter, one of the preeminent artists of his time, cut his first etchings as a copperplate map engraver for the U.S. Coast Survey or that he received formal training in cartography. Whistler came from a family of mapmakers. His grandfather, U.S. Army Capt. John Whistler, was called upon to survey, map, and construct frontier military posts. His 1808 plan of Fort Dearborn, on the present site of downtown Chicago, is now in the National Archives. His surveying and mapping skills were learned in the field.

George Washington Whistler, whose portrait his son James painted from memory, was the first to receive formal training in mapmaking. George was trained at the Military Academy at West Point, then the only professional school for engineers in the nation. His 1818 plan of West Point was probably prepared under the tutelage of Claudius Crozet, a French émigré who introduced technical texts and teaching methods from the Royal School of Bridges and Roads in Paris and the Military School of Engineers and Artillerists at Metz, France. Whistler's map reveals a talent for cartography. After graduation, he made maps during field assignments, including the Army's first topographic contour map. Later, he became a top railroad engineer. He died at age 48 in St. Petersburg, where he spent his last years building railroads for the tsar.

James McNeil Whistler's association with cartography was briefer but included both formal training and an apprenticeship as a copperplate map engraver. At West Point for three and a half years, he took newly designed courses in topography and landscape drawing. Lt. Seth Eastman's *Treatise on Topographical Drawing* (New York, 1837), the first cartographic manual published in America, was available. After his "little difference with the Professor of Chemistry," the artist was appointed a map engraver at the U.S. Coast Survey. After three months he left for Europe to seek his fame; but by that time he had etched at least two copperplates. Criticized by his supervisor for including a flock of birds on his chart of Anacapa Island, he replied, "Surely the birds don't detract from the sketch. Anacapa Island couldn't look as blank as that map did before I added the birds." The flock of birds remains on the completed copperplate, adding an artistic touch.

1857-59 PORTRAIT OF MAJOR WHISTLER BY JAMES MCNEILL WHISTLER
OIL ON WOOD PANEL
FREER GALLERY OF ART, SMITHSONIAN INSTITUTION, WASHINGTON, D.C.

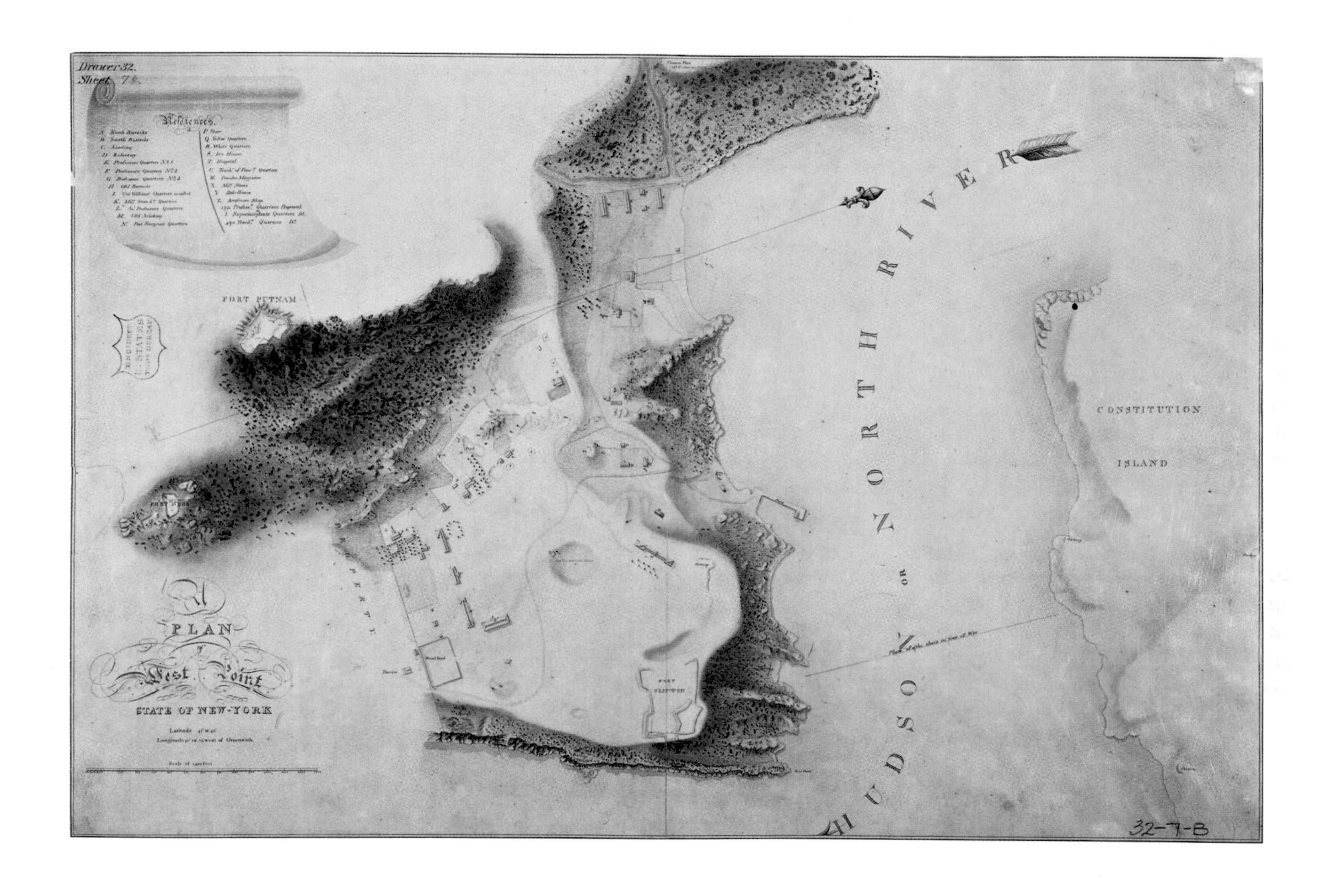

1818 Plan of West Point, State of New-York by Cadet George Washington Whistler
Manuscript with Watercolor Wash, 25 x 38.75 in.
U.S. National Archives and Records Administration, Washington, D.C.

1833 Detail of Topographic Map of Nilgiri Mountains of Southern India Surveyed by Capt. B. S. Ward and Col. Colin McKenzie
Copperplate Engraving by J. & C. Walker, 25 x 38 in.
Library of Congress, Washington, D.C.

The Great Survey of India

IN A FASCINATING EXAMPLE OF THE POWER OF MAPS, geographer Matthew H. Edney asserts that modern India as a geographical entity was the creation of 19th-century British mapmakers. Basic surveying and mapping of the subcontinent took place during British dominion over India from 1757 to 1947, by officials of the British East India Company and, after the Sepoy Mutiny of 1858, by the Crown Colony of India. The East India Company, a commercial enterprise established by Queen Elizabeth I in 1600, consolidated political sovereignty over much of the subcontinent by the late 18th century.

The systematic mapping of the subcontinent began in 1802 with the Great Trigonometrical Survey of India. William Lambton, and his successors, including George Everest, for whom the world's highest mountain is named, established a network of geodetically measured triangles stretching from the tip of India 1,400 miles north to the Himalaya, and across India from Bombay to Calcutta. Before this triangulation survey was completed in 1866, the British East India Company, anxious for updated maps, initiated the first topographic survey of British India, the largest connected map survey of the 19th century.

Royal Military Engineers compiled the maps. Native Hindu and Muslim explorers were sent surreptitiously to map northern regions beyond British contol. They mapped much of Kashmir and Tibet, posing as mullahs or traders. Nain Singh, on a secret two-year trip to the "Forbidden City" of Lhasa, Tibet's capital, mapped some 1,300 miles, determined 276 latitude sightings with his sextant, measured nearly 500 elevations, and discovered "the existence of a vast snowy range." Recipient of a Royal Geographical Society gold medal, he was considered to have "added a greater amount of positive knowledge to the map of Asia than any individual of our time."

The topographic survey lasted from 1825 to 1906, producing 358 printed maps at a scale of one inch to four miles. The maps were reduced and engraved in London until 1867, when British copperplate engravers were established in Calcutta. New printing methods such as lithography and photozincography reduced printing costs, but none matched the quality of the copperplate engravings, as illustrated in this 1833 detail of the Nilgiri Mountains of southern India. The most obvious feature of this map is the hachure, a symbol consisting of a series of short lines drawn to follow the direction of the slope of the land. Engraved correctly, hachures provide an almost three-dimensional image of the terrain.

Early Thematic Atlas

NEW MAPS OF PHYSICAL AND CULTURAL geographies burst upon the public in 1845 with the publication in Germany of the first volume of Heinrich Berghaus's *Physikalischer Atlas*. In addition to geology, Berghaus treated his readers to the latest discoveries in volcanism, meteorology, tidal forces, and magnetism, and to detailed distribution maps of flora, fauna, and ethnography. The atlas was a tour de force with 90 thematic maps that expanded the definition of cartography to include natural and human activities not formerly mapped. Before the second volume appeared in 1848, the Edinburgh publisher Alexander Keith Johnston issued a condensed version with the assistance of Berghaus that is even more beautifully illustrated. "[T]he physical atlas of Mr. Berghaus has nothing equal to it in any country," observed the president of the Royal Geographical Society in 1849. A second edition appeared in 1852, and later a six-volume version was issued.

A gifted cartographer, Berghaus published his first atlas after leaving school at the age of 14. He served as a cavalry officer in campaigns against Napoleon and briefly with the Prussian Land Survey, but compiled maps in his spare time. Later he established Berghaus's Engravers School for Geography in Potsdam, where students worked on his atlases. One student, August Petermann, went on to fame in England and Germany, first as "physical geographer and engraver on stone to the Queen of England," then as founder and editor of the noted geographical journal *Petermanns Geographische Mitteilungen*.

Berghaus's atlas coincided with the rise of scientific geography in Germany, initiated by Carl Ritter and Alexander von Humboldt. The three geographers were closely associated; later they established the Geographic Society of Berlin, now the world's second oldest geographic association. Von Humboldt first suggested to Berghaus an atlas to accompany *Kosmos*, his six-volume work on physical geography. Von Humboldt believed that the study of geography required measuring and classifying nature's phenomena; so he made field trips throughout Western Europe and Russia, then took a five-year trip to South America, Cuba, and Mexico with French botanist Aimé Bonplan. Returning in 1804 via Washington, D.C., Von Humboldt briefed President Thomas Jefferson on the American Southwest.

Although not a trained cartographer, the Prussian geographer prepared important maps to illustrate his books. He devised two innovative cartographic devices: the isotherm, a line that links equal values of temperature, still used by cartographers to map surface temperatures; and an altitude profile to illustrate plant and animal variations at different elevations. Berghaus's world map of plant geography combines von Humboldt's altitude profile of equatorial plants with a plant distribution map by the Danish botanist Joachim Frederik Schouw.

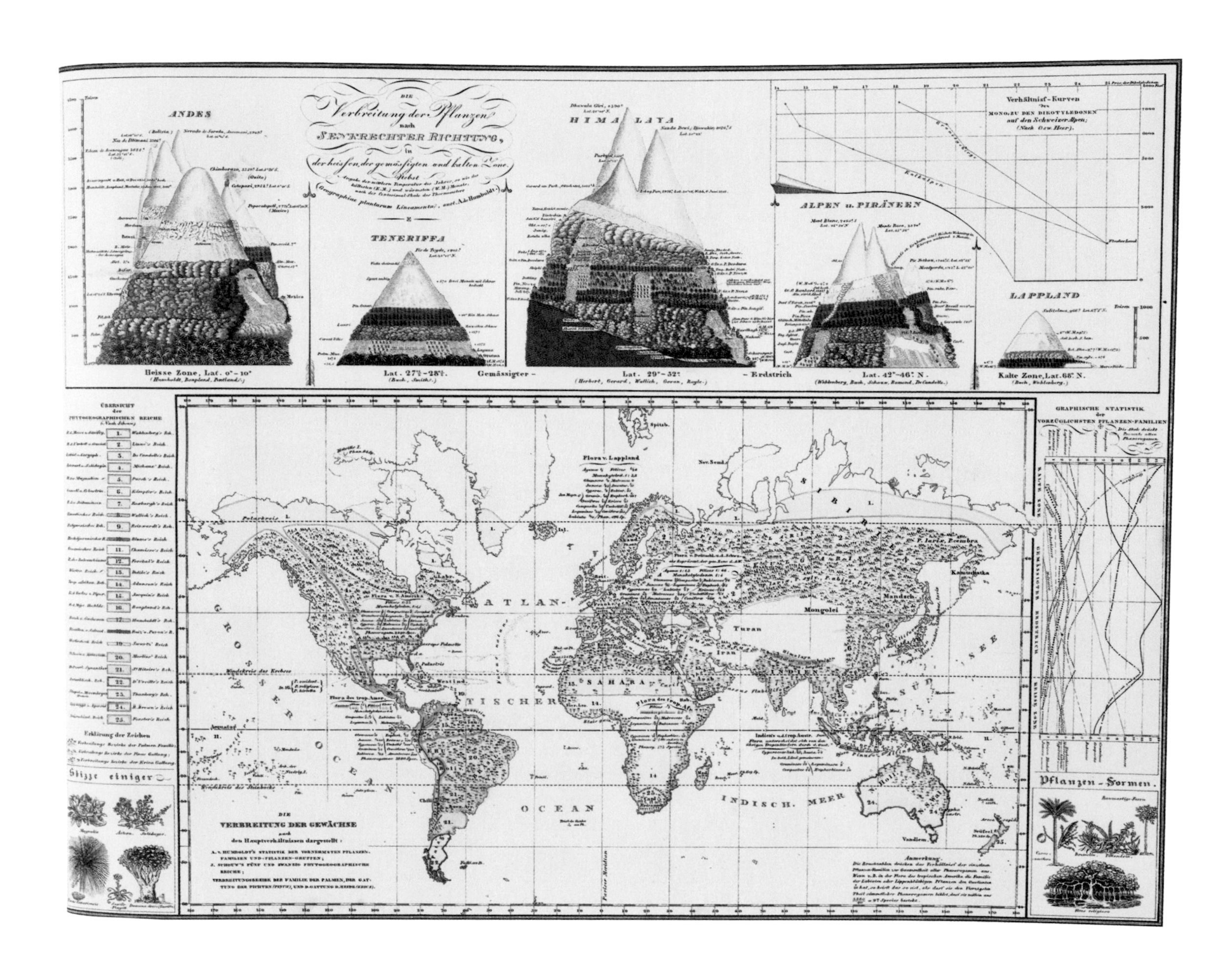

1845 Umrisse der Pflanzengeographie (Outline of Plant Geography) by Alexander von Humboldt and Joachim Frederik Schouw
Copperplate Engraving, Hand-Colored, 14.25 x 18.3 in.
Library of Congress, Washington, D.C.

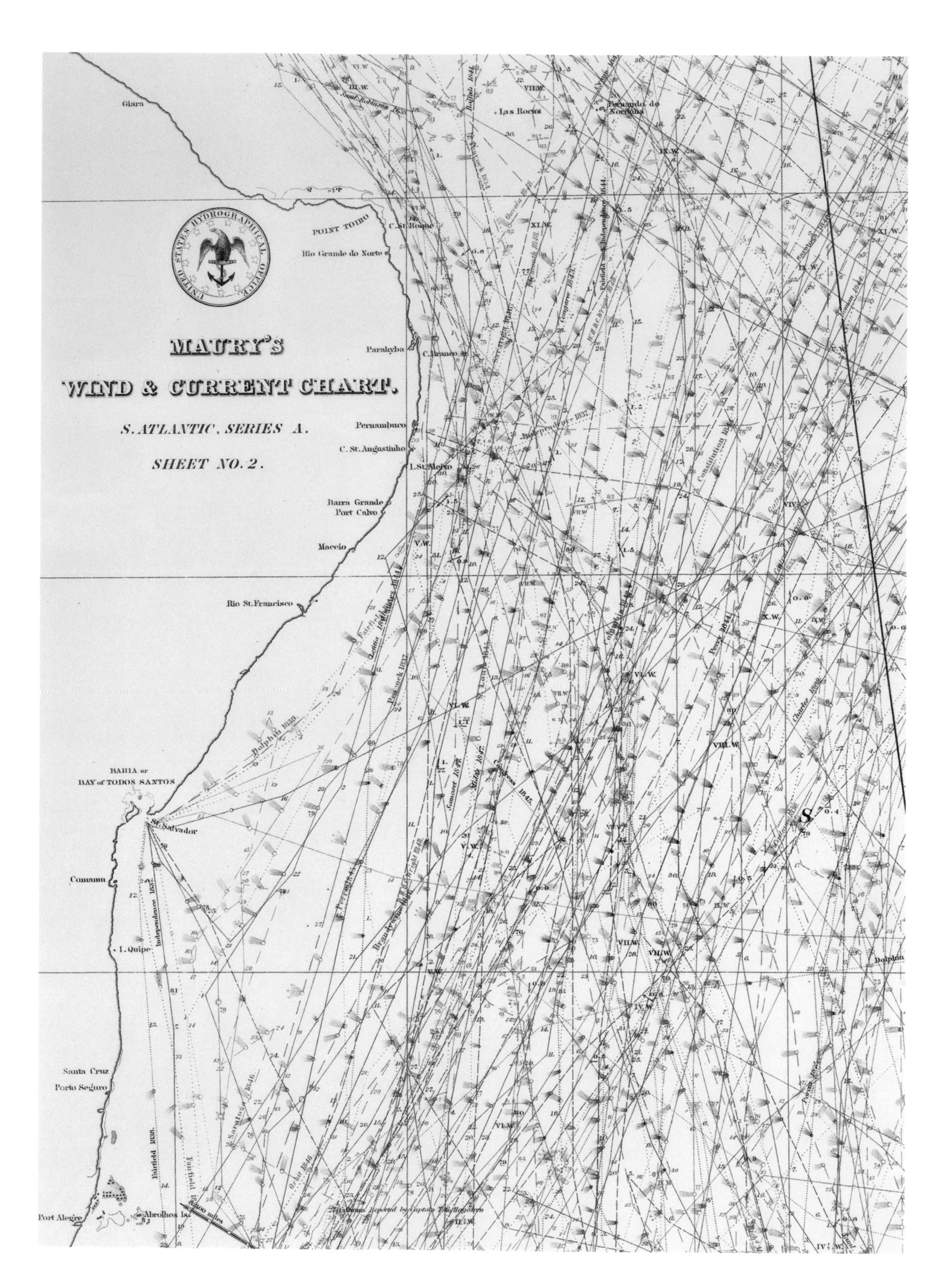

1847 Maury's Wind & Current Chart S[outh] Atlantic by Lt. Matthew Fontaine Maury
Chromolithograph by William Endicott, 35 x 49 in.
Library of Congress, Washington, D.C.

Maury's Wind and Current Charts

WIND AND WAVES! DURING THE AGE OF SAILING ships these two elements of nature determined the speed of ocean passages. In 1847, Matthew Fontaine Maury, a U.S. naval officer, developed a radically new map type that provided sailors with critical information. These maps, called "wind and current charts," shaved days from sailing times, saving shipping companies millions of dollars and delivering passengers in record time. They became indispensable components of a ship captain's navigational kit.

Despite some four centuries of maritime navigation, little was known about the true character of ocean currents and winds in the mid-19th century. No science of the sea existed until Maury turned his attention to this field. Appointed a midshipman in the U.S. Navy with the help of Sam Houston, then a Tennessee congressman, Maury spent 12 years at sea, mastering navigation, learning the finer techniques of sailing, and observing the Pacific and Atlantic Oceans. A leg broken in a stagecoach accident ended further sea duty. In 1842, he was placed in charge of the Navy's Depot of Charts and Instruments, which held ships' logs and journals dating back to the founding of the U.S. naval services. These relics of the past contained detailed information from all the routes sailed by Navy vessels, including weather conditions, winds, and currents, which Maury and his staff converted into maps. Additional information was obtained from new forms devised by Maury and filled out by ships' captains during cruises. These "abstract logs" added even more detailed daily weather observations.

Lt. William B. Whiting prepared the first set of eight wind and current charts, printed in 1847, covering the Atlantic Ocean. Small brushlike symbols indicated wind intensity and direction. Arrows denoted currents. Roman numerals signified the degree of magnetic variation. Ships' tracks were shown by dotted lines that were color-coded, a process made possible by the development of chromolithography.

Reluctant at first to accept these new charts, the sailing fraternity was persuaded when a veteran Baltimore sea captain cut 34 days from his usual 110-day coffee run to Rio de Janeiro. "Every day saved at sea lessened the danger to seafarers, reduced by that much the risk of shipwreck or storm damage to cargo, and thus . . . cut the cost of shipping," noted Maury biographer Frances Leigh Williams. These charts received further acclaim during the California gold rush of 1849, when their use shortened passage from New York to San Francisco by more than 40 days. Eventually wind and current charts were issued for all of the world's oceans. They were continuously updated and accompanied by "Explanations and Sailing Directions" until 1861—when Maury resigned from the U.S. Navy to join the Confederacy.

After the Civil War and some years in Mexico and England, Maury taught meteorology at the Virginia Military Institute at Lexington, until his death in 1873.

Jewel of the South Sea

TRADITIONAL PICTORIAL MAPS CONTINUED TO BE produced in China well into the 19th century alongside Western-style mathematical maps. This representative scroll map portrays tropical Hainan Island off the southern coast of China, known as the "Jewel of the South Sea." At 13,000 square miles, it is the second largest island in the China Sea and the southernmost point of Chinese territory. An unknown government official prepared this map for administrative purposes, despite its appearance as a work of art. He brought to it his training in traditional Chinese mapmaking. Based on an examination of many maps, the respected cartographic historian Cordell Yee has listed these skills as calligraphy, draftsmanship, literary and historical textual study of the place mapped, and "sometimes even poetic composition." All are reflected in this map of Hainan Island.

In an intriguing transformation from geographical feature to map, the mapmaker altered the island's oval shape into a rectangle, oriented to the four cardinal points, with south at the top. "According to the principles of feng shui ('wind and water'), the traditional Chinese science of siting or 'geonomy,'" Yee notes, "the rectangle is an auspicious shape." Geographer Li Xiaocong proposes that the squared shape also may have been selected to conform to the Chinese traditional belief that Earth was flat. This transformation continued with the placement of the island's two major ports, the ancient cities of Yazhou (now Sanya City) and Haikou, at the top and bottom respectively. Their locations, indicated by pictorial drawings of Chinese vessels, provided a north-south axis "about which other towns are symmetrically placed."

Traditional Chinese pictorial maps should be viewed as landscape paintings, suggests Yee. The viewer imagines being "in the landscape, moving oneself to take in different views, or turning one's head to orient oneself properly to another part of the terrain." The map image should be viewed sequentially, by following a coastline, road, or river valley.

As with most traditional Asian maps, scale and perspective vary throughout the map. Waterways are portrayed in two dimensions, but vegetation, terrain, and man-made features such as city walls and buildings are displayed in variable perspective.

Place-names and legends provide information about locations and the activities of the Li people, the ethnic group that originally settled the island and occupied the mountainous interior during the 19th century. Everyday life, various economic activities, and tribal festivities are depicted: Men fish with bows and arrows and construct rafts on lakes; women flay rice; oxen stir soil and water in preparation for planting; and two men pay tribute. Two thatched bamboo buildings on stilts represent Li dwellings. The walled cities, inhabited primarily by mainland Chinese, reflect the numerous uprisings and rebellions of the Li people during the Ming and Qing dynasties.

Circa 1850 Qiong Jun Di Yu Quan Tu (Map of Hainan Island)
Ink and Color Scroll Painting, 72.5 x 36.6 in.
Library of Congress, Washington, D.C.

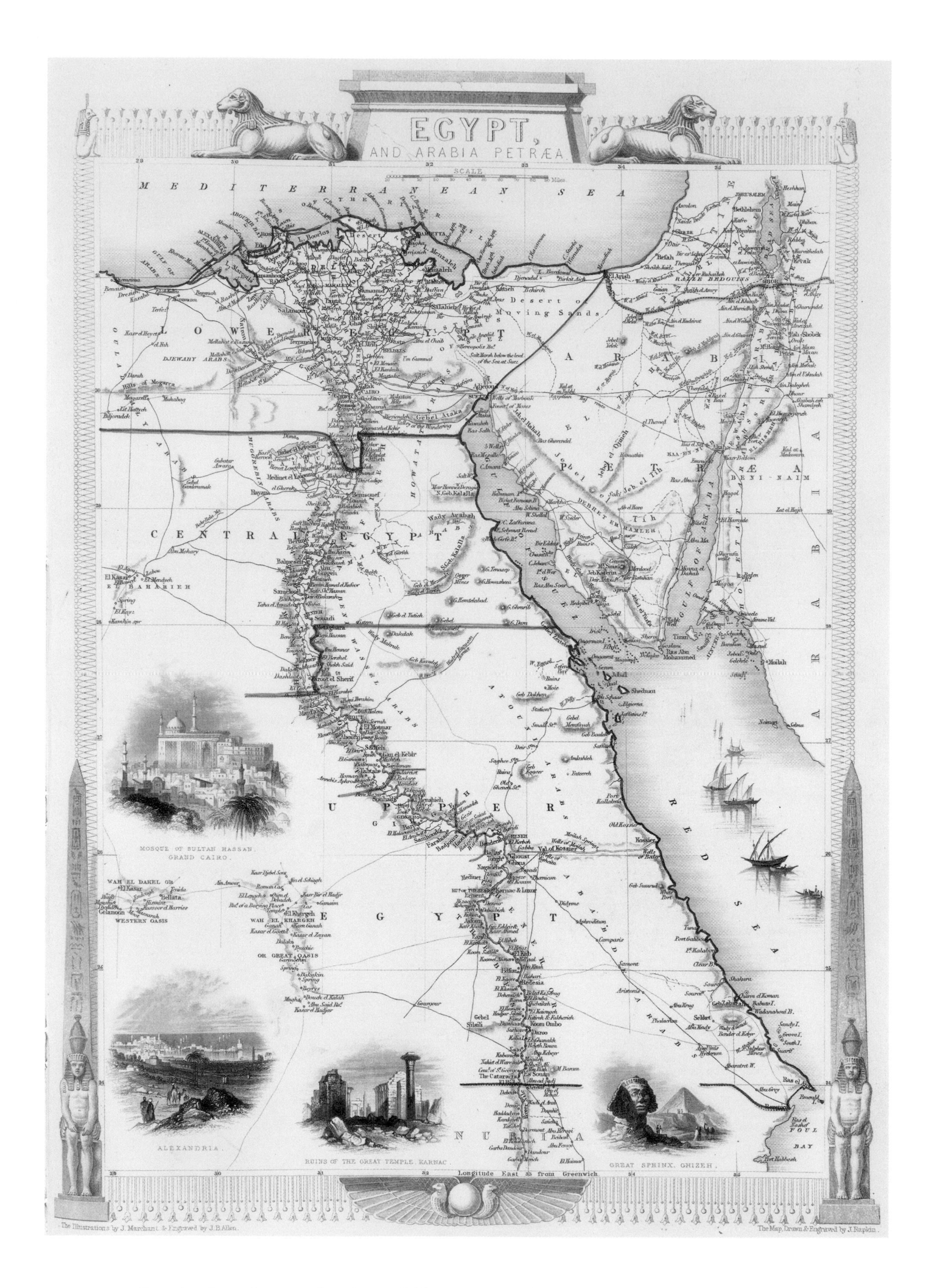

1851 Egypt And Arabia Petraea by John Rapkin, J. Marchant, and J. B. Allen, from John Tallis's Illustrated Atlas
Steel Engraving, with Outline Hand-Coloring, 14.25 x 10.5 in.
Library of Congress, Washington, D.C.

Illustrated Atlas in a Handy Size

THE 19TH CENTURY WITNESSED A RESURGENCE IN the production of geographical world atlases in Europe and America. The development of modern geography—established in Germany by Alexander von Humboldt and Carl Ritter in the first decades of the century and popularized by professional societies such as the Société de Géographie of Paris, the Royal Geographical Society of London, and the National Geographic Society in Washington, D.C.—provided a framework for acquiring, analyzing, and disseminating data. The official topographic surveys initiated in Europe and its colonies, and the final surge of exploration in the Pacific and elsewhere, continued to furnish atlas publishers with new information. An increase in literacy, the adoption of compulsory education, and a broadening interest in the world, stimulated by exploration, colonial spheres of influence, and emigration, ensured a growing market for these works. Improved transportation and communication systems such as the steamship, railroad, transoceanic cable, and telegraph aided in the transmission of information. Finally, new printing processes such as steel engraving and lithography reduced production costs.

Nineteenth-century geographical world atlases differ from earlier atlases in both format and content. The basic model evolved in Germany, with the introduction of Adolph Stieler's *Hand-Atlas* in 1816. Stieler's objective was to provide "a general usefulness . . . as well as instruction . . . for the daily use for people from all walks of life." To this end, his atlases were produced in a "handy size" and sold at moderate prices. Maps were geographical and topographical in content, based on the latest information, and prepared at relatively small scale on uniform projections. Text described the history and geography of the region portrayed. Place-names were important, since the atlases often were used to help newspaper readers follow current events. They were enhanced with color and illustrations to appeal to the eye.

John Tallis's *Illustrated Atlas* was one of the more popular general atlases of this period. Tallis was a successful and innovative London publisher with offices in New York, Edinburgh, and Dublin. He entered the map business in 1838 with his *London Street Views*, a popular series of engraved street plans that included illustrated profiles of buildings. These views were published in 1847 as *Tallis' Street Views & Pictorial Directory.* Following its success, Tallis began work on his *Illustrated Atlas*, the maps of which initially were distributed in serial form on a subscription basis.

A complete volume was published in 1851 to commemorate the Great London Exhibition of that year. Text was written by Montgomery Martin, known for his work on the history of the British colonies. Maps were drawn and steel-engraved by John Rapkin. Today, Tallis's atlases are prized for their beautiful vignettes, drawn and engraved by skilled artists.

Charting the Coastline

President Thomas Jefferson created the Survey of the Coast in 1807 as a scientific organization charged with preparing coastal charts of the United States, "in which shall be designated the islands and shoals and places of anchorage." Still in operation as the Office of Coast Survey, a component of the National Ocean Service, it is unique among national nautical charting organizations. Since its inception, it has been administered as a civilian agency, although Army and Navy topographers were occasionally assigned to carry out surveys. Ferdinand Rudolph Hassler, a Swiss-born geodesist experienced in trigonometric surveying, was appointed the agency's first superintendent. He placed the organization on a firm scientific footing, laying the groundwork for triangulation networks along the East Coast, but little surveying and charting was accomplished prior to his death in 1843.

Alexander Dallas Bache, the great-grandson of Benjamin Franklin, succeeded Hassler. A graduate of the U.S. Military Academy at West Point and a first-rate scientist, Bache was also a superb administrator. Bache ably implemented Hassler's basic plan, which he adapted to meet the needs of an expanding nation. He further determined that the agency's charts should be made available to everyone having an interest in commerce, navigation, geography, or science. A flatbed, single-roller, copperplate engraving press was acquired in 1842, and the first engraving to be printed was a chart of New York Bay and Harbor.

Bache divided the Atlantic and Gulf coasts into eight sections, in which topographic and hydrographic surveys were conducted simultaneously. As Texas was added to the Union in 1845 and California in 1850, new survey teams were established. Discovery of gold in California at Sutter's Mill in 1849 and the subsequent gold rush provided an additional impetus for charting the Pacific coast, and led to the only mutiny in Coast Survey history. The five-man crew of the survey gig *Ewing* tossed their commanding officer, Midshipman William Gibson, overboard on September 13, 1849, and headed for the gold fields. Later convicted of "Mutiny, Desertion, and running away with a boat," two of the crewmen were "hanged from the yard arm." The hydrographic survey of the West Coast was conducted by James Alden, an accomplished artist and hydrographer. Headland sketches were added to these charts to aid navigators, a device that dated from the earliest European charts.

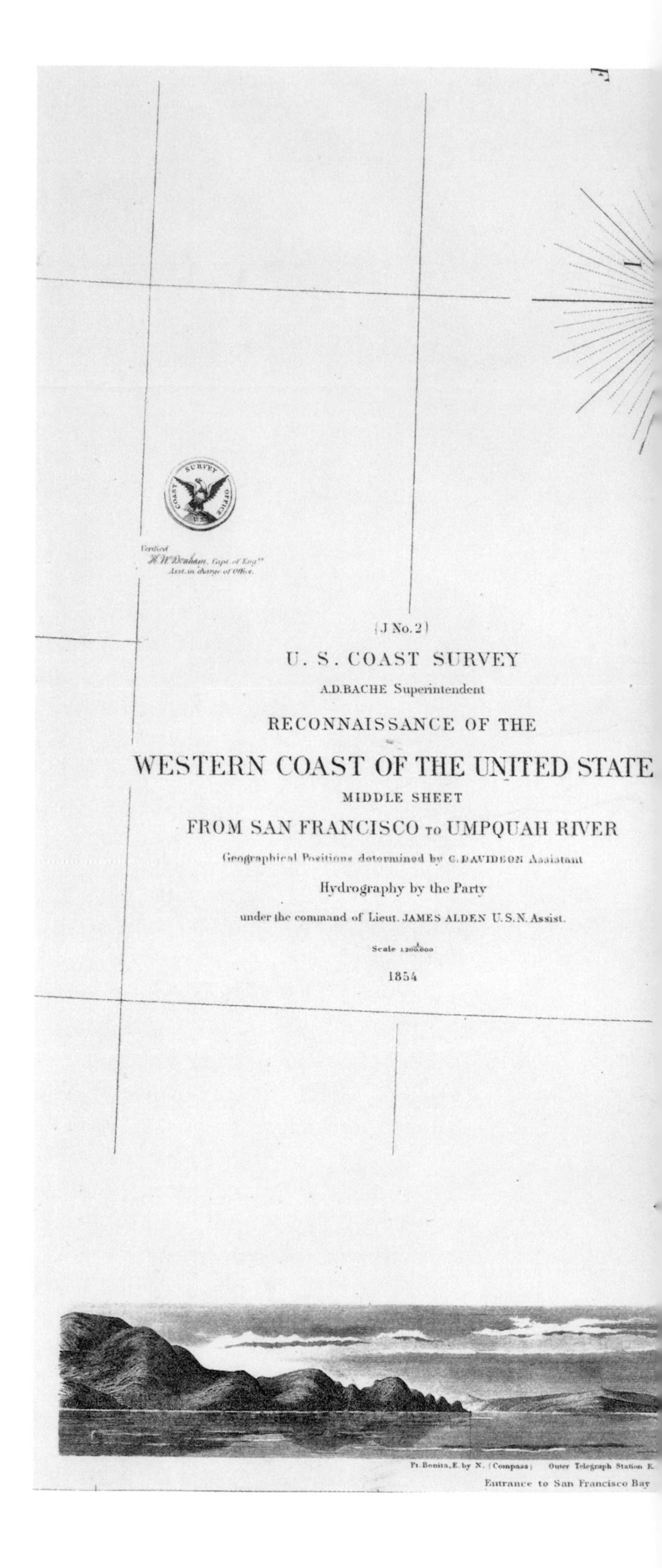

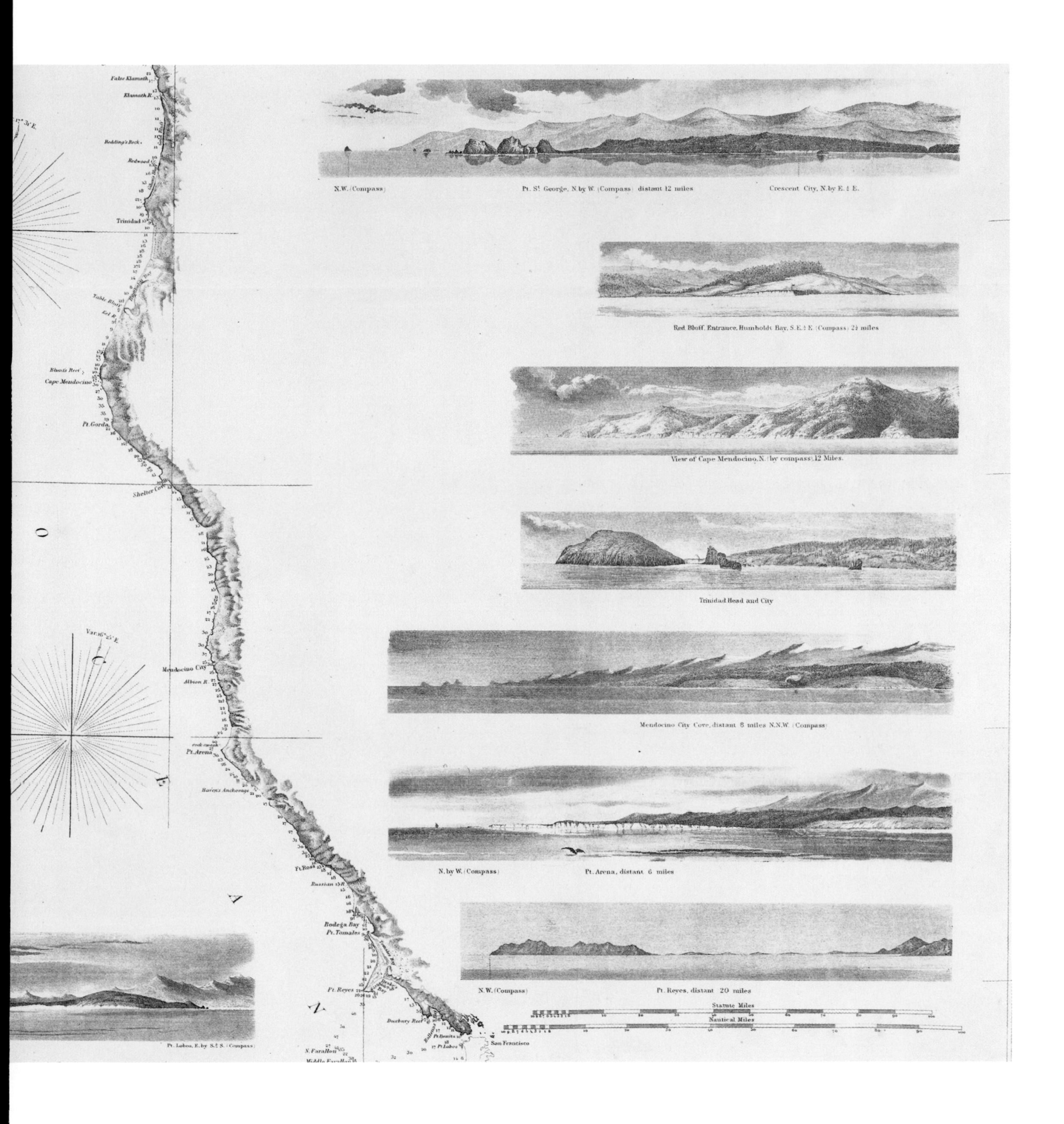

1854 Reconnaissance of the Western Coast of the United States by Lt. James Alden and George Davidson
Copperplate Engraving, 23 x 22.5 in.
Library of Congress, Washington, D.C.

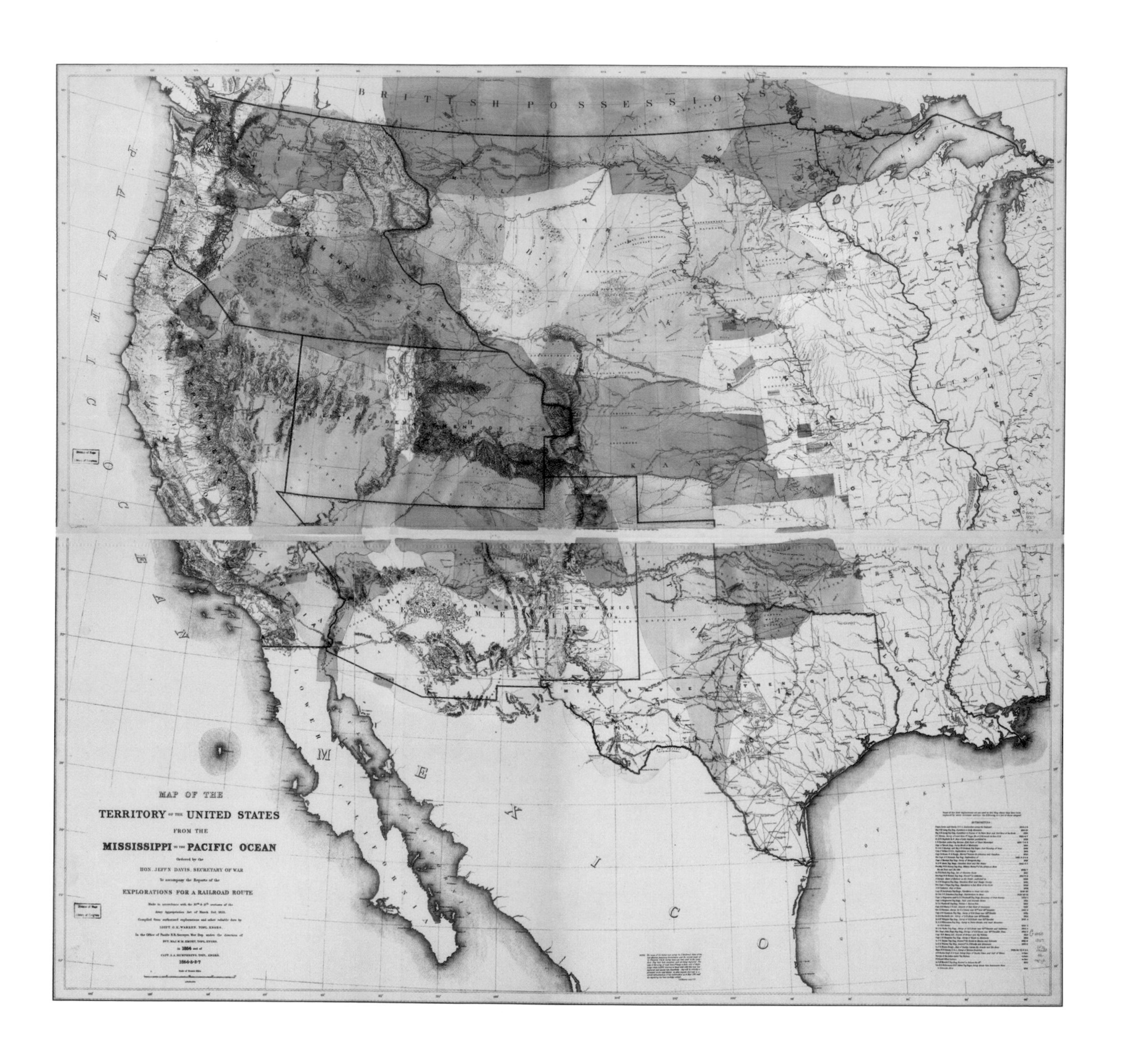

1857 Map of the Territory of the United States from the Mississippi to the Pacific Ocean by Lt. Gouverneur K. Warren
Lithograph, Hand-Colored, 45.5 x 42.25 in.
Library of Congress, Washington, D.C.

The Army's Master Map of the West

THE WAR DEPARTMENT'S MAP OF THE TERRITORY of the United States from the Mississippi to the Pacific Ocean is one of the great foundation maps of the American West. Published by Congress in 1857, it was the first map of the entire trans-Mississippi West to show geographical positions and topographic relief with reasonable accuracy. Measuring nearly four feet in both directions, it depicts information one might expect to find on a military map that might be used for strategic and commercial purposes: existing and projected roads, military fortifications and towns, and the hunting grounds of western Indians.

This map was prepared in the Office of Explorations and Surveys, a small unit established by Secretary of War Jefferson Davis to pursue special mapping projects. It was designed by Maj. William H. Emory, compiled by Lt. Gouverneur K. Warren, and drafted principally by civilian cartographers Edward Freyhold and Baron Frederick W. von Egloffstein, a Prussian topographic artist. Revisions made in 1858, 1868, and 1879 incorporated data from later Army expeditions to the headwaters of the Yellowstone and Colorado Rivers. Areas lacking data were left blank; areas where the quality of the data was questionable were depicted with faint lines.

The primary data used to compile this map were based on the Pacific Railroad surveys of 1853 to 1855, a congressionally directed effort to determine the best railroad route to link the Mississippi Valley with the West Coast. While the routes chosen reflected regional political interests rather than practical considerations, their distribution ensured that much of the trans-Mississippi West would be surveyed and mapped. These included a northern route following the 47th and 49th parallels from Minnesota Territory to Puget Sound, Washington; a central route extending from Fort Leavenworth, Kansas, to California through Colorado, Utah, and Nevada; and two southwestern routes, one stretching from Fort Smith, Arkansas, westward along the 35th parallel to Los Angeles through present-day Oklahoma, the Texas Panhandle, New Mexico, and Arizona, and one from East Texas westward along the 32nd parallel to San Diego. This last route, which closely tracks the Mexico-U.S. border, was added later at the insistence of southern congressmen who hoped to expand slavery and commerce along its course. Several other surveys running parallel to the Pacific coast were undertaken in California, Oregon, and Washington.

This three-year effort culminated in a massive 13-volume work entitled *Reports of Exploration and Surveys to Ascertain the Most Practicable and Economic Route for a Railroad from the Mississippi River to the Pacific Coast.* Volume 11 was devoted exclusively to maps. Published by Congress between 1855 and 1861, this set constitutes the single most important contemporary source of knowledge on western geography and cartography for the mid-19th century.

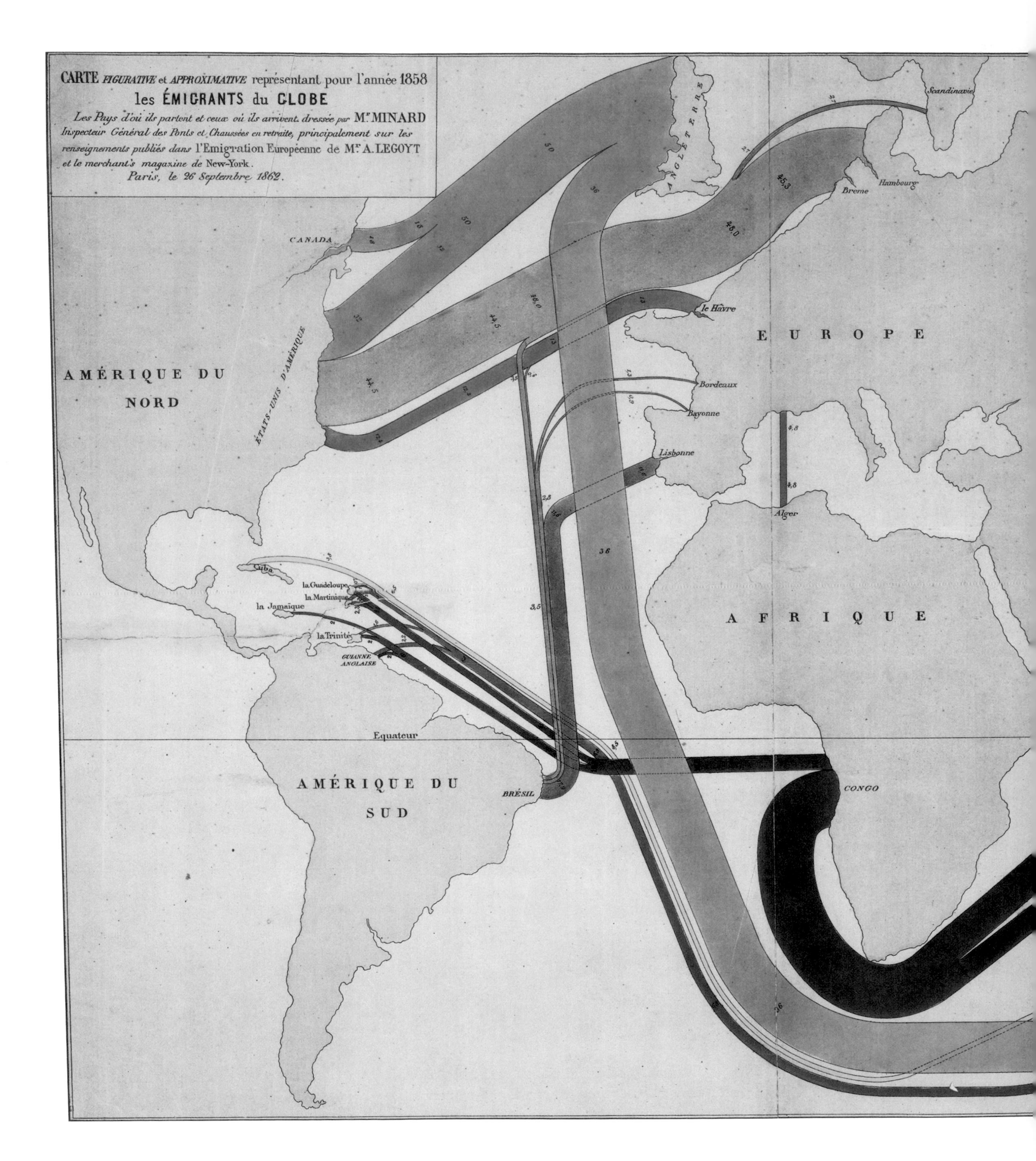

1862 Carte figurative et approximative représentant pour l'année 1858 les émigrants du globe (Map of World Emigration in 1858)
Hand-Colored Lithograph by Charles Joseph Minard, 25 x 28.5 in.
Library of Congress, Washington, D.C.

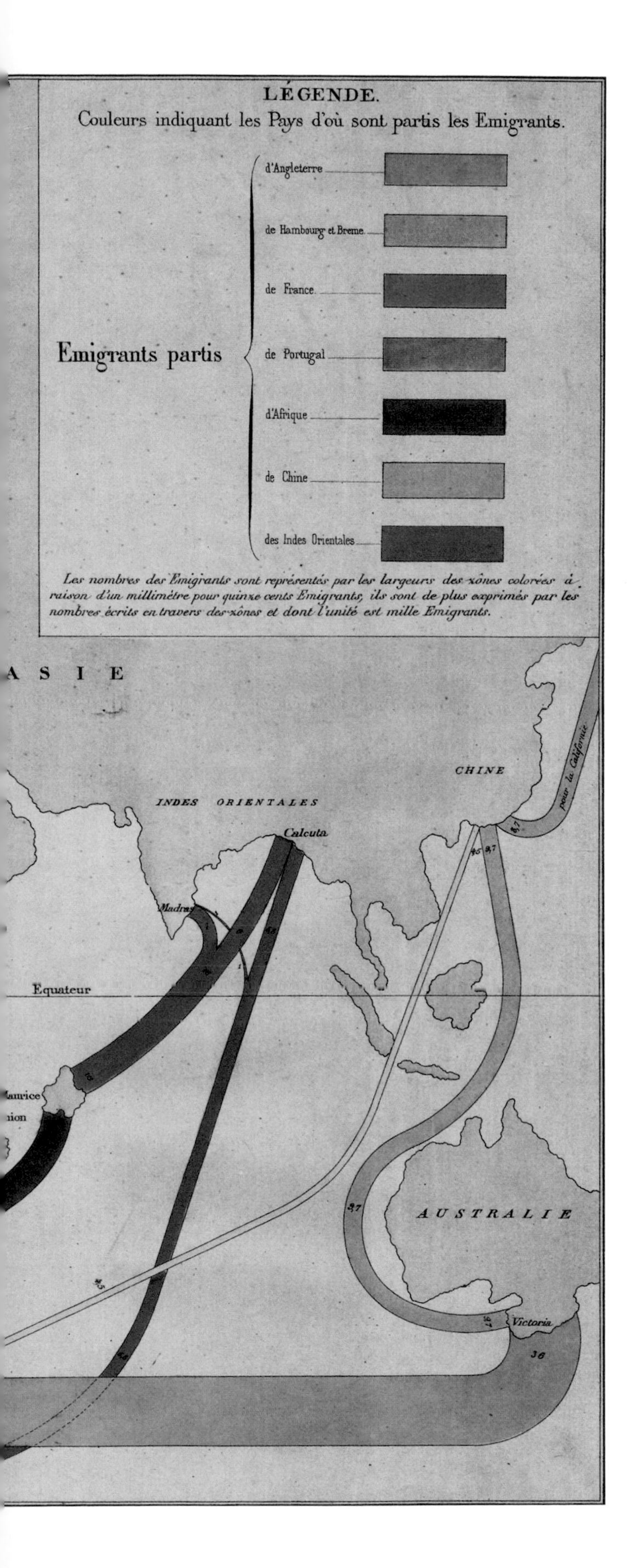

Emigration Flow Maps

MODERN FLOW MAPS, SEEN DAILY ON TELEVISION, in newspapers, and on computer screens, first made their appearance in the late 1830s. They allow mapmakers and graphic artists to display the movements of complex data through the use of lines of varying widths. Flow maps have a narrative power, according to the eminent statistician Edward Tufte, author of the groundbreaking work *The Visual Display of Quantitative Information.* They tell dynamic, visual stories with variable data.

The flow map was developed and popularized primarily by Charles Joseph Minard, a student and instructor associated with the School of Bridges and Roads in Paris, and a practicing civil engineer focusing on maritime port facilities, canals and inland waters, and, later, railroads. Retiring in 1851 at the mandatory age of 70, Minard devoted the remaining 19 years of his life to his second career as a thematic cartographer, developing new ways to visualize social and economic data. Between 1844 and his death in 1870, he created and published 69 unique maps that portrayed numerical data in various graphic forms. More than half of these works were flow maps.

Minard's *Carte figurative* depicting European, African, Indian, and Chinese emigration in 1858 is representative of this genre. The statistical data were obtained from publications by Alfred Legoyt, head statistician with the Ministry of Trade in Paris, and from Freeman Hunt's *Merchants' Magazine and Commercial Review,* a New York encyclopedia of commercial subjects. Minard designed his map "to convey promptly to the eye the relation not given quickly by numbers."

By using varying line widths and colors, Minard was able to portray four different data sets simultaneously. These include the *size* of emigration, which is illustrated by the width of the lines and by accompanying numerical figures expressed in thousands; the *location* of the various population movements over the surface of the map; the *direction* of emigration, which is indicated by narrowing widths of the flow lines; and the different *nationalities and racial groups* participating voluntarily or involuntarily in emigration, as in the case of Africans from the Congo kidnapped and forced into slavery. These groups are indicated by different colors.

It is interesting to note that the large majority of slaves in 1858 were sent to two Indian Ocean islands, Réunion and Maurice (Mauritius), rather than the West Indies or South America.

Sun Prints

ARMED CONFLICT STIMULATES MAPMAKING. MAPS are required for all aspects of modern warfare, from the battlefield to the home front. The American Civil War was no exception. During the struggle from 1861 to 1865, Union and Confederate topographical engineers surveyed and mapped thousands of miles of the South, many never before charted. At the same time, commercial map publishers and newspapers provided the public with the latest maps of military campaigns so that they could follow the course of the war and track the movements of family members.

Topographical field surveying and mapping units generally were assigned to each field army. The best-known Rebel cartographers were Maj. Albert Campbell, head of Robert E. Lee's Topographical Department, which functioned as the map bureau for the Confederacy, and Jedediah Hotchkiss, a civilian schoolteacher and brilliant amateur mapmaker who was chief cartographer for Stonewall Jackson's Army of Northern Virginia. Hotchkiss's 1862 manuscript map of the Shenandoah Valley, sketched from horseback, is a classic work, and was recently digitized by the National Park Service to help identify Civil War sites for preservation and interpretation.

Union mapmakers of note were generally from the Army's two elite units, the Corps of Topographical Engineers and the Corps of Engineers. Graduates of the U.S. Military Academy at West Point, they were well trained in surveying and mapmaking. Capt. William E. Merrill, for example, was chief topographical officer with the Army of the Cumberland, commanded by Gen. George H. Thomas. He operated the most sophisticated field mapping unit in the Civil War, with its own printing and lithographic presses, photographic equipment, draftsmen, lithographers, and map mounters. One of his unit's innovations was the use of photographic techniques to reproduce maps quickly for tactical purposes. Developed by Capt. William C. Margedant, this process involved placing a map drawn on tracing paper over a sheet of paper coated with silver nitrate. The sheets were then exposed to the sun, which "printed" a negative image on the photographic paper. Such maps could be updated quickly and often.

Another innovation was cloth maps. Ater the compilation of a map of Georgia for Gen. William Tecumseh Sherman's Atlanta campaign, copies were lithographed on muslin for cavalry officers. "Many officers sent handkerchiefs to the office and had maps printed on them," noted topographical engineer Thomas Van Horne in his *History of the Army of the Cumberland*. The maps held up and could be washed.

One of the more interesting Union topographers was the legendary George Armstrong Custer. Perhaps the Army's greatest cavalry officer, Custer, incredibly, initially was assigned to Prof. Thaddeus Lowe's experimental balloon corps, where he made some of history's first aerial-sketched maps.

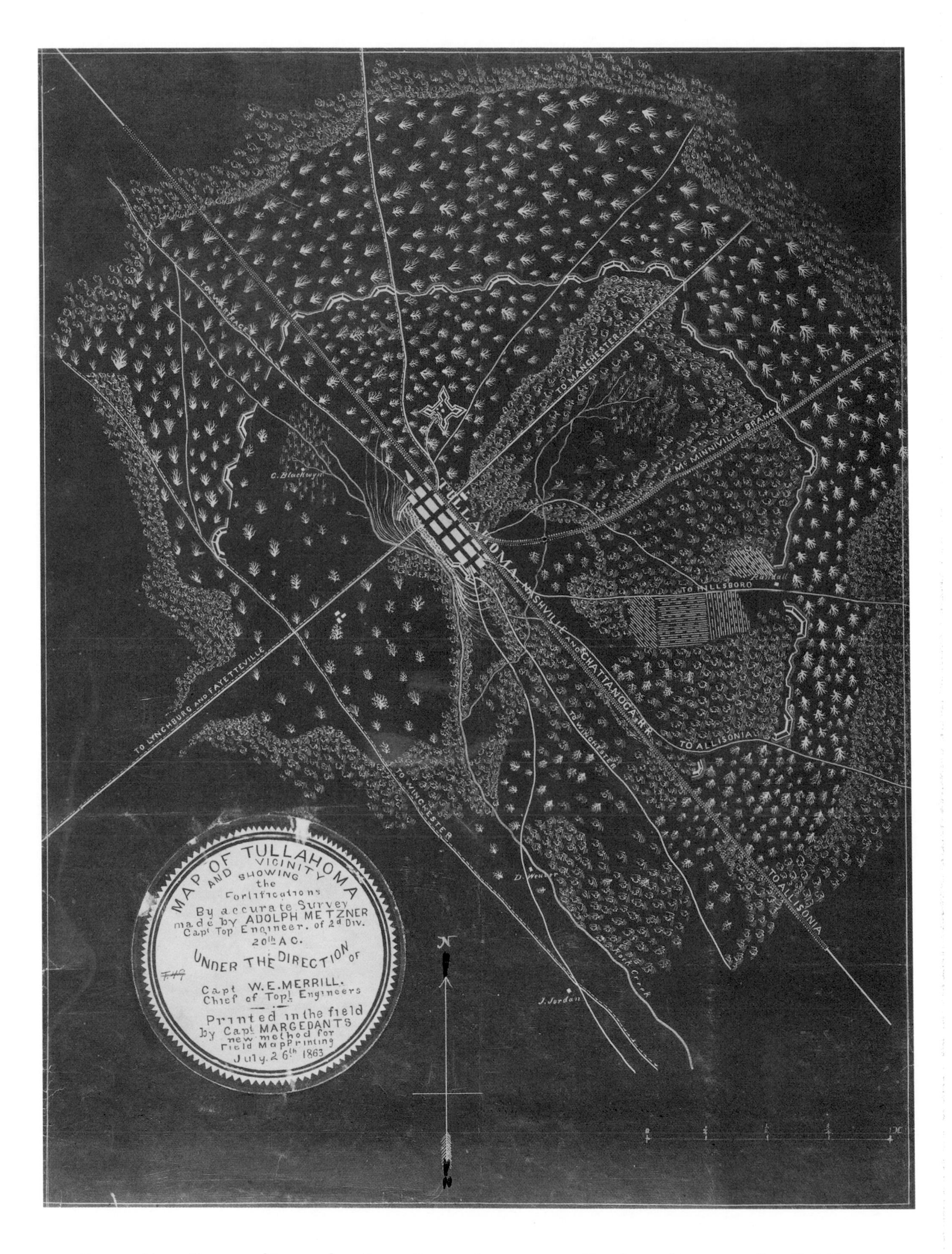

1863 Detail of Map of Tullahoma [Tennessee] and Vicinity Showing Fortifications by Capt. Adolph Metzner and Capt. W. C. Margedant
Photo-Processed, Hand-Colored, 18.6 x 14 in.
U.S. National Archives and Records Administration, Washington, D.C.

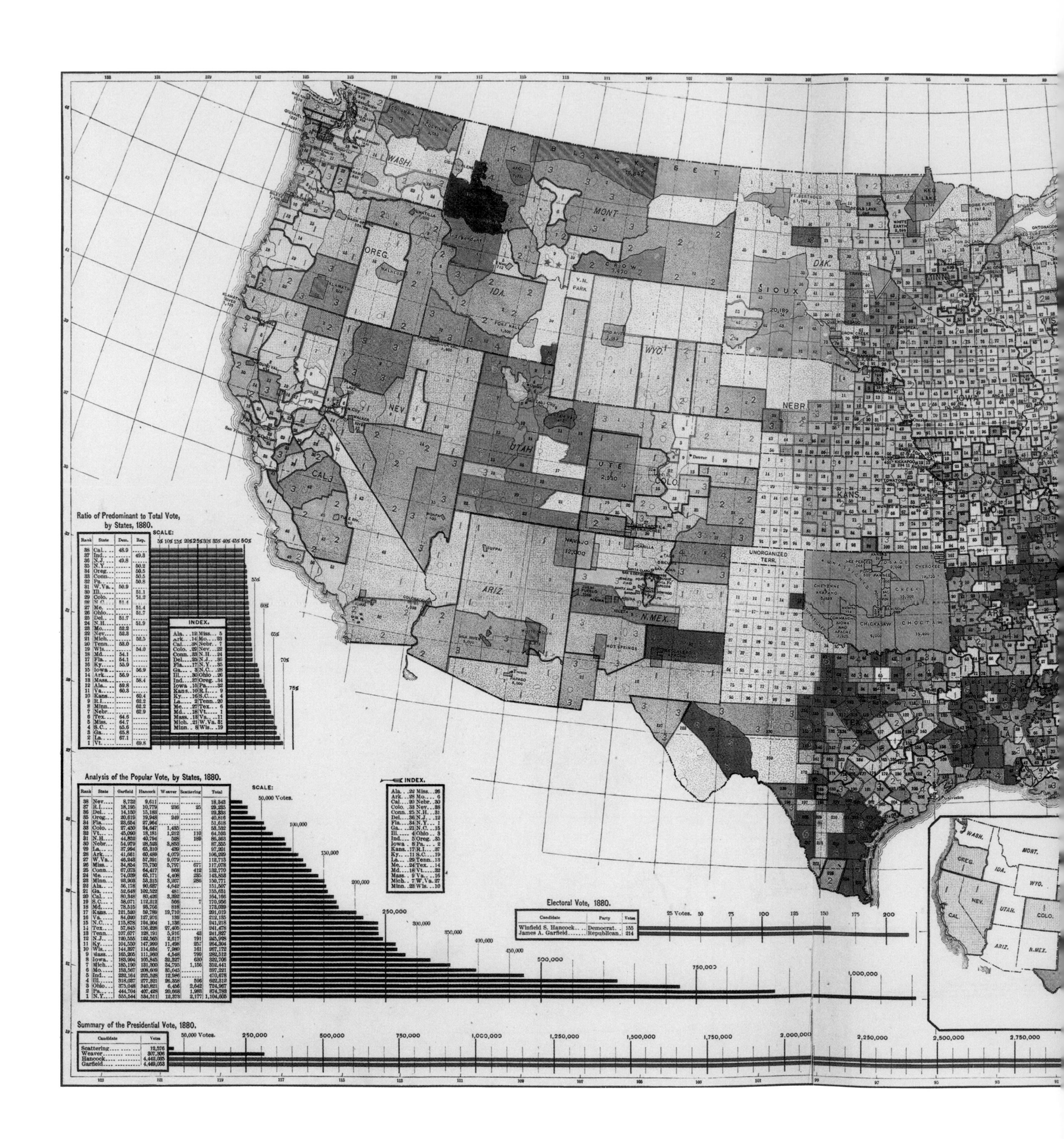

Circa 1883 Popular Vote Ratio of Predominant to Total Vote by Counties in 1880 by Fletcher W. Hewes and Henry Gannett
Chromolithograph, 18.75 x 26.75 in.
Library of Congress, Washington, D.C.

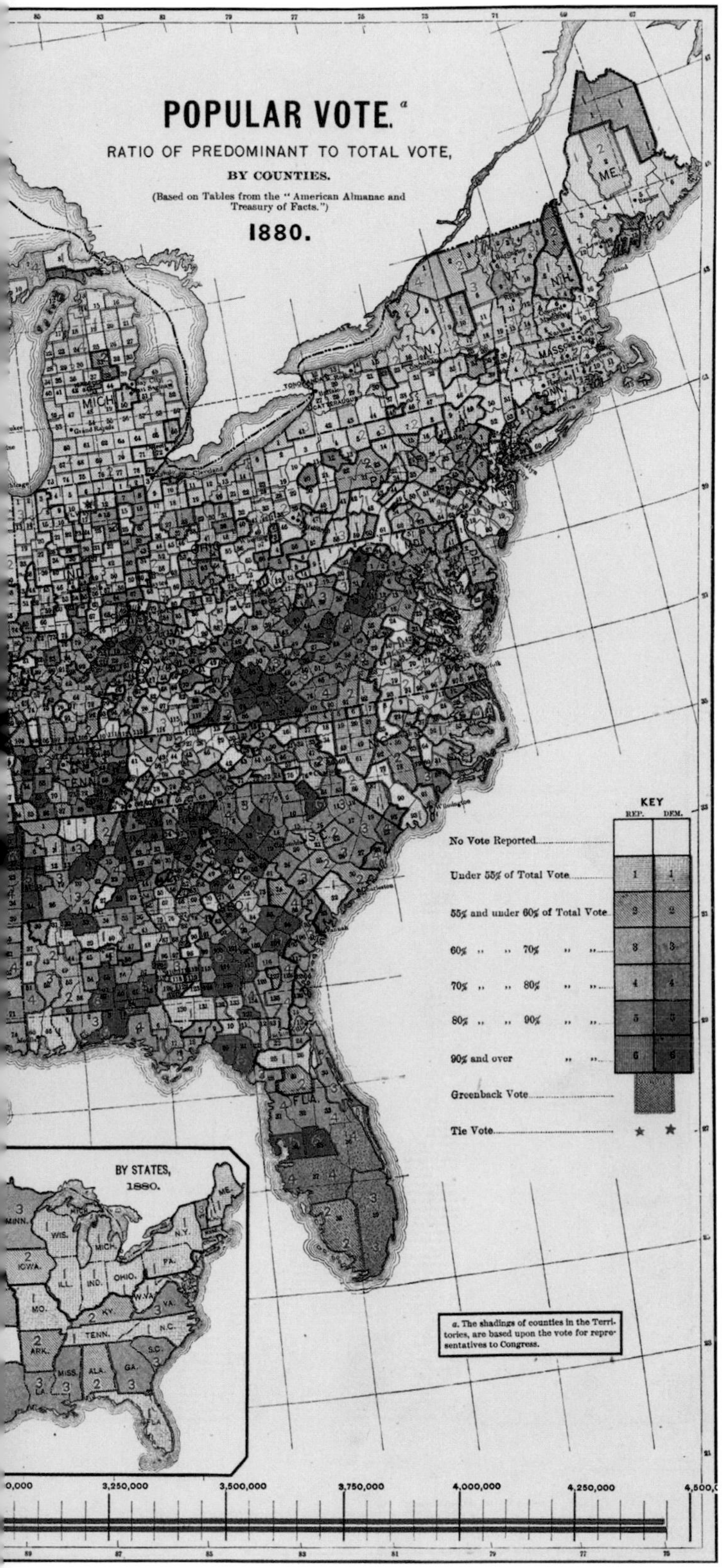

Mapping the Red and Blue States

THE 2004 ELECTION HIGHLIGHTED THE CONCEPT OF "red and blue states" in the American lexicon, where "red" denotes Republicans and "blue" Democrats. While the term has become highly charged in the current political climate, the choice of these colors dates from the earliest presidential elections that were mapped in color. The 1880 electoral map, for example, depicted the popular presidential vote by county for Republican James A. Garfield, Democrat Winfield Scott Hancock, and Greenback Labor candidate James B. Weaver. The contest, which was won by Garfield by a margin of fewer than 10,000 votes out of more than nine million cast, was illustrated with red and blue colors. The 2004 and 1880 electoral maps look similar; in fact, they are practically mirror images of each other, with southern counties mostly red and northern counties mostly blue. However, in 1880, the blue states were Republican and the red states Democrat!

This 1880 electoral map was one of 151 colored lithograph map plates prepared for *Scribner's Statistical Atlas of the United States*, published in 1883, which was based on the returns of the Tenth Decennial Census. The United States conducted its first official national decennial census in 1790, but for the next eight decades census results were used only sporadically to create statistical maps. The 1850 census, for instance, was gleaned by Dr. Edward H. Barton, a pioneer surgeon, for the preparation of several maps in 1853 that would show sanitary conditions for his medical reports on epidemics. German geographer August Petermann used returns from the 1850 and 1860 censuses for several articles appearing in his journal *Geographischen Mitteilungen*.

The major breakthrough, however, came in 1870, when Francis Amasa Walker, superintendent of the Ninth Decennial Census, published a statistical atlas based on census returns. This pioneering work, printed in color by the New York lithographer Julius Bien, contained 54 maps that provided a variety of new images of the physical, economic, and social geography of the United States. It was followed by Fletcher W. Hewes and Henry Gannett's more sophisticated work, which appeared in *Scribner's Atlas* and is reproduced here, "showing by graphic methods [the counties'] present condition and their political, social and industrial development."

Gannett, a co-founder of the National Geographic Society and chief geographer of the United States Geological Survey, was detailed to the Census Bureau in 1879 to provide geographic expertise.

POPULAR VOTE.[a]
RATIO OF PREDOMINANT TO TOTAL VOTE,
BY COUNTIES.
(Based on Tables from the "American Almanac and Treasury of Facts.")
1880.

KEY
REP. DEM.
No Vote Reported
Under 55% of Total Vote
55% and under 60% of Total Vote
60% " " 70% " "
70% " " 80% " "
80% " " 90% " "
90% and over " "
Greenback Vote
Tie Vote

BY STATES, 1880.

a. The shadings of counties in the Territories, are based upon the vote for representatives to Congress.

SECTION FIVE

Maps for Everyone

Aerial Photograph of Laketown Township, Minnesota, 1937
Prepared by Soil Conservation Service

INTRODUCTION

MAPS BECAME READILY AVAILABLE IN THE 20TH CENTURY, PARTICULARLY IN EUROPE AND NORTH AMERICA. They were no longer restricted to the specialist but widely accessible to people in all walks of life. New methods of data acquisition, map printing, and map distribution drove down costs and increased press runs. Factors as diverse as global wars, new modes of transportation, and the rise of an affluent and well-educated middle class spurred the development of new markets. A professional class of mapmaking specialists emerged, developing standards that were adopted worldwide. Mathematical cartography supplanted the traditional humanistic and artistic forms favored by earlier Asian mapmakers. By mid-century, the format and content of maps were similar throughout the world. The standardized symbols and colors of Western mathematical cartography served as a universal language.

NEW TECHNOLOGIES HAD A PROFOUND IMPACT ON HOW MAP DATA WERE GATHERED. AERIAL PHOTOGRAPHY and photogrammetry—the use of photographs, especially aerial photographs, to make accurate measurements—were the most dramatic of these innovations. No longer was it necessary to send large numbers of surveyors and mapmakers into the countryside to prepare basic topographic maps. The early use of these instruments and associated processes is illustrated here by the photograph of the U.S. Army aerial photographer and the 1921 Michigan Schoolcraft quadrangle. The role of aerial photography in the mapping process expanded greatly during World War II, which is touched upon with William Bostick's 1941 D-Day map. The use of conventional, high-altitude, and space photography for mountain cartography is represented by Bradford Washburn's 1988 map of Mount Everest.

Other instruments made possible the acquisition of mappable data previously unattainable. Data for Alvaro Espinosa and Marie Tharp's 1981 world map of earthquakes, for example, were obtained from seismic monitoring stations. Tharp's base map of the ocean floor was derived from data obtained by echo-sounding devices developed during World War II. Government geodesists established a worldwide geodetic reference base that greatly improved map accuracy, represented by the 1962 Russian military map.

The introduction of offset lithography and, later, computer technologies, made possible the widespread distribution of maps during the 20th century. The offset press, on which a zinc or aluminum printing plate is attached to a rotating cylinder, increased production speed in comparison to flatbed presses with copper engravings or lithographic stones. In the United States, map publishers turned to offset lithography soon after its introduction in 1904; their European counterparts did not fully adopt it until after World War II because their print runs were smaller and they held large inventories of original stones. The first offset presses could produce about 1,500 two-color maps every hour. By 1959, the General Drafting Company was printing 5,000 sheets in four colors each hour. The majority of the maps in this section were produced by offset lithography.

The perfection of color printing during this period made possible the portrayal of complex concepts, such as shaded relief and ethnography, which can be seen in George Blum's 1895 map of California and the 1919 ethnographic map of Macedonia.

The National Geographic Society was one of the earliest mapping organizations to adopt computer-assisted mapmaking, first using an Israeli-produced computer originally developed to design fabric patterns for the textile industry. The Society's 1988 map of aircraft over the New York metropolitan region represents an early example of this technology.

As maps became less costly and easier to reproduce, new markets of map users emerged. Among the first were farmers and tradesmen in rural areas and small towns in the United States and Canada at the close of the 19th century. With pride they purchased landownership maps and atlases that depicted their farms and homes, and panoramic maps of their towns, such as Henry Wellge's 1882 view of Santa Fe, New Mexico.

The inventions of the bicycle, the automobile, and the airplane, like the railroad of the 19th century, spurred travel to unfamiliar places, creating demand by travelers and advertisers for special maps. New map forms were created—the free oil company road map and the airline souvenir map. Representative examples include George Blum's 1895 California map for cyclers; Texaco Oil Company's 1927 map of Florida; and Britain's Imperial Airways' 1934 map of its London-to-Paris route. Closely related was the tourist map, which is illustrated by Heinrich Berann's magnificent 1989 panoramic view of Yellowstone National Park.

THE GLOBAL REACH OF TWO WORLD WARS, KOREA AND VIETNAM, AND THE COLD WAR FUELED interest in maps and world geography. Demand was filled in part by freelance cartographers such as Richard Edes Harrison with his 1943 perspective view of the PacificTheater. Like other wartime journalists, cartographers used maps to persuade as well as inform. A fascinating example is a map of the German socialist revolution and Russian Civil War, issued by the Moscow State Publishing House of the Red Proletariat in 1928.

The wars of the 20th century introduced servicemen and -women to mapmaking and map reading. Probably a billion maps were produced during World War II! Front-line topographic battalions made many of them. Some Navy ships carried their own offset lithographic printers. William A. Bostick's 1944 D-Day map of Omaha Beach is reproduced here. He used such a press to print maps for George Patton's invasion of Sicily. Special-purpose maps were introduced in response to unique requirements. The aeronautical chart, for instance, which first appeared during World War I, is represented by a strip map from the British Royal Navy in 1915. A manuscript map by the Japanese naval aviator who led the December 7, 1941, strike on Pearl Harbor is an example of an early damage assessment map. One of the most unique is a 1944 escape and evasion map of southern Germany that was sent to American POWs in decks of playing cards.

Circa 1918 U. S. Army Air Service Aerial Photographer, Aerial Photo Recon School, Langley Field, Virginia
U.S. National Archives and Records Administration, Washington, D.C.

Mapping Takes to the Air

MAPMAKING WAS RADICALLY CHANGED BY AERIAL photography. Initially, its images were used for "interpretation," or intelligence gathering of various kinds. Then, with the aid of optical analog plotting instruments and the applications of photogrammetry, aerial photographs provided the raw material for compiling topographic maps. Measurements were taken directly from these photographs, and they became substitutes for costly and time-consuming ground surveys. Later, with the introduction of sensing devices beyond the normal visual range of film, many different kinds of maps were produced.

Photographers were intrigued with adopting their techniques for remotely sensed imaging almost from the beginning. They experimented with a variety of flying platforms. Paris was photographed from a free-flying balloon in 1858, Boston a year later. After the development of sensitized gelatin emulsions in 1870 and roll film, carrier pigeons were harnessed with small, breast-mounted cameras with built-in timers. G. R. Lawrence photographed the San Francisco earthquake and fire in 1906 from a height of 3,280 feet with a kite-supported camera. The invention of the airplane provided a camera platform that could be flown from place to place. Wilbur Wright took the first aerial photographs from an airplane during a flight over Centocelli, Italy, in 1909, but it was World War I that revealed their value. By 1918, French aerial units were taking 10,000 aerial photos daily.

While the majority of World War I aerial photographs were used for interpretation, the British Expeditionary Force in Egypt experimented with using them to prepare maps of Palestinian towns behind Turkish lines.

Photogrammetry is the art, science, and technology of obtaining information about the environment through the technical processes of recording, measuring, and interpreting photographs. The basic principles were developed initially in France and Austria for mapping towns and mountainous regions using photographs taken from roofs and mountains. More sophisticated stereo-plotting instruments, mounted on portable drafting boards called plane tables, were developed to analyze aerial images in Germany after World War I. Multi-lens aerial cameras, 3-D plotters designed for mapping continuous strips of overlapping vertical photographs, and color films were introduced in the 1930s and 1940s. By mid-century, aerial photographers and cartographers were preparing basic topographic map coverage for much of the world.

Bird's-Eye View of Santa Fe

BIRD'S-EYE VIEWS, OR PANORAMIC MAPS, WERE popular from the 1870s to the 1920s. They were similar to perspective maps of European cities produced during the Renaissance but depicted their subjects as if seen from a higher angle. They were also larger and more detailed. Although the total number printed is unknown, it was substantial. Panoramic artist Oakley H. Bailey stated that he personally had prepared maps of nearly 600 places during a career that spanned 55 years. Prussian emigrant Albert Ruger produced 60 panoramic views in one year. More than 2,000 survive in the Library of Congress and the Boston Public Library.

Panoramic artists spent most of their time on the road, traveling from city to city and from door to door to line up subscribers. City officials, real estate agencies, and local chambers of commerce were likely customers, often commissioning maps for promotional purposes. Proud home owners purchased the town views as wall hangings and enjoyed pointing out their houses to visitors.

When a sufficient number of subscribers had signed up, the mapmaker walked the streets, sketching individual buildings, houses, and major landscape features. Sometimes problems ensued. Panoramic artist Thaddeus Mortimer Fowler was jailed briefly during the height of World War I spy mania while sketching Allentown, Pennsylvania. City officials arrested the mapmaker on suspicion of being a German agent. Fowler died four years later at the age of 80 after falling on an icy street while drawing a map of Middleton, New York.

Improvements in printing methods, such as lithography, photoengraving, and chromolithography, increased print runs and reduced prices. Print runs of 250 to 2,000 maps were common. In 1872, one publisher sold panoramic views for $3.00 and charged an additional $2.50 for varnishing and mounting on a "plain frame."

Henry Wellge's work is representative. Like a number of panoramic mapmakers, he began his career in Madison, Wisconsin, home of the leading panoramic map publisher, Joseph J. Stoner. During a prolific career, Wellge mapped towns in 24 states, including making this view of Santa Fe. In typical fashion, it portrays the street layout, prominent public buildings, banks, hotels, churches, and homes. The central plaza, with the Palace of the Governors, is at left center. Fort Marcy, headquarters of the Army District of New Mexico, and the post's horse corral are to its left.

1882 Bird's-Eye View of the City of Santa Fe, New Mexico, by Henry Wellge
Chromolithograph, 9.5 x 19 in.
Library of Congress, Washington, D.C.

1895 Map of California Roads for Cyclers by George W. Blum
Photoengraving, 19 x 13.4 in.
Library of Congress, Washington, D.C.

Boneshakers

TODAY'S AUTOMOBILE ROADMAPS EVOLVED FROM BICYCLE maps. The bicycle craze of the late 19th century literally paved the way for the automobile age, according to transportation historians. Bicyclists promoted better roads, developed travel clubs, and encouraged manufacturing skills and technological improvements later associated with the automobile. A foot crank attached to the front wheel propelled the first pedal-driven bicycles. Invented in Paris in the late 1860s, these "boneshakers" and "mechanical horses" were easily recognized by their large front wheels, which measured up to five feet in diameter. In an essay titled "Taming the Bicycle," Mark Twain encouraged his readers: "Get a bicycle. You will not regret it, if you live."

The development of the safety bicycle with chain-drive and pneumatic tires in the late 1880s led to a bicycle boom that lasted until the introduction of motorcycles and automobiles. By 1897 the United States had produced more than two million bicycles, one for every 30 people. One million cyclists competed with horses for road space in England. Cycling clubs formed in France, England, and North America. The League of American Wheelmen (LAW), for example, established in 1880, numbered 102,000 members. Even a new fashion, bloomers, was designed to give women cyclists freedom from long skirts.

Cyclists needed good maps that showed the location and condition of roads. Cycling maps began to appear in England as early as 1880. In the United States, LAW encouraged local clubs to maintain logs of bike routes, or "road books." From them, route maps were constructed.

One of the most famous was George W. Blum's 80-page *Cyclers' Guide and Road Book of California,* which sold for a dollar in 1896. It contained itineraries for trips throughout the state, and three maps: one of the state of California, one of Golden Gate Park, and a larger-scale map in seven sections of the route from Chico to San Diego. Blum's *Map of California Roads for Cyclers* provided coverage from Mendocino south to San Diego, and inland to Lake Tahoe and Yosemite. Bicycle roads were colored red and rated by general surface condition and grade of slope. The road from Sacramento to Lake Tahoe, for example, was rated "F. H." (Fair, Hilly) to Placerville, then "P. M." (Poor, Mountainous), and finally "F. M." (Fair, Mountainous).

Terrain, an important feature of cycling maps, was portrayed by shaded relief, a new process that combined several printing technologies, some not available until the late 1880s—photography, metal engraving, halftone screens, and color printing.

Advertisers covered initial production costs. Their advertisements provide a unique look at the culture and the goods and services that encompassed the bicycle trade at the turn of the century. Portrayed in formal wear with bow tie and boutonniere, Joe E. Shearer, the inventor of Si-ko-lene, "the best lubricant on Earth for cycle chain," encouraged cyclists to "Try it!" The Premier Cycle Company of San Francisco appealed to patriotism. Its ad boasted that the Helical Tube, "American made throughout" was "the only wheel backed by impartial tests of the Ordinance Department U.S. Government."

Flying Map With Chart Holder

In contested skies over Europe during World War I, racing a mile a minute in "flying coffins" of wood and paper, daring pilots were guided to their destinations (and destinies) by strip maps, mounted on mechanical scrolling devices strapped to their legs to prevent loss from gusts of wind in open cockpits.

Wilbur Wright's epoch flight at Kitty Hawk in 1903 created the need for this new type of map—a chart to assist aviators navigating from one point to another high above the Earth's surface. Hermann Moedebeck, a Prussian artillery officer and balloonist, first advanced the idea for these maps. He prepared a series of aerial maps for Count Zeppelin's airship line in 1909. Working tirelessly to promote aviation cartography, Moedebeck published widely, held forums, and inspired aeronautical clubs to establish committees on flying maps. In an effort to coordinate and standardize this work, the International Commission on Aeronautical Maps first met in Brussels in 1911. Delegates recommended the creation of an international aeronautical map, but no agreement was reached with respect to conventional symbols except for high-tension electric overhead wires, to be indicated appropriately by "strings of red crosses."

Air navigation maps based on the commission's recommendations were first produced in France and England, motivated by large-scale military maneuvers in 1911, perhaps in anticipation of the war to come. In France, which claimed 352 licensed pilots in 1911—more than the rest of the world combined—these maps were prepared under sponsorship of the Aéro-Club de France, and in England by the Geographical Section of the Army's general staff.

The onset of the war dramatically increased the need for aviation maps. "Entire squadrons have landed behind the German lines, due to misinterpretation of their maps," noted one advocate. All combatants issued them, usually overprinting their national topographic map series with information relevant to aviators. Strip maps were used in Italy and England, including this 1915 map of the Kaiser Wilhelm, or Kiel, Canal from the British Hydrographic Office. A strategic waterway in north-central Germany connecting the North and Baltic Seas, the canal was a target for reconnaissance or bombing missions.

In the United States, Gen. William "Billy" Mitchell was a strong proponent of the aeronautical chart, but little progress was made during the war years, forcing American pilots on the western front to use French maps. At home they relied on Rand McNally road maps. It was not until the early 1920s—when U.S. Air Mail pilots (the first to fly on a daily basis) and Army Air Service aviators teamed up to demand better aviation maps—that adequate aeronautical charts were tested and produced. By the end of the decade, the United States' airways were linked by a series of strip charts and sectional maps that changed little in basic format and content until World War II.

1915 The Kaiser Wilhelm (Kiel) Canal by the Hydrographic Department of the Admiralty
Chromolithograph, 6.25 x 56 in.
1914 Aviation Chart Holder Manufactured by Houghton-Butcher for the British War Department
Private Collection

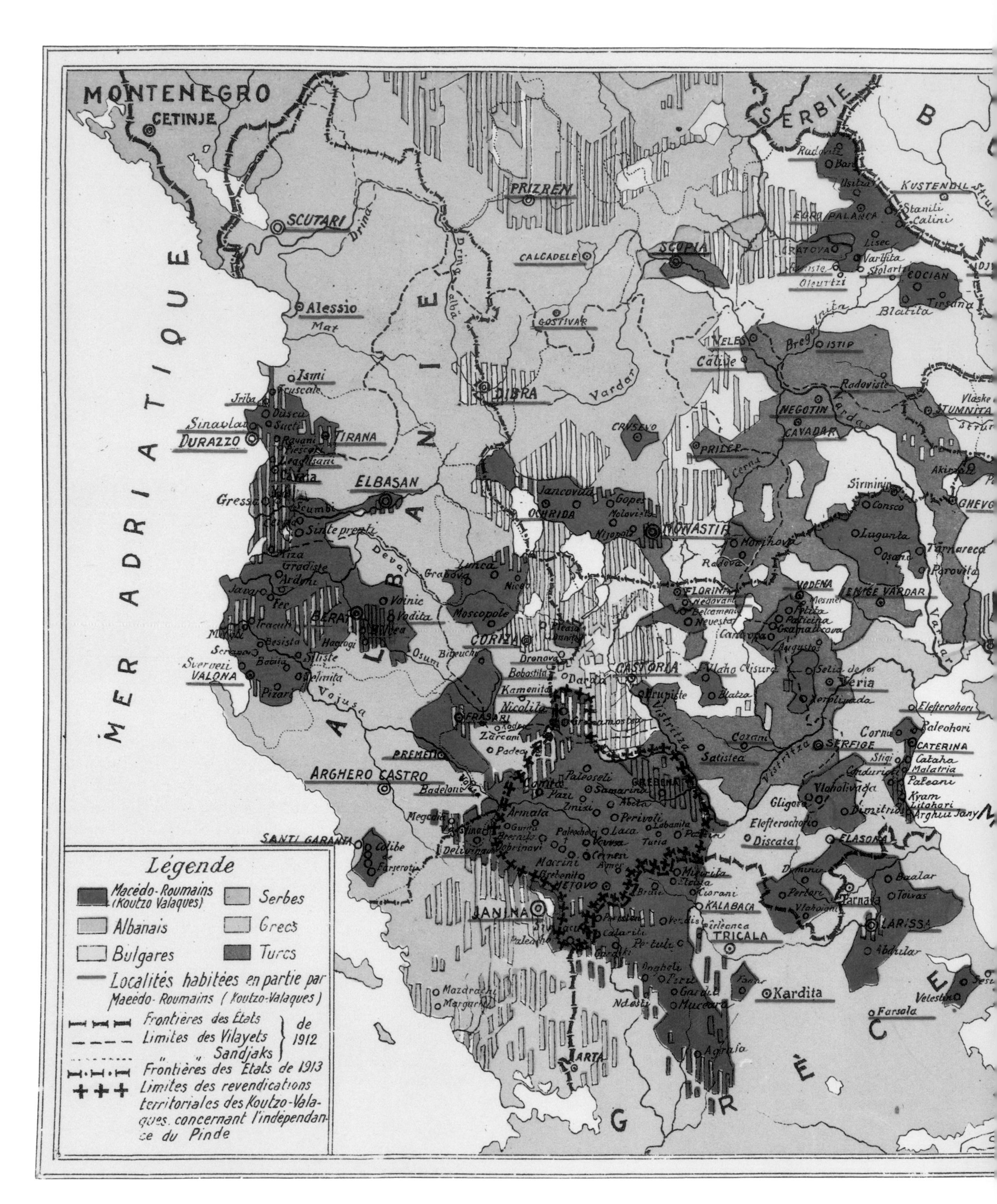

1919 Carte Ethnographique des Macédo-Roumains (Ethnographic map of Macedonians and Romanians) by A.D. Atanasiu
Chromolithograph, 9 x 11.5 in.
Library of Congress, Washington, D.C.

Mapping the Peace

MAPPING ETHNIC GROUPS HAS ALWAYS POSED problems. The cartographic limitations were resolved in the 19th century with the introduction of lithography and color printing, but the technical issues of gathering and deducing meaningful statistical data from incompatible census returns that cover disparate districts, countries, or regions challenge cartographers to this day. There are also political and economic considerations, and the threat of charges of racism associated with characterizing people by ethnic affiliation, traditions, and language. The earliest ethnographic maps date from the 1840s. Gustav Kombst's *Ethnographic Map of Europe*, which appeared in A. K. Johnston's *National Atlas* (Edinburgh, 1843), was based on "the principle of the natural physical differences of the different varieties of the Caucasian species inhabiting Europe." Kombst and Johnston used different colors to portray these variations, a technique followed by later cartographers.

A region of Europe that generated numerous ethnographic maps was the Balkan Peninsula, beginning in 1847 with Ami Boué's *Ethnographische Karte Osmanischen Reichs Europäischen Theils und von Griechenland (Ethnographic Map of the European Part of the Ottoman Empire and Greece)*. Numerous maps by ethnographers and linguists followed, attempting to sort out settlement patterns in regions of clashing cultures and constantly changing political boundaries. These maps were little more than academic curiosities until the end of World War I and the Paris Peace Conference at which diplomats, particularly the U.S. delegation, were looking for a "scientific" peace, one based on factual data. "Each of the central European nationalities had their own bagful of statistical tricks," according to participant Mark Jefferson, a U.S. cartographer. "When statistics failed, use was made of maps in color."

A. D. Atanasiu's 1919 map is a classic example. Centered on the region of modern Macedonia and the adjoining countries of Serbia and Montenegro, Bulgaria, Greece, and Albania, it was designed by a Romanian nationalist who supported his country's program of expansion. Six ethnic groups are portrayed by different colors, most prominently the Koutzo-Valaques (Aromani Vlachs), highlighted in red. These nomadic peoples spoke a language similar to Romanian, and for that reason were claimed by Romania. Provincial and district boundaries are depicted for the region along with the territorial boundaries for 1913, which were created after Turkey's defeat in the Balkan War of 1912.

Schoolcraft Quadrangle

IN EARLY 1920, THE ARMY AIR SERVICE DIRECTED ITS best aerial photographer to proceed to Michigan to photograph the Schoolcraft Quadrangle, a 225-square-mile area near Kalamazoo, for the U.S. Geological Survey (USGS). In seven hours flying time, Capt. Albert W. Stevens snapped 274 overlapping vertical aerial photographs using a single-lens camera. Survey mapmakers used the images to make the 1921 *Michigan Schoolcraft Quadrangle*, the first printed map in the U.S. constructed from aerial photographs. A new phase in American mapmaking had begun.

Created by Congress some 40 years earlier to consolidate several scientific and mapping programs, the USGS began a topographic mapping program, still ongoing. They were to produce an atlas conceived in 1897 as a set of map sheets of uniform size, which, when conjoined, would provide a map of the country. For mapping purposes, the country was divided into quadrilateral tracts bounded by parallels of latitude and meridians of longitude. The basic "quadrangle" or "quad" map, representing one map sheet in topographic atlases of the U.S., covers 15 minutes of latitude and longitude (about 197 to 282 square miles) at a scale of one inch to one mile (1:62,500).

Before aerial photography, survey field parties spread across the United States each season plotting manuscript quads on plane tables. Parties generally consisted of a topographical engineer, a recorder, two rodmen (who held the surveyor's rod level), and "sometimes an umbrella man." Packers, teamsters, drivers, and cooks provided support. Additional men might be employed to clear timber from summits or to construct observation towers and signal stations. They traveled by horse or "four-mule wagon" and later by automobile. One topographer working in the West estimated that he traveled some 10,000 miles on horseback between 1900 and 1914. Another used a motorcycle, towing his rodman on a bicycle at the end of a 15-foot rope. Alaskan mapmakers traveled by dogsled, canoe, and pack train. Aerial photography reduced this kind of fieldwork, particularly with the invention of stereographic plotters by which terrain could be copied directly from the photographs.

Standard symbols, colors, and place-names were adopted by the agency to simplify map interpretation and to encourage map reading by the public. Four basic colors were used to convey information: black for cultural features such as place-names, roads, railroads, and building symbols; brown for terrain, indicated by contours (lines showing equal elevation) and elevation figures; blue for lakes, rivers, and swamps; and green for woodlands. Most daring was the use of contours to depict an area's topographic character. It was the first time that an entire country was mapped with this symbol on a uniform series of maps. Survey officials used the term "topographic expression" to describe this process, recognizing that contouring remained as much art as science.

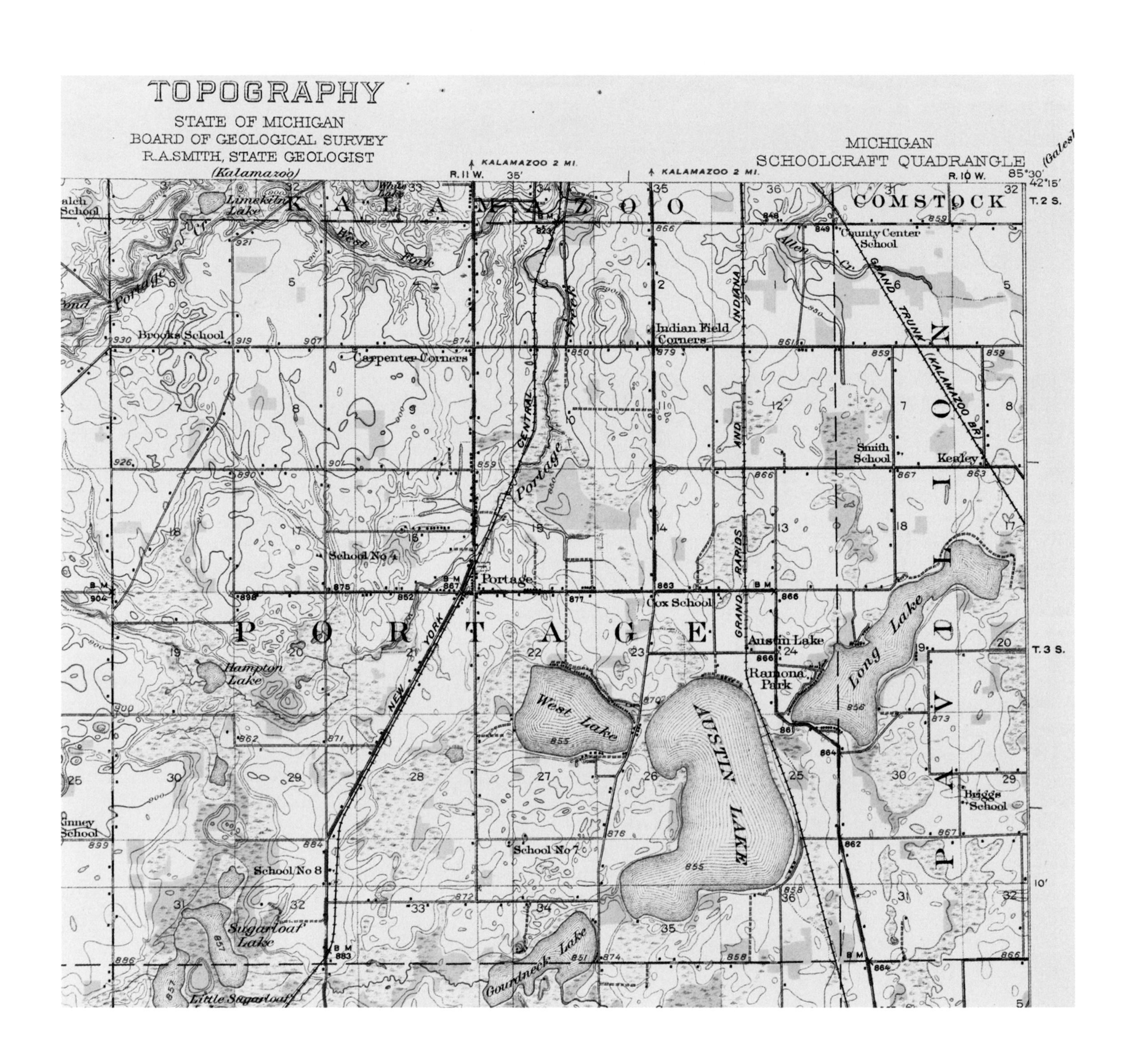

1921 Michigan Schoolcraft Quadrangle
by George T. H. Hawkins, J. H. Wilson, H. S. Senseney, J. R. Ellis, R. B. Steele, E. E. Harris, and E. C. Biobee
Topography by C. L. Sadler and G. R. Richardson
1922 Chromolithograph, 19.75 x 16 in.
Library of Congress, Washington, D.C.

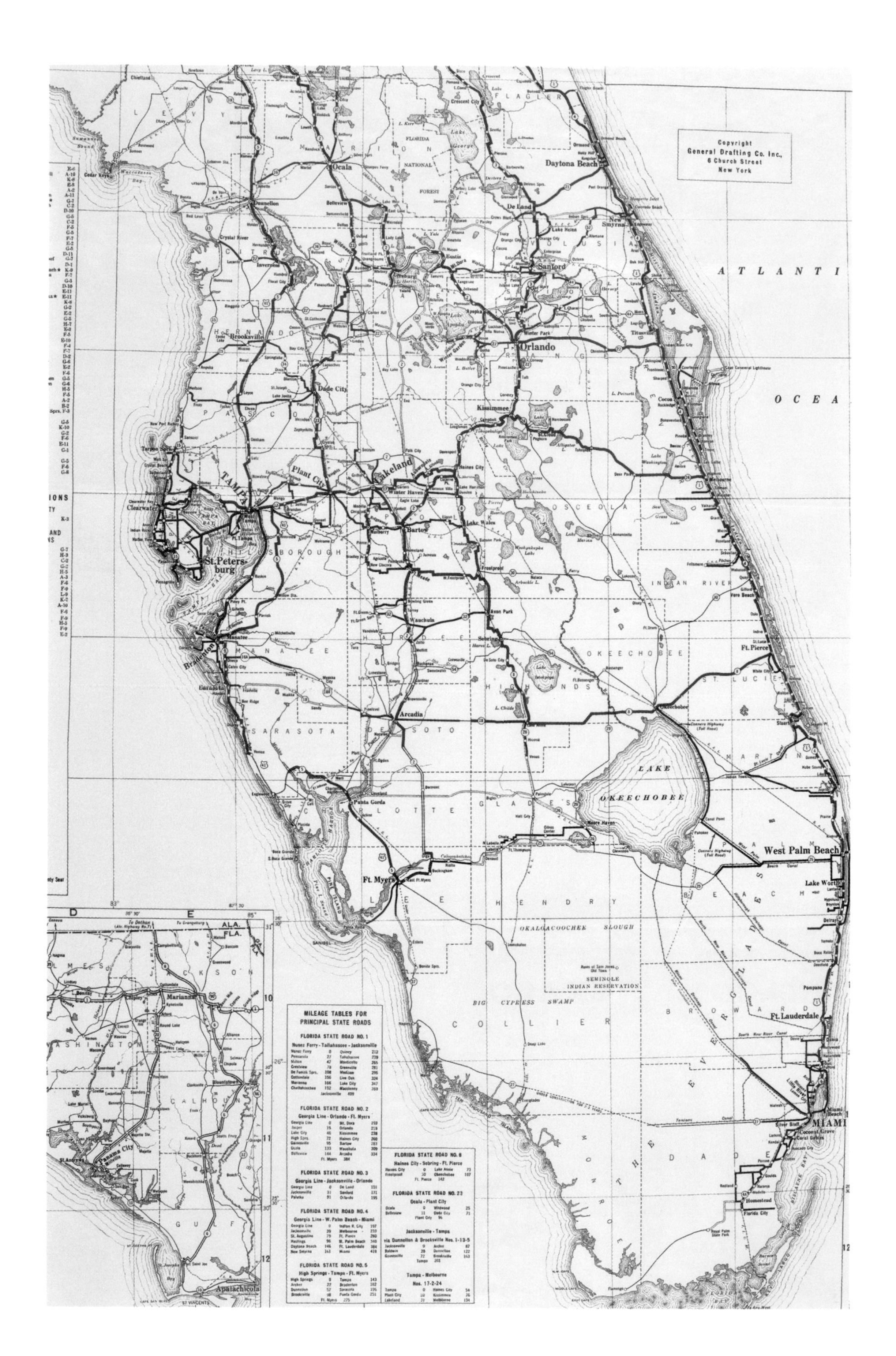

1927 Texaco Road Map of Florida by General Drafting Company, New York
Offset Lithograph, 27 x 21 in.
Library of Congress, Washington, D.C.

Gas Maps

U.S. MOTORISTS ENJOYED FREE ROAD MAPS FOR much of the 20th century. A uniquely American phenomenon, the maps were distributed by touring companies, automobile clubs, oil companies, and drive-in service stations to promote their businesses and products. Popular with the driving public, they came to be taken for granted. It was easier to ask for a new map at the gas station than to look for one crumpled on the floor. Walter R. Ristow, former chief of the Geography and Map Division of the Library of Congress, has estimated that more than ten billion road maps were made and distributed in the six decades prior to the first oil embargoes in the 1970s, which signaled the end to free maps. This staggering number reflected an expanding transportation system, an emerging consumer society, and aggressive marketing that catered to the automobile culture.

Road maps evolved from the promotional maps produced by railroad companies and bicycle clubs. Cyclists, in fact, marked and mapped some of the first auto roads. The National Highways Association (NHA) formed in 1911, and many local groups that sprang up were advocates of road maps and guidebooks.

Major map-publishing companies were quick to enter this market. General Drafting, founded by Finnish immigrant Otto G. Lindberg in 1908, prepared a road map of Vermont in 1912 for the recently established American Automobile Association. Rand McNally issued a small black-and-white strip road map of the New York City area as early as 1904, but its major contribution was a national series of sectional maps known as *Auto Trails Maps—A Guide to the Blazed Trails*, initiated in 1917. These maps included the names of all roads marked by auto clubs and associations. At the urging of map publishers, Congress eventually required that major highways be numbered rather than named, simplifying mapping and motor travel.

In 1923, the General Drafting Company began publishing road maps in two colors (red and black) by offset lithography, a new printing process that was cheaper and faster than conventional lithography. This innovative firm also introduced varied line widths to indicate six or seven road classifications, and point symbols of different sizes to designate relative sizes of cities and towns. By the 1950s, four-color offset lithographic presses were producing up to 5,000 map sheets per hour. One of the most appealing features of the gas map was its vibrant artwork, which still captivates collectors.

1927 COVER FROM STANDARD OIL COMPANY FLORIDA ROAD MAP
BY GENERAL DRAFTING COMPANY, NEW YORK

Soviet-Style Propaganda

"THE MAP IS THE PERFECT SYMBOL OF THE STATE," observed Mark Monmonier in his classic work, *How to Lie With Maps*. From the time of the first national atlases of England and France in the 16th century, monarchs and ministers have used maps to legitimize their rule and control, none more so than the totalitarian regimes of the 20th century.

Although most countries and cultures have used map propaganda to advance their well-being, Nazi Germany and the Soviet Union were clearly different. Conflict was the major cartographic theme displayed in their small-scale political and historical maps, with National Socialists stressing ethnology, race, and geographic space, and Communists emphasizing class struggle, the proletariat, and the inevitable rise of Marxist ideology. Each became masters at using the graphic images of maps to promote their political identity and spread their ideology.

The Division of Military Literature of the State Publishing House of the Red Proletariat provides a fascinating example of this genre. It prepared a set of ten dramatic historical maps in 1928 that illustrated the Bolshevik Revolution of October 1917, the following cvil war, and, last, the destruction by the Red Army of the Cossacks and the counterrevolutionary forces.

Communist cartographers used bold colors, action symbols, and didactic prose to dramatize their story. A vivid red designated the centers of Communist control in Russia, Bavaria, and Hungary in this map of socialist uprisings and foreign intervention during the winter of 1918-19. Moscow, capital of the new Bolshevik government, is highlighted by a series of star outlines and a huge red flag. Red flames denote Communist uprisings in German cities. Wartime "was the age of the verb," journalist John McDonough wrote in another context. The arrow symbol represented the strongest verb in the visual language of cartography. It was used here to portray attacks on Communist forces by foreign armies from England, Serbia, France, and Greece. Finally, figures of menacing armed men and galloping horses delineate the boundaries of conflict.

During his leadership from 1918 to 1924, Vladimir Lenin encouraged the use of maps and other visual aids to convey Soviet political propaganda, partly in recognition of the low level of Russian literacy in the 1920s. Each map in this set reproduced a different quote from Lenin, exhorting his followers to support the Red Army and his Communist revolution.

1928 Germanskaia Revoliutsiia I Interventsiia Antanty (The German Revolution and the Intervention of the Entente)
by N. N. Lesevitskiy, D. N. Kravcehenko, and A. N. De-Lazari
Chromolithograph, 27.5 x 41.5 in.
Library of Congress, Washington, D.C.

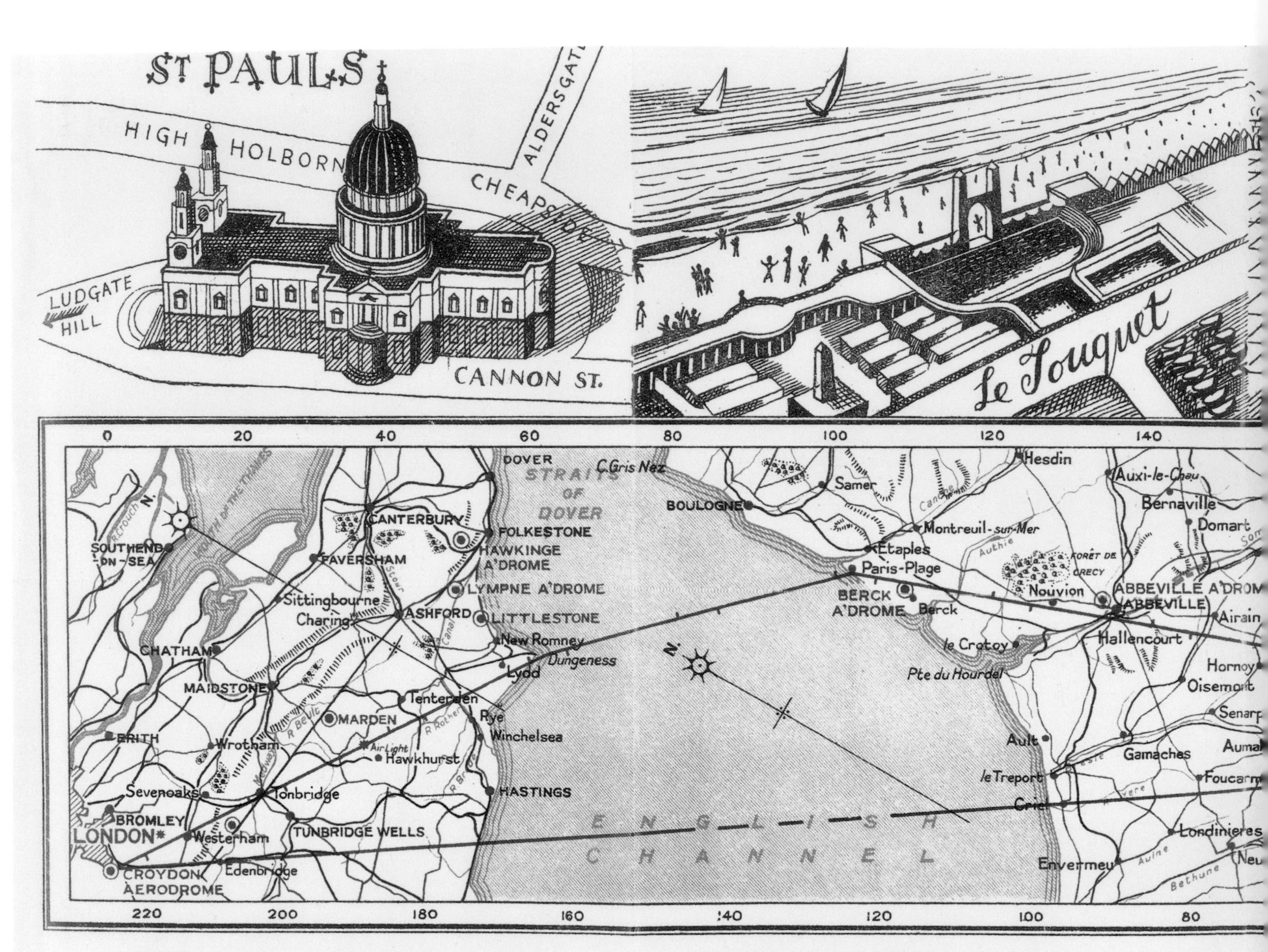

1934 Detail from Map of the European Air Routes of Imperial Airways by Raynoil Maps, Ltd.
Chromolithograph, Strip Map, 8 x 57.75 in.
Private Collection

Flying Maps for Passengers

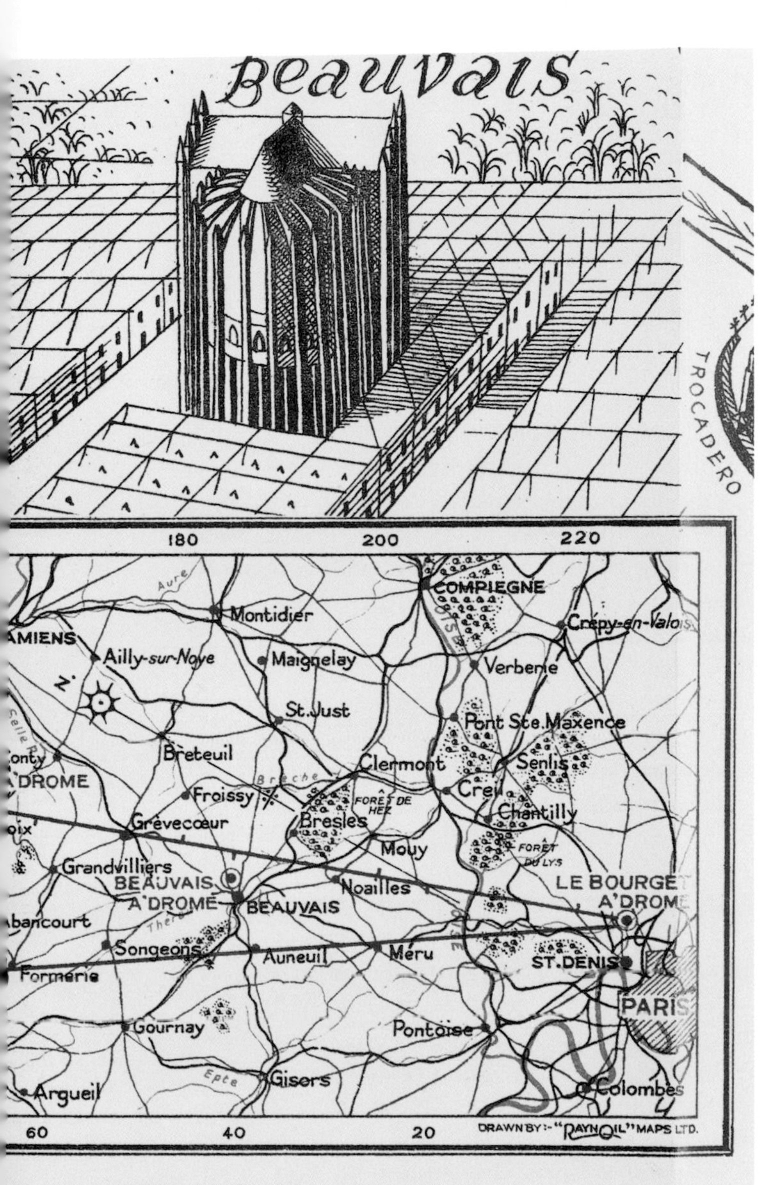

THE LAST LAP TO PARIS

old cathedral stands out at Beauvais a little to the of the course. At the winding River Oise, just north 'aris, the pilot prepares to land, and in a few minutes red-roofed houses of the northern suburbs of the tal pass beneath the wings, and there ahead are the e hangars of Le Bourget

AIRLINE SOUVENIR MAPS INFORMED, RELAXED, AND amused airline passengers during long flights from the beginning of scheduled passenger service following World War I until about the 1970s, when on-board movies diminished their appeal and cost-cutting reduced their value. During the late 1920s and early 1930s, the number of commercial airlines increased dramatically in response to various forces: the creation and maintenance of national airway infrastructures in Europe and the United States; subsidized air-mail contracts; and the introduction of multi-engine, all-metal passenger airplanes that carried six to ten passengers.

For these passengers, attractively illustrated maps were designed to promote the emerging airline industry as well as individual airlines. To alleviate fear of flying, additional information was provided on the dynamics of flight, the impact of weather, and the elements of air navigation. "Either pilot can handle the plane single-handed," a Colonial Air Transport 1930 strip map noted. "Before them is a map of the route, and instruments, which tell accurately the height, speed and direction of the plane."

The first airline souvenir maps were associated with cross-Channel air service between London and Paris. Handley Page Transport issued a large, foldout, pictorial map of the region in 1921, with the route indicated by colored dots. "The actual route flown may vary slightly at the discretion of the pilot," passengers were warned. A similar foldout appeared in a timetable issued by the French *Compagnie Des Grands Express Aériens* (1920-1923).

By the mid-1920s, separately published foldout "flying maps," modeled after the navigational strip maps used by pilots and navigators, made their appearance. These maps were frequently pasted or printed in passenger guidebooks and generally were accompanied by aerial photographs or drawings of significant landmarks that could be seen from the air along the flight route. This Imperial Airways 1934 flying map of its London-to-Paris route portrays St. Paul's Cathedral, a French seaside resort, and the old cathedral at Beauvais. It was prepared by the Raynoil Map Company, which also produced air navigational charts. A dark red line marks the air route. Red circles indicate the main aerodromes (airfields) in London and Paris and emergency landing sites along the route. Unscheduled landings were common in an era of low-quality aviation fuel, unreliable aircraft engines, and propellers that splintered during rainstorms.

"Tora! Tora! Tora!"

On December 27, 1941, Comdr. Mitsuo Fuchida stood before Emperor Hirohito in the Imperial Palace in Tokyo. Twenty days earlier, circling high above Pearl Harbor in a three-crew bomber, he had sent his famous coded radio message, *"Tora! Tora! Tora!"* (Tiger! Tiger! Tiger!), signaling to his superiors that the Imperial Japanese Navy's aerial strike force had achieved complete surprise. Now Fuchida described the successful dawn attack and illustrated his presentation with several maps.

Among them was this after-action damage assessment map, prepared by Fuchida himself aboard the aircraft carrier *Akagi* from reports provided by all the flight officers. It was drawn in traditional Japanese style, with varying scales and pictorial renderings. Three large Japanese characters, painted in bright red and bracketed within a rectangle, notified the reader that this map was "Top Secret." Adjoining characters described its purpose: "Estimated damage report against surface ships on the air attack of Pearl Harbor, December 8, 1941 [the date in Japan]." An English-speaking Japanese native probably made the transcription of the characters in red ink, as the names of several ships are misspelled.

Fuchida portrayed seven layers of information with pictorial renderings, colors, and symbols. Outlines were used to show the number and location of U.S. ships. The instruments of their destruction were illustrated by three symbols: A red X represented one 250-kilogram bomb; a solid red dot, an 800-kilogram bomb; and an arrow, one torpedo. Ship types were classified by color, with blue denoting warships such as battleships, cruisers, or destroyers, and yellow signifying support ships, such as oil tankers. Damage assessment was illustrated qualitatively by red slash marks, ranging from minor (one slash) to serious (three slashes). Crossed slashes indicated that a ship was sunk. Billowing, reddish-yellow flames indicated burning vessels. The light-blue lines surrounding ships represented oil slicks or burning oil.

After-action maps and their grim statistics and images remained as important tools in strategic and tactical planning throughout the war. All combatants used similar maps to evaluate bomb strikes. As World War II progressed, Allied air forces turned to aerial photography for damage assessments, but no subsequent flight leaders found themselves reporting to a president, prime minister, or chancellor with map or aerial photographs in hand, as Fuchida had.

One of Japan's top pilots, Fuchida eventually became the air officer of the Combined Japanese Fleet and was promoted to captain. After Japan's surrender, and inspired by a former American prisoner of war, he became a Christian missionary and lived briefly in America. His map was given to Gordon W. Prange, chief of the historical section in Japan under Gen. Douglas MacArthur. After Prange's death, the map was sold to a private collector.

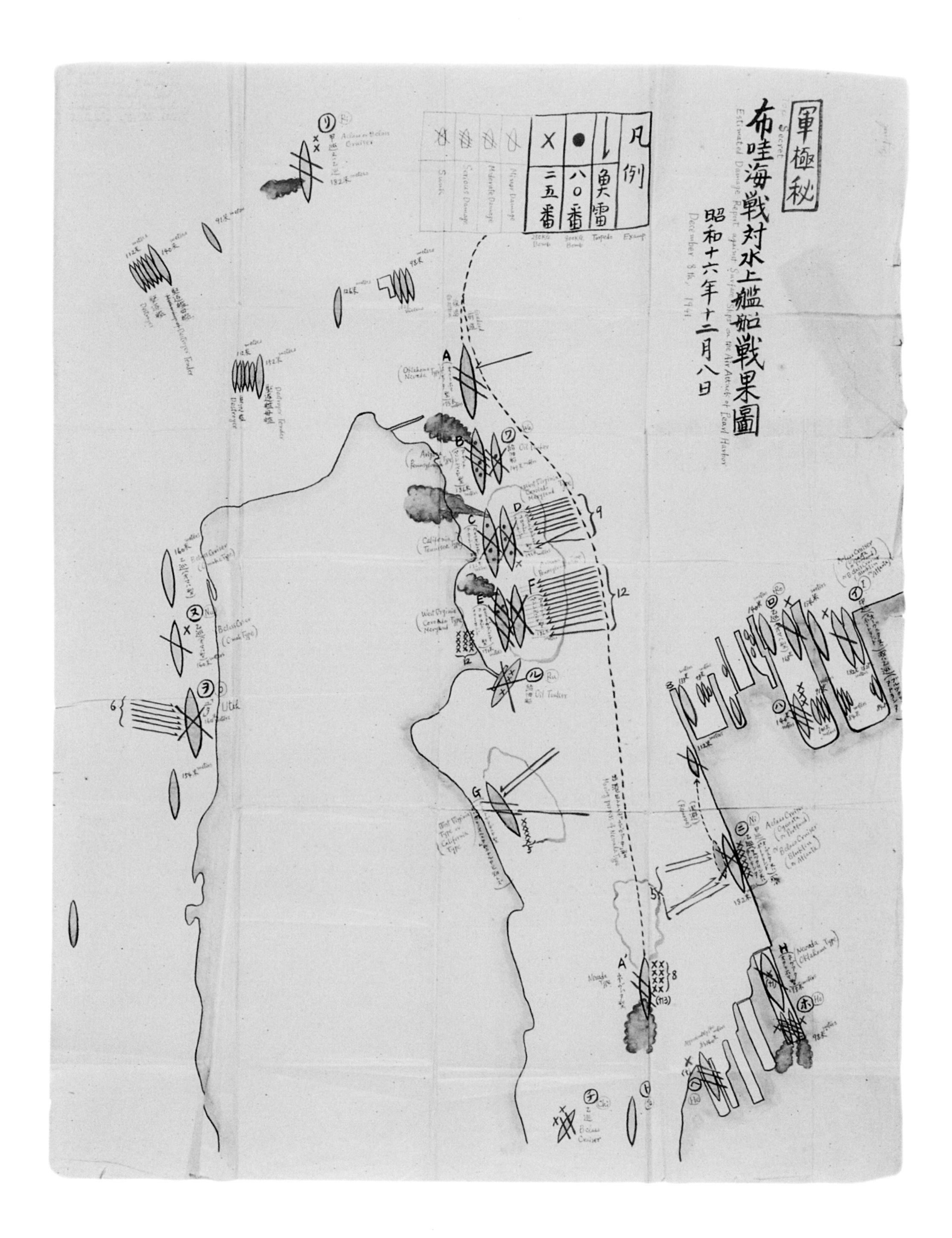

1941 After-Action damage Assessment Map of Air Attack on Pearl Harbor, Dec. 7, 1941, by Comdr. Mitsuo Fuchida
Ink and Watercolor on Paper, 31.75 x 23.6 in.
Private Collection

1943 From the Solomons, One of A Set of Four Maps by Richard Edes Harrison
Offset Lithography, 13 x 20.75 in.
Library of Congress, Washington, D.C.

"There's Nothing But Water Here"

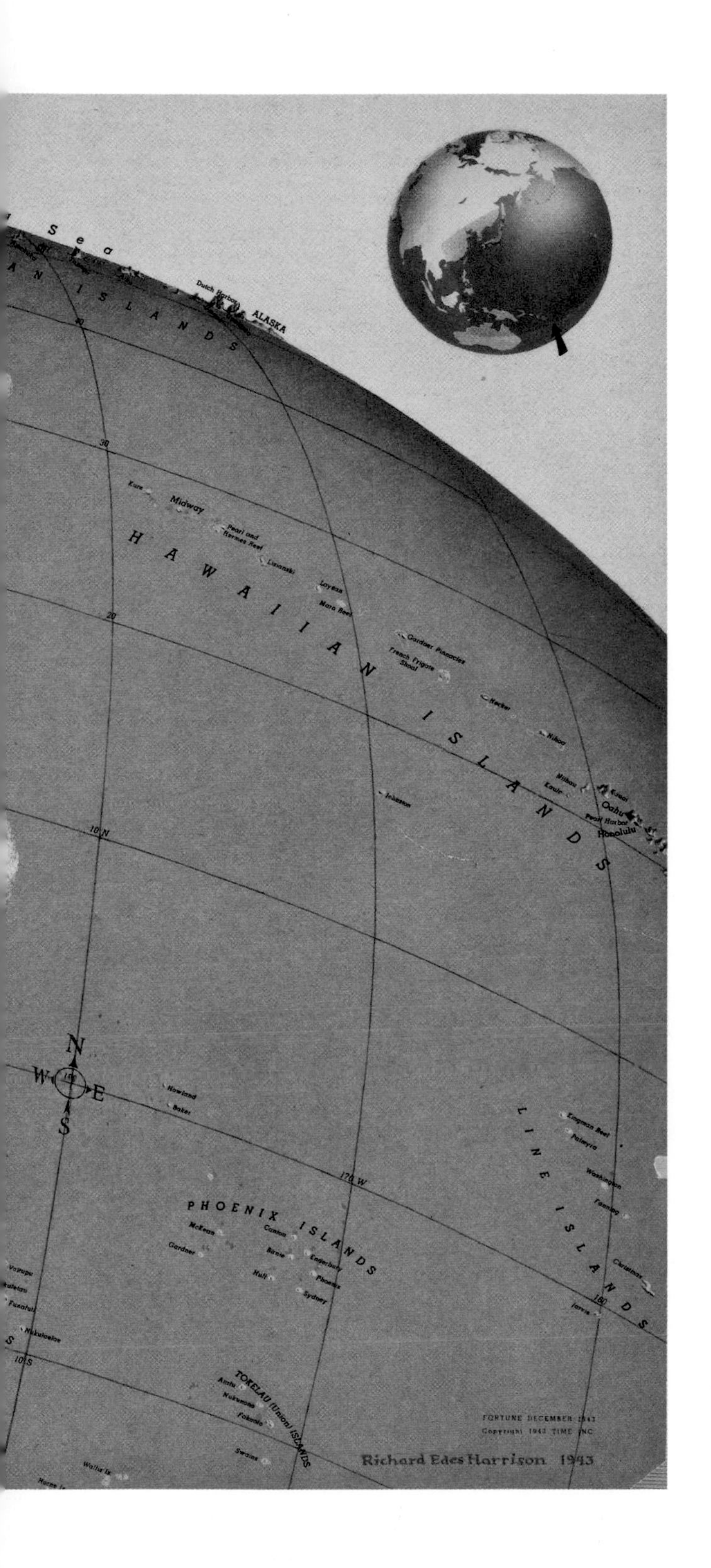

When Richard Edes Harrison submitted his manuscript map of the Pacific Ocean, *From the Solomons,* to the editor of *Fortune* magazine late in 1943, it was rejected, with the comment, "There's nothing but water here." "Yes, but that's where the war is going to be fought," Harrison shot back. "End of complaint," he later scribbled on his map.

Harrison's manuscript map was part of a series of innovative maps that the cartographic artist designed at the beginning of the U.S. Pacific Campaign of World War II. "The maps, which show the approaches to Japan from Alaska, from the Solomons, and from Burma and China, indicate not only the relationships of land and water, distances and directions, but also the shape of mountains and plains," wrote the *Fortune* magazine editor. "It is in and around East Asia that the war has finally to be won."

Harrison's objective was to illustrate World War II as a global conflict fought in a new age defined by the airplane. This was a new kind of war, and it required fresh ways of looking at the world's geography, particularly the spatial relationships of continents and oceans.

Harrison reintroduced old map projections in novel ways to produce "air age" maps to generations raised on the limitations and distortions of the Mercator map projection. With a brilliant series of wartime maps in *Fortune,* Harrison taught the American public (and his magazine editor) to visualize geography in perspective from a single viewpoint high above Earth, rather than from the infinite viewpoints of a conventional map. Harrison's magazine maps were reissued in atlas format under the title *Look at the World: The Fortune Atlas for World Strategy.* While such perspectives are common today, they were revolutionary in the early 1940s. As Harrison had predicted, the war in the Pacific raged through the summer of 1945, well after Germany's surrender in May. And that is where it was won, with aircraft and two atomic bombs.

Harrison was self-taught, as were most newspaper and magazine mapmakers. Born in Baltimore in 1901, he earned degrees in both zoology and architecture at Yale before turning to mapmaking in 1932, when he substituted for the mapmaker at *Time* magazine. During a 60-year career, he taught cartography at Syracuse University and produced maps for many leading publishers as a freelancer. Not restrained by traditional academic and official government cartographic conventions, he introduced new ways of displaying spatial information.

Charting D-Day

THE INVASION OF CONTINENTAL EUROPE ON JUNE 6, 1944 (D-Day) required special maps for the British, Canadian, and American assault forces landing on the beaches of Normandy, code-named Sword, Juno, Gold, Omaha, and Utah. Prepared under the highest security classification, code-named BIGOT, the maps provided detailed information on the lay of the land, roads, hedgerows, buildings, and German installations. This Omaha Beach amphibious chart/map, used by beachmaster Joe Vaghi to direct troops onto dry land, also included details about underwater obstacles and beach gradients. A "panoramic shoreline sketch" below the map was designed to aid Navy coxswains in navigating the choppy inshore waters of the English Channel. Sunlight and moonlight tables, and data on inshore currents and tidal stages, were provided on the back of the chart/map.

Designed by Navy Lt. William A. Bostick, a graphic artist trained at the Carnegie Institute of Technology, the map covered a two-and-one-half-mile section of beach. Assigned to Task Force 122, the naval command that carried the troops ashore at both Sicily and Normandy, Bostick and his small staff of enlisted men, selected especially for their mapmaking and drafting ability, prepared the chart/maps for the two landing sites assigned to the Americans: Omaha and Utah.

The mapmakers used a variety of sources, including a nautical chart that dated "from the time of Napoleon." Their most important sources, however, were depth soundings and sketch maps obtained by Navy frogmen and oblique panoramic aerial photographs of the shore taken by low-flying twin-engine P-38 Lightnings. "I used these photos to make watercolors of the beaches as landing craft skippers would see them as they approached from the channel," Bostick later wrote. "They needed every navigational assistance we could give them."

Unlike his landing maps for the Sicily campaign, which were printed with an offset lithographic press carried aboard the command ship, the U.S.S. *Ancon*, Bostick's chart/maps of both Omaha and Utah were printed by the United States Army Corps of Engineers Printing Plant at Cheltenham, England, which could handle the larger print run that was required.

On the 50th anniversary celebration of D-Day at Normandy, Bostick "was proud to discover" that the chart/map that he had designed was displayed at two French museums commemorating the event.

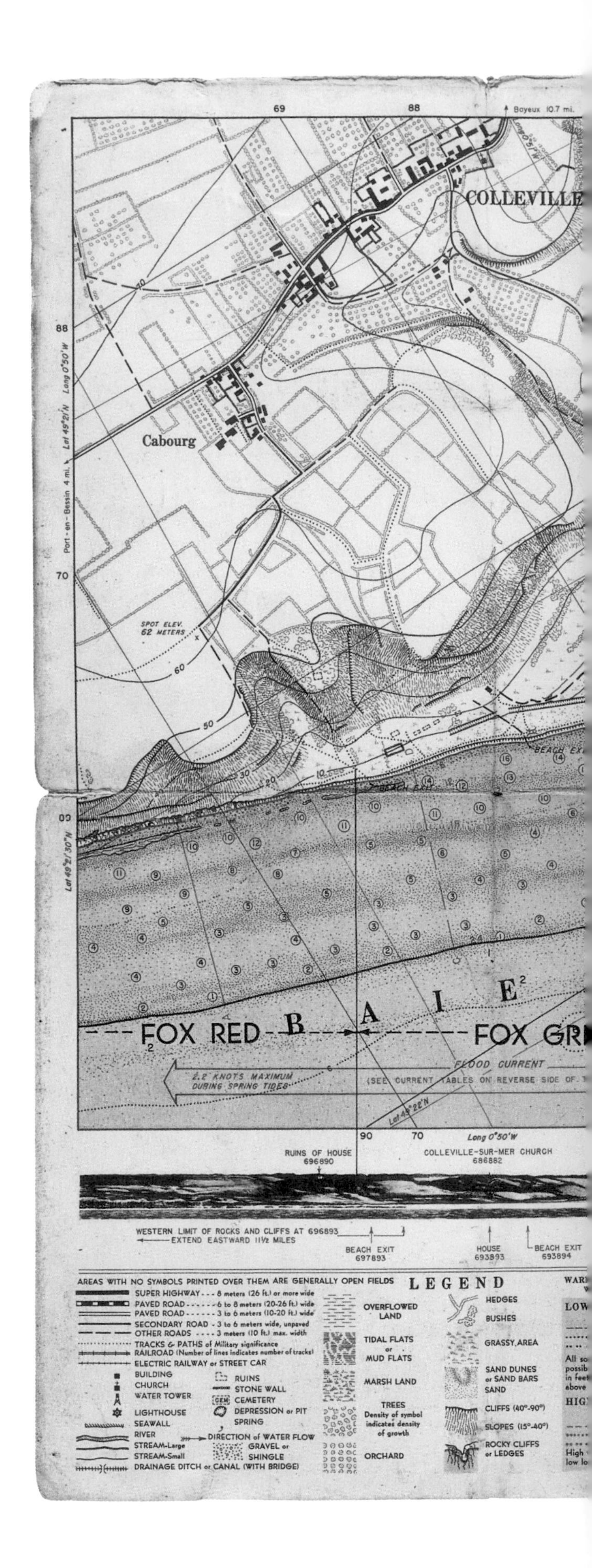

1944 Omaha Beach-East (Colleville-Sur-Mer) by Lieut. William A. Bostick, U.S. Navy
Offset Lithography in Five Colors, 17 x 22 in.
Joe Vaghi Collection

SECRET

(approx.) :
Lat : 47° 47′ 30″ N.
Long : 8° 45′ E.
and the Hohentwil, (688 m.high), position (approx.) :
Lat : 47° 46′ N.
Long : 8° 49′ E.

These two hills rise from the surrounding plain, and form landmarks visible for a distance of 60 kilometres in clear weather from a north eastern or north western direction. They are however still in Germany, and they themselves must be avoided as German O.P.s are stationed there.

6. Having found these hills, the next poir
the two parallel chimneys of the brickyards
They are inside Swiss territory and would be

7. Should the two volcanic hills be sightec
of his position, he would of course have to pa
would be able to sight the factory chimneys.
chance of avoiding detection would be to pass

1944 Escape and Evasion Map of Southwestern Baden-Württemberg concealed within deck of Bicycle Playing Cards
Map Reproduction by United States Playing Card Company, 1990, 19 x 25.5 in.
Library of Congress, Washington, D.C.

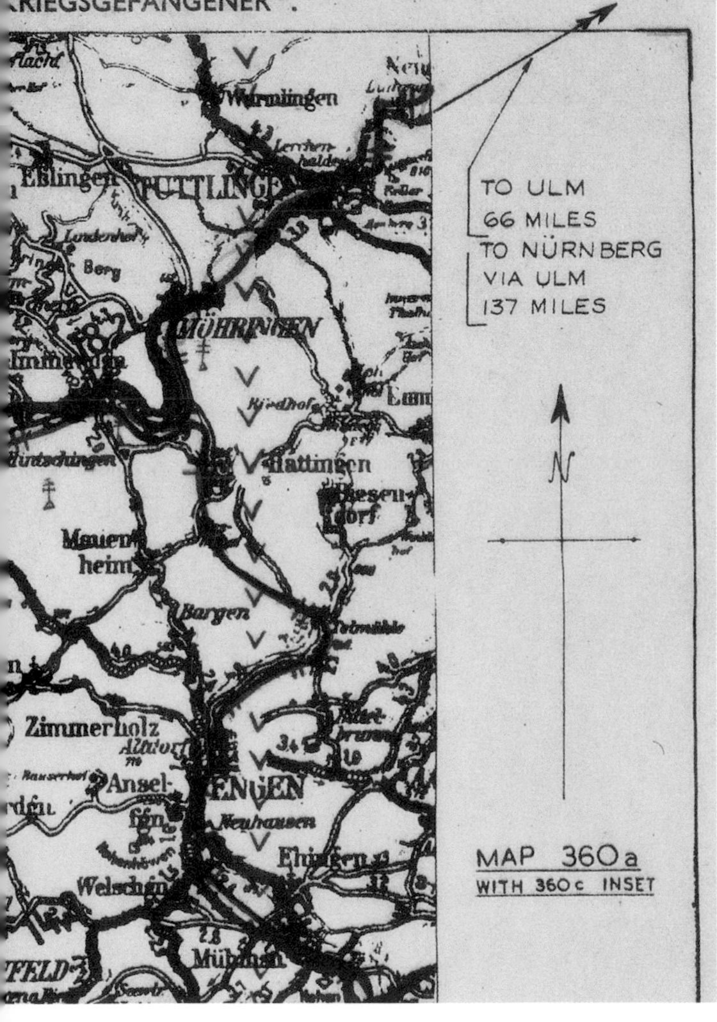
tween the two hills, and then try to identify and aim for
chimneys.
no account should the railway line Singen-Schaffhausen
th be crossed, as the course of the frontier then becomes
d, and it would be possible to cross into Switzerland and
ediately back into Germany through ignorance of the

re are also two salients of German territory, Busingen
s, completely surrounded by Swiss territory inside the
haffhausen. As soon as a fugitive is reasonably sure he is
land, he should make himself known to Swiss peasants.
ion will almost certainly be good, and the danger of
back into German territory would be avoided.
stretch of frontier round the Ramsen salient, and also
from Erzingen westwards to the Rhine must be avoided,
wire charged with electric current has been erected.
sh prisoners are now working on many farms and roads
Western Germany. They wear a yellow arm band marked
KRIEGSGEFANGENER".

Cartographic Card Trick

SOME 35,000 WORLD WAR II ALLIED COMBATANTS escaped from prisoner of war (POW) camps or evaded capture behind enemy lines after their airplanes were shot down. Many were aided by clandestine maps produced and distributed by British Military Intelligence-9 (MI9) and its American counterpart, MIS-X. These two units developed a variety of maps, "the escaper's most important accessory," according to Christopher Clayton Hutton, one of MI9's most creative officers.

The primary map was printed on fabric such as silk or rayon, or on tissue paper. Such maps have an ancient lineage but were conceived and developed independently by Hutton, who initially used maps of Europe obtained from the famous Edinburgh publisher John Bartholomew. These were reprinted in three colors. Subsequent fabric maps were published at larger scales for the European and Far Eastern theaters in four, eight, and nine colors. Some 1.3 million escape and evasion maps were eventually issued. The majority were carried into battle by air crews and paratroopers, but many were sent to POW camps with aid packages.

Maps destined for POWs were stuffed into hollow handles of shaving and shoe brushes, hidden in cigarette tins, folded into the spines of books, or placed in cavities under the top surface of table tennis paddles and board games. The most ingenious of these furtive devices were playing cards.

The United States Playing Card Company of Cincinnati prepared several hundred decks of cards that were shipped to POW camps in 1944, according to MIS-X operative Lloyd R. Shoemaker. Before the card sheets were printed and cut, a map was substituted for the black security filament normally placed between the front and back layers of cards to prevent light from revealing the numbers on the card's face. These three layers were held together by a soft adhesive that could easily be pulled apart. The encapsulated sheets were then cut into 52 cards, two jokers, and two company logos, with each card containing a small section of the map. In addition to the map, detailed instructions gave potential escape routes, landmarks, and hazards, such as places where enemy troops were known to be stationed. Instructions with a map for Baden-Württemberg warned escapees to avoid an area guarded by electrified wire.

While it is not known how many POWs escaped using these playing cards, former POW David Pollak told the *Cincinnati Enquirer* in 1990 that the "prisoners knew which decks had maps."

Mapping the Cold War

THE COLD WAR ERA WAS A PERIOD OF INTENSE cartographic competition and innovation. The United States and its NATO allies, on the one hand, and the Soviet Union and the Warsaw Pact countries, on the other, each initiated or improved large-scale topographic map series to support their policies of containment in the face of mutual annihilation through nuclear war. Cartography during this period, historian John Cloud has demonstrated, briefly converged with geodesy and photogrammetry (the use of photos for measuring) to produce new generations of precise military maps covering great spans of Earth's surface. Basic to these map series was the creation by the mid-1960s of a world geodetic datum, or frame of reference, in three dimensions by the U.S. Department of Defense. Devised initially to send satellites into Earth orbit and to target intercontinental ballistic missiles, a consolidated global datum dramatically improved the accuracy of maps. At the same time, Central Intelligence Agency and Air Force mapmakers and photogrammetrists conducted classified space reconnaissance imagery programs with exotic coded acronyms such as CORONA that provided the first detailed remotely sensed images of Eurasia. RACOMS, or Rapid Combat Operations Mapping System, was designed to produce four large-scale topographic quadrangles at a scale of 1:50,000 from new photographs within 48 hours!

After World War II, Joseph Stalin initiated a major topographic mapping program of the newly reconfigured Soviet Union. He directed his military and civil topographic services to survey the entire nation at a scale of 1:100,000. Much of the mapping was done from aerial photographic surveys and aerial geodetic triangulation, but field teams were still sent to establish geodetic control points in poorly surveyed areas such as Siberia. A Communist Party political officer accompanied each field crew "to conduct political and security surveillance." This project was completed in 1954, then extended to other countries, particularly the United States.

This detail of Northern Virginia and the District of Columbia is taken from a Soviet map of the national capital region on the eve of the Cuban Missile Crisis, October 18-29, 1962. The Potomac River, National Airport, the Pentagon, Arlington Cemetery, the Mall, and the Capitol are clearly seen. This map was based on larger-scale compilations made from 1951 to 1959. After 1979, editions were updated by high-resolution satellite images.

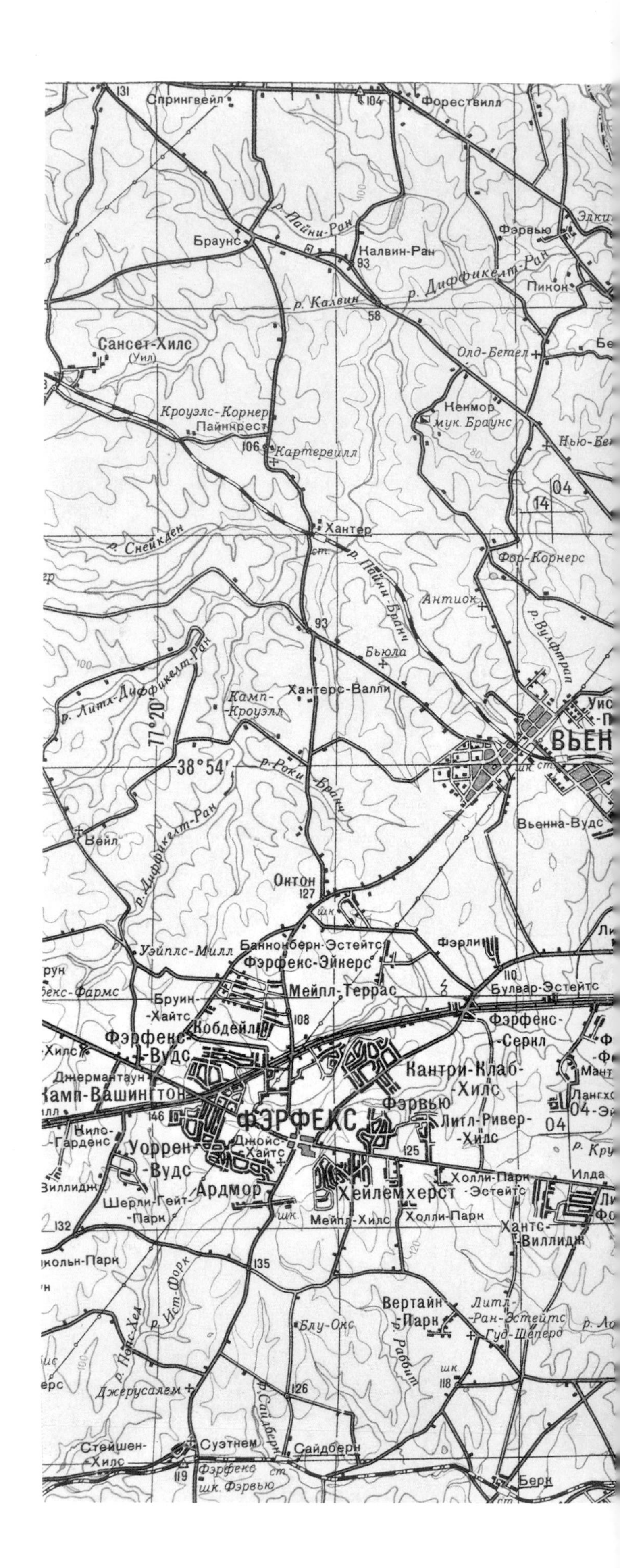

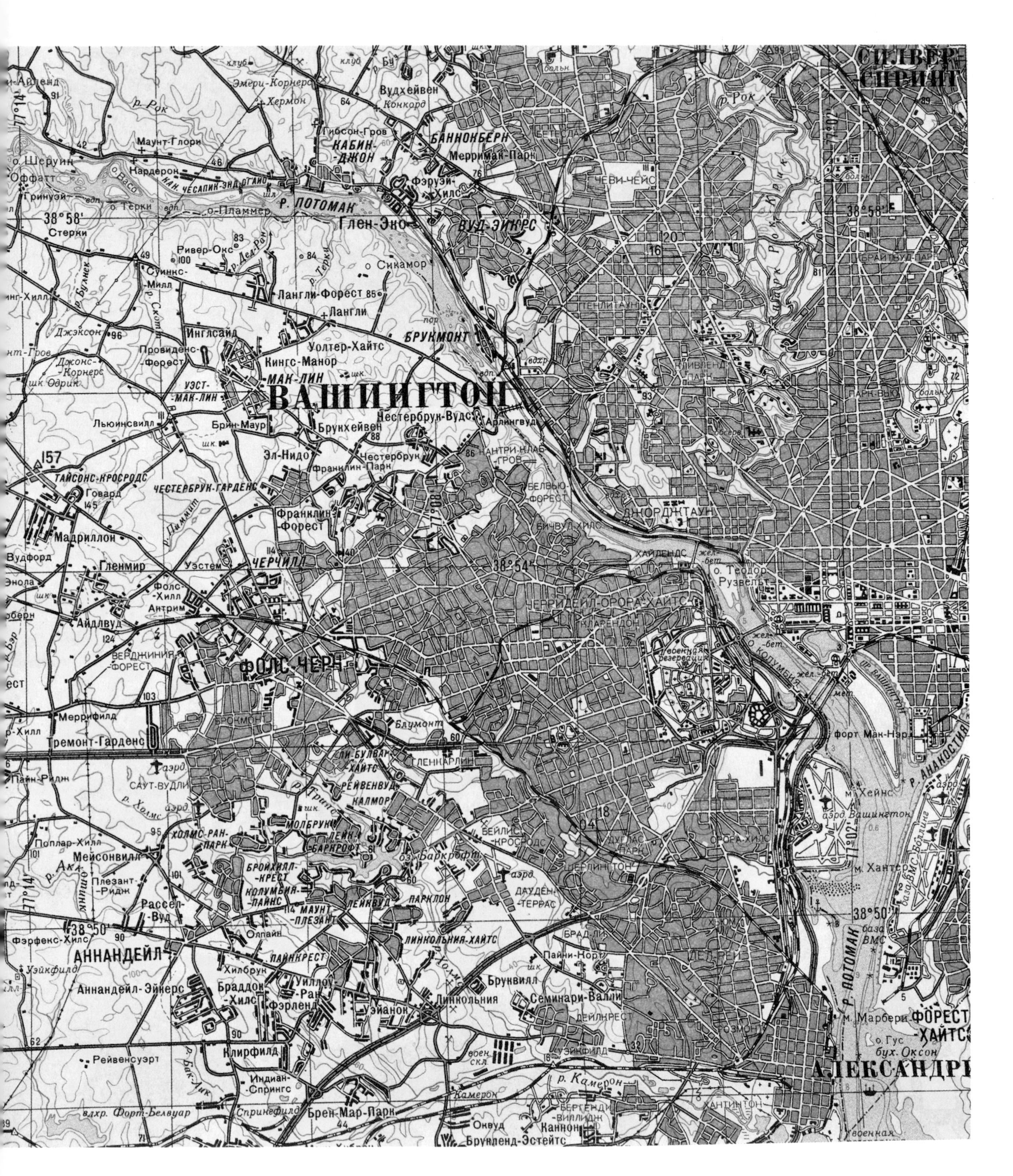

1962 SShA shtaty Virginiyai Merilend, Okrug Kolumbiya (USA States of Virginia, Maryland, and the District of Columbia)
by Military Topography Department of General Staff, Russian Military Forces
Chromolithograph, Scale 1:100,000, 18 x 20.25 in
Library of Congress, Washington, D.C.

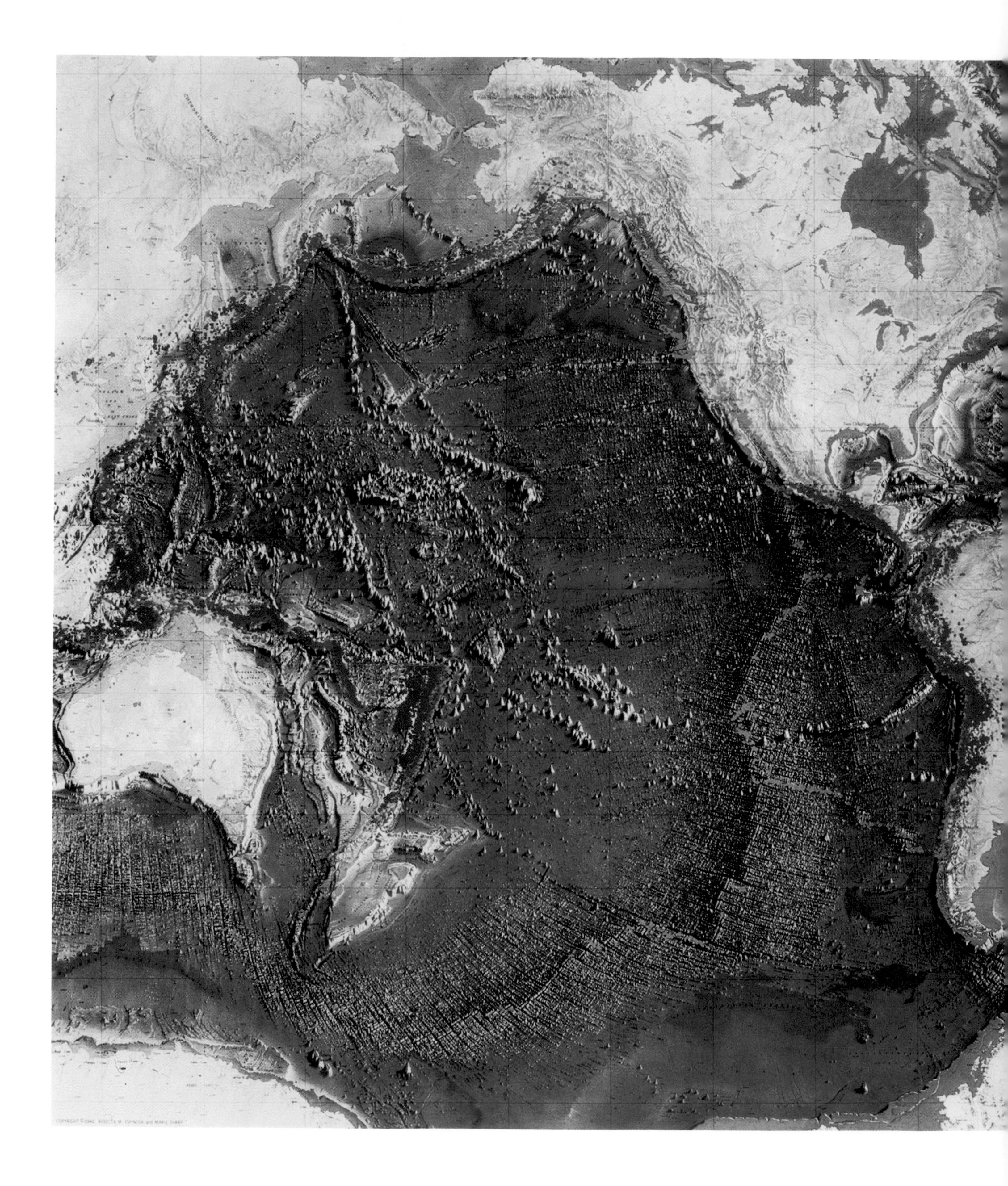

1981 Seismicity of the Earth 1960-1980
Oceanfloor Topography by Bruce Heezen and Marie Tharp; Seismicity by Alvaro F. Espinosa, Wilbur Rinehart, and Marie Tharp
Chromolithograph, 21.25 x 37.25 in.
Library of Congress, Washington, D.C.

SEISMICITY OF THE EARTH
1960-1980
BY ALVARO F. ESPINOSA
U. S. Geological Survey
WILBUR RINEHART
World Data Center A, for Solid Earth Geophysics
and MARIE THARP
Lamont-Doherty Geological Observatory of Columbia University
1981
WORLD OCEAN FLOOR PANORAMA
BY BRUCE C. HEEZEN AND MARIE THARP, ©1977
Sponsored by
UNITED STATES NAVY THROUGH THE OFFICE OF NAVAL RESEARCH
Mercator Projection
SUBMARINE DEPTHS IN CORRECTED METERS
Bruce C. Heezen (1924-1977)
IN MEMORIAM
Magnitude ≥4.5 ≥7.5
Depth (km)
Number of Earthquakes
0-33
27,788
34-100
17,585
101-300
7,329
301-700
3,167

Ring of Fire

EARTHQUAKES ARE POWERFUL EXPRESSIONS OF Earth's energy and help scientists better understand the mysteries of the planet's formation and structure. Maps of earthquakes have contributed to a better understanding of plate tectonics, the theory that Earth's outer shell is a mosaic of large and small plates that ride on, and move over, more fluid materials. The margins of these massive slabs, imbedded with continents and islands, are marked by volcano and earthquake zones where new crustal material is constantly generated, causing the plates to expand and collide. At collision points, edges of the plates may be folded and compressed, creating mountains, or destroyed when part of one plate slides under another, forming submarine troughs such as the 36,000-foot-deep Marianas Trench in the Pacific, southeast of Japan.

The world map of seismic geography on pages 216-17 dramatically portrays the earthquake zone that snakes around the planet like a baseball seam. Fifty-six thousand earthquakes are depicted, collected over a 20-year period by geologists Alvaro Espinosa and Wilbur Rinehart. Red circles mark the epicenter of each earthquake. The more powerful the earthquake, the larger and brighter the circle. The data reveal the close correlation of earthquakes, volcanoes, mid-oceanic ridges, and submarine trenches. Most spectacular are the Ring of Fire, the belt of volcanic and seismic activity that encircles the Pacific Ocean, and the Mid-Atlantic Ridge, a submerged mountain range that zigzags from the Arctic Ocean south to Antarctica. Another intriguing feature of this map are the thermal centers, or "hot spots," located in the center of the Pacific Ocean. Hot spots are fixed sites under the tectonic plates that periodically emit lava, building chains of islands as a plate passes over them. The Hawaiian Islands are a classic example. The orientation of the island chain indicates the direction of the plate's movement, with the oldest islands in a chain located the greatest distance from the hot spot.

Cartographer Marie Tharp, with marine geologist Bruce Heezen, pioneered ocean floor mapping. Tharp drew this map based on their 1977 *World Ocean Floor Panorama*, the first map to display in detail the mid-oceanic ridges and related fracture zones. Tharp and Heezen began their joint effort to map Earth's submarine topography in 1947 at Columbia University's Lamont-Doherty Geological Laboratory.

Tharp prepared the maps while Heezen collected ocean floor data from echo-sounding profiles of the sea bottom made aboard the university's ship, *Vema*. Trained in geology and mathematics, Tharp contributed to the discovery of the Mid-Atlantic Ridge through her study of earthquakes. "Earthquakes helped more than anything to map the ocean," she said. Tharp and Heezen provided geologists with the evidence to confirm the theory of plate tectonics. Their cartography contributed significantly to the current integrated geologic picture of Earth.

Greater Yellowstone: Map as Art

HEINRICH CAESAR BERANN WAS KNOWN AS "THE father of the modern panorama map." Tom Patterson, a leading American practitioner of the art of landscape mapping, considered Berann, who died in 1999, one of the world's "most gifted" mapmakers. A native of Innsbruck, Austria, Berann studied art and design before turning to cartography in 1934 when he won a competition for his panoramic map of the Grossglockner Hochalpenstrasse, a new alpine highway. During the next six decades, until retiring in 1994 at the age of 79, he painted nearly 600 panoramic maps. His clients included tourist resorts, map-publishing firms, and government agencies. From 1956 to 1998 he prepared the official maps for the Olympic Games held in Cortina, Rome, Innsbruck, Sarajevo, and Nagano, Japan.

A long association with the National Geographic Society began in 1963 when his young daughter, after looking at one of the magazine's maps of a mountain region, sent a letter to the Society, writing in effect, "My father can draw a better map." After seeing a sample of his work, the Society agreed. He was hired to paint panoramic views of Khumbu Himal and Mount Everest in the Himalaya. Later Berann collaborated with Bruce Heezen and Marie Tharp when the Society published a number of their pioneering ocean floor maps.

Berann's maps resembled traditional panoramic maps in that they portrayed a portion of the countryside from a point high above the scene with varying emphasis in different parts of the map. But they differed because of his unique ability to evoke "a sense of place—some essence or even spirit of the landscape," as one admirer observed. This genius is demonstrated in *Greater Yellowstone*, one of a series of four maps he prepared for the U.S. National Park Service between 1986 and 1995. The original map was hand-drawn and painted with the aid of existing topographic maps, aerial photographs, and sketch maps Berann prepared during several flights over the region in a small aircraft. The painting on pages 220-21 required more than six months to produce.

To achieve the most realistic re-creation of a landscape, Berann changed some standard cartographic conventions. Like a medieval mapmaker, he oriented his map toward the south to focus the reader on the most important features of the park: Yellowstone Lake, the Grand Canyon of the Yellowstone River and its Lower Falls, Mammoth Hot Springs, and the Lower Geyser Basin. Selected features were exaggerated to add interest and geographical context, including the great Teton Range at the top of the map, and Old Faithful Lodge at upper right. Most dramatically, the north-south-trending Tetons were turned in an east-west direction to reveal the better-known east face. Other signature characteristics of Berann's panoramic maps are his misty mountain lakes and deep-blue cloudscapes.

1989 Greater Yellowstone by Heinrich C. Berann
29 x 38.25 in.
Library of Congress, Washington, D.C.

BERANN

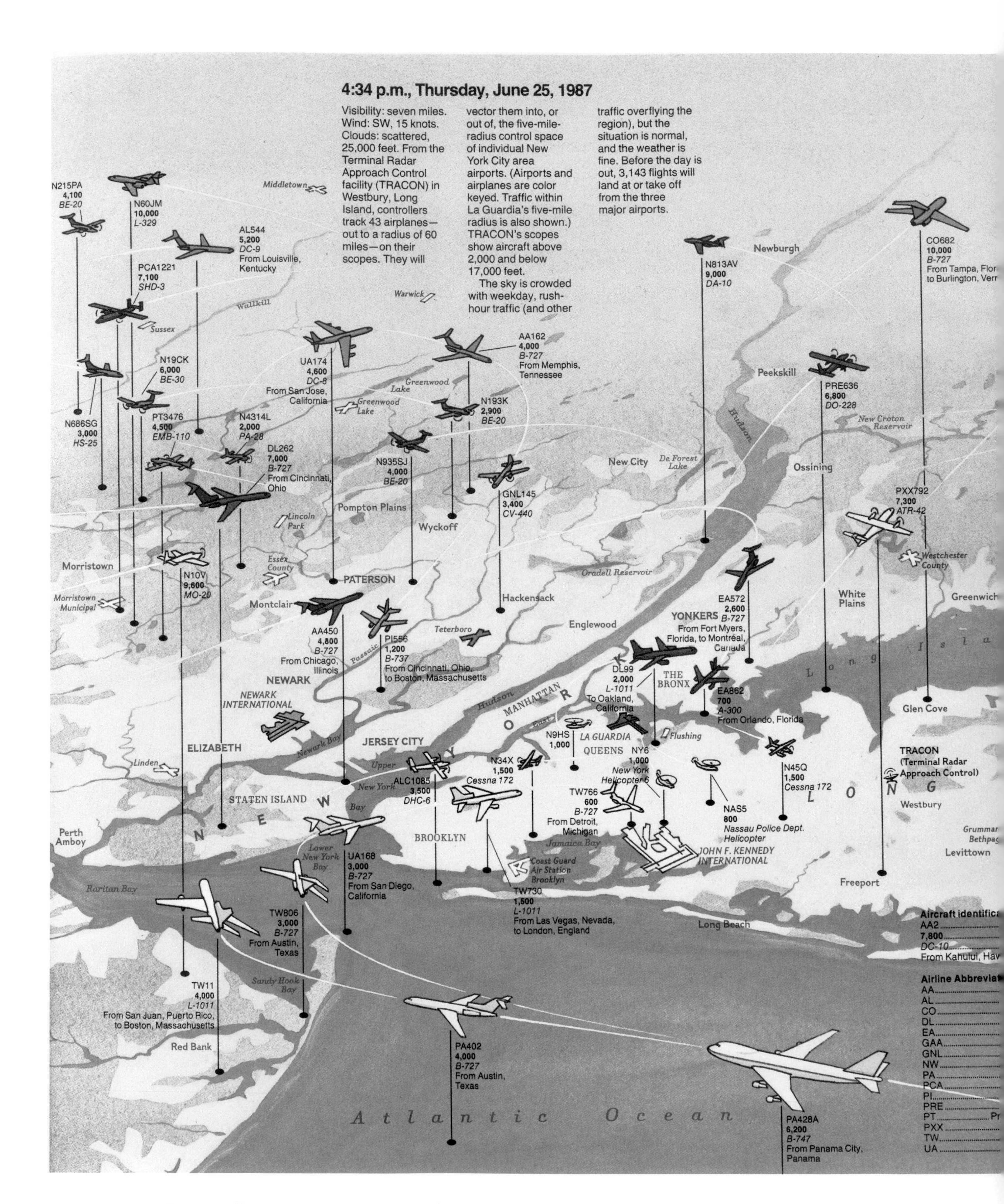

4:34 p.m., Thursday, June 25, 1987
Visibility: seven miles. Wind: SW, 15 knots. Clouds: scattered, 25,000 feet. From the Terminal Radar Approach Control facility (TRACON) in Westbury, Long Island, controllers track 43 airplanes—out to a radius of 60 miles—on their scopes. They will vector them into, or out of, the five-mile-radius control space of individual New York City area airports. (Airports and airplanes are color keyed. Traffic within La Guardia's five-mile radius is also shown.) TRACON's scopes show aircraft above 2,000 and below 17,000 feet.
The sky is crowded with weekday, rush-hour traffic (and other traffic overflying the region), but the situation is normal, and the weather is fine. Before the day is out, 3,143 flights will land at or take off from the three major airports.
Atlantic Ocean
Aircraft identifica
Airline Abbrevia

1988 Flying Late, Flying Nervous: 4:34 p.m., Thursday, June 25, 1987, by National Geographic Society Cartographic Division
Computer-Generated Map, 11.25 x 16 in.
Library of Congress, Washington, D.C.

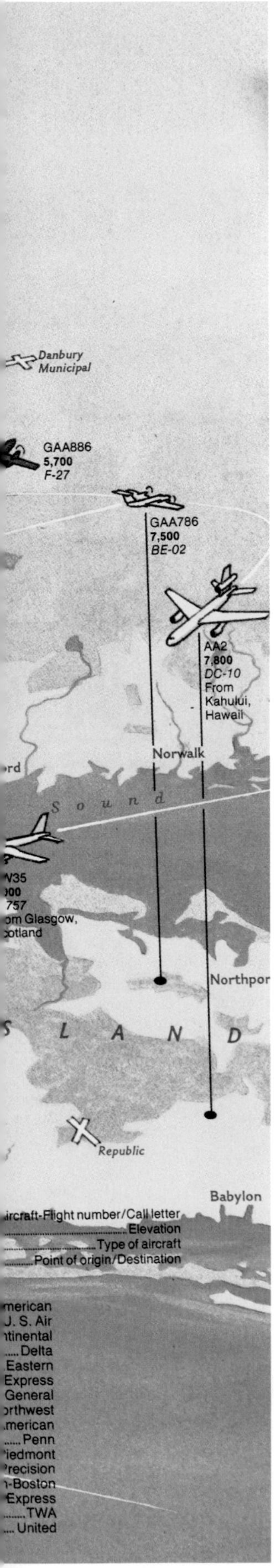

Flying Late, Flying Nervous

ONLY A SKILLED CARTOGRAPHER COULD REPLICATE the dramatic images that air traffic controllers view constantly while sitting before their radar screens. With this pictorial map a National Geographic cartographer visually captured an instant in time—4:34 p.m. on June 25, 1987—from air traffic control monitors at the Terminal Radar Approach Control facility at Westbury, Long Island, in New York. Forty-three aircraft are shown taking off or preparing to land at the three major airports of the New York metropolitan area: Newark International (color-coded green), La Guardia (red), and John F. Kennedy International (yellow). Name of airline, flight number, altitude, and point of origin or destination further identify each aircraft.

This is one of 380 maps specifically designed and compiled for the Society's *Historical Atlas of the United States*, a massive work published in 1988 to celebrate the Society's centennial. Edited by Wilbur E. Garrett and John B. Garver, Jr., director of the cartographic division, this atlas contained an additional 600 photographs, graphs, and historical maps. More than 160 staff writers, editors, illustrators, cartographers, researchers, indexers, computer specialists, and production staff contributed.

Maps have been associated with the Society since its founding in 1888, when four foldout charts of the great blizzard of that year were included in the first issue of NATIONAL GEOGRAPHIC magazine. For the first 27 years of the Society's existence, however, other firms prepared the maps.

In 1915 Albert H. Bumstead, a veteran mapmaker who had participated in several National Geographic Society exploring expeditions, established the Society's cartographic division. Since then, this division has produced the Society's maps. These include the famous single-sheet supplement maps, first inserted in the magazine in 1899 by Society President Gilbert H. Grosvenor; page maps illustrating magazine stories; general reference wall maps; and atlases. From its beginning, the aim of the division was to produce maps that combined "visual appeal, accuracy, and originality," Garver said.

The Society turned to computer-generated maps in 1983, one of the first major mapping firms to do so. By 1987, the cartographic division's yearly production normally included six large double-sided supplemental maps and about 60 magazine page maps. All told, some 660 million supplements and page maps were printed that year—an astronomical increase from the print run of 1,417 in 1899.

Mount Everest, A Worldwide Effort

In 1988, Bradford Washburn produced the most accurate and detailed map ever made of Mount Everest. One of America's most accomplished mountaineers and a pioneer mountain photographer, Washburn was, in addition, a skilled cartographer, photogrammetrist, and aerial photographer. He was the founding director of the New England Museum of Natural History, now the Boston Museum of Science, a position he held for 40 years.

After graduating from Harvard in 1933 and doing graduate work in geology, surveying, and aerial photography at Harvard's Institute of Geographical Exploration, he conducted a two-year National Geographic Society expedition to the Canadian Yukon to map one of the world's last uncharted wildernesses. Similar Society projects followed, producing classic maps of Mount McKinley, Mount Hubbard, and the Grand Canyon. For his 1960 map of Mount McKinley, Washburn managed to have a U-2 high-altitude spy-plane photograph the mountain from 68,000 feet.

Washburn's interest in surveying and mapping Sagarmatha or Chomolungma, as the Nepalese and Tibetans call Everest, was inspired by one of his graduate school instructors, Army Capt. Albert W. Stevens, a pioneer aerial photographer. The first step in his four-year project involved Swiss aerial photographer Werner Altherr photographing some 300 square miles of the Mount Everest region from a Learjet flying at 40,000 feet. Ground control was established by a comprehensive computer evaluation of existing British, Chinese, and Austrian topographic maps and infrared pictures of the same area taken from U.S. space shuttle *Columbia*, 156 miles above Earth. With the Learjet photographs and the newly established network of control points, Swiss Air Survey photogrammetrists prepared a large-scale manuscript contour map. Next, cartographers from the Swiss Federal Office of Topography, noted for its terrain mapping, meticulously depicted Everest's cliffs, crevasses, and moraines. The three-dimensional appearance of the map was made by airbrushing shades of gray to simulate topographic contours. Known as shaded relief or hill shading, this method of representing terrain was perfected in Switzerland in the 1880s. The final step in the production of this map was the selection and addition of place-names, an effort that took two years. It was carried out by National Geographic Society place-name experts in consultation with their counterparts in Nepal and China.

Washburn's Everest map was a remarkable achievement in cooperative mapmaking among different countries, traditions, and cultures. Specialists from nine countries participated. "Truly this was a project of people, working in concert and respect," Washburn wrote in the piece accompanying his map in the November 1988 issue of National Geographic. More than ten million copies of the map were printed and distributed by the National Geographic Society.

1988 Detail of Mount Everest = Sagarmatha = chu-mo-lung-ma feng (qomolangma) by Bradford Washburn
Computer-generated Chromolithograph, 36.5 x 23.5 in.
Library of Congress, Washington, D.C.

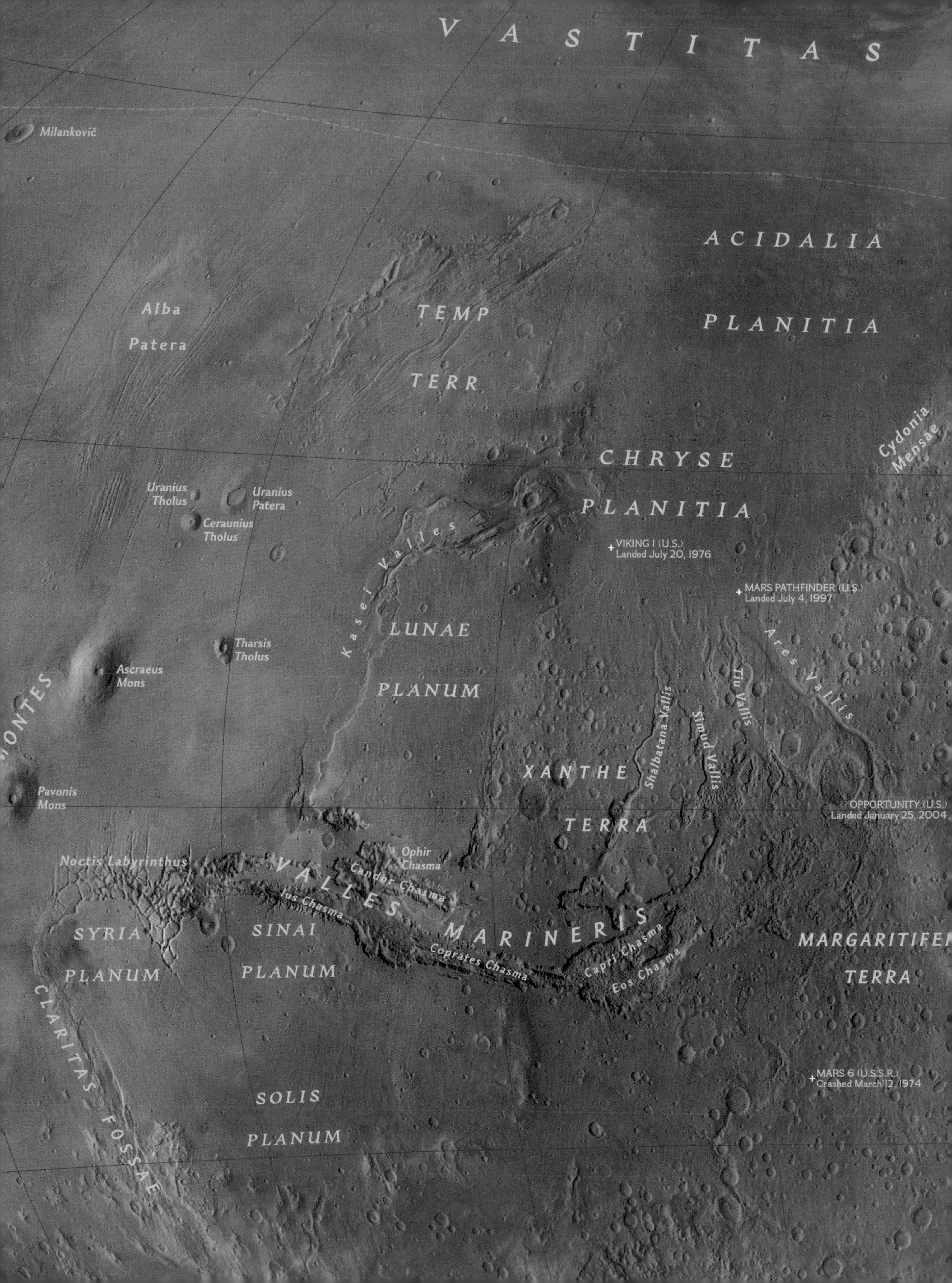

VASTITAS

Milanković

ACIDALIA

PLANITIA

Alba

Patera

TEMP

TERR

Cydonia Mensae

CHRYSE

PLANITIA

Uranius Tholus

Uranius Patera

Ceraunius Tholus

Kasei Valles

VIKING I (U.S.)
Landed July 20, 1976

MARS PATHFINDER (U.S.)
Landed July 4, 1997

LUNAE

PLANUM

Tharsis Tholus

Ascraeus Mons

ONTES

Ares Vallis

Tiu Vallis

Simud Vallis

Shalbatana Vallis

XANTHE

TERRA

Pavonis Mons

OPPORTUNITY (U.S.)
Landed January 25, 2004

Noctis Labyrinthus

Ophir Chasma

Candor Chasma

VALLES MARINERIS

Ius Chasma

SYRIA

PLANUM

SINAI

PLANUM

Coprates Chasma

Capri Chasma

Eos Chasma

MARGARITIFER

TERRA

CLARITAS FOSSAE

MARS 6 (U.S.S.R.)
Crashed March 12, 1974

SOLIS

PLANUM

SECTION SIX

Satellite Imaging, Digital Mapping, and Virtual Reality

Image of Mars compiled from NASA Spacecraft Data, 2000
by Michael Caplinger and Michael Malin

Introduction

Cartography was transformed, and the map redefined, over the past 20 years. The widespread adoption of the personal computer, the introduction of remote-sensing satellites for public and commercial use, and the development of a massive networking infrastructure known as the Internet, with its various languages for disseminating data, such as the World Wide Web, have changed the way map data are collected, manipulated, and disseminated.

Geographical information systems, or GIS, have changed the way maps are prepared. GIS is a technology that uses the computer, mapping software, and geo-referenced data to create, integrate, analyze, and distribute maps and mappable data at various scales. While the traditional map serves as both storage and archival medium for spatial information, and as a medium for communicating this information, in GIS the storage medium, the data files, are maintained separately from the communication medium, the mapping software. Because of the ease by which computer-generated maps can be produced, most electronic maps today are created by non-cartographers. The electronic map is challenging the cartographic standards that evolved during the past two centuries and that gave rise to the modern map with its universal look and language.

When displayed on computer screens, GIS maps become virtual maps: They are unique, temporary map images that exist only briefly in time. When the computer is turned off, they disappear unless saved in the computer's memory or on a printout. Virtual maps are the predominant map form today, with a potential viewing audience of one-half billion Internet users worldwide. National surveys in 1998 and 2002 revealed that nearly half of all Internet users, or about 87 million in the United States, consulted a map at least once a month. Various GIS software packages helped generate the 2005 map of HIV in Africa, the 2002 intertidal elevation model of Willapa Bay, in Washington, and the 2002 map of lower Manhattan.

Because computer images are being constantly refreshed, virtual maps also can be interactive. Any element of an electronic map can be enlarged or reduced in size, scales can be changed, and different layers of geo-referenced data can be viewed with ease. Map users have become mapmakers. Numerous Internet sites provide interactive maps, including traditional map publishers such as the National Geographic Society, with its popular *MapMachine.*

Complementing GIS is the global positioning system (GPS), which is replacing traditional methods of determining latitude, longitude, and elevation. GPS gives surveyors and mapmakers precise geographic coordinates for Earth's surface features through a worldwide network of orbiting satellites and receiving units. It is the primary tool for land and field surveying, and is being adopted for navigation in cars, boats, and aircraft. Integrated with remote sensing devices and aircraft navigation systems, it contributes to the production of highly accurate maps such as the map of Willapa Bay and NASA's 2004 Synthetic Vision Chart.

The 1972 launch by the National Atmospheric and Space Administration (NASA) of the first civilian remote-sensing vehicle into Earth orbit inaugurated satellite mapping. Satellites substantially speeded up data collection and dramatically expanded the range of mappable information. What once took years or months to survey can now be done in hours or minutes. The surface of the Earth is now mapped daily by numerous remote-sensing satellites, producing vast archives of mappable data that is received, analyzed, and maintained by cartographers, scientists, and technicians scattered across the globe.

Mapmakers are overwhelmed with data. NASA's Terra satellite, illustrated at the beginning of this section, collects nearly 20 terabytes of data—an amount equivalent to the entire book collection of the Library of Congress—every three months from five sensors. The millions of satellite images that have been acquired and archived since the introduction of remote-sensing satellites have been used to produce millions of maps on such topics as agriculture, global change, and regional planning. The 1991 infrared map of China and "The Blue Marble" (2004) are representative.

Powerful new technologies have also enhanced the value of conventional aerial photography. One of these is LIDAR (Light Detection and Ranging), which is used for producing highly accurate three-dimensional maps. The 2002 elevation map of Willapa Bay, Washington, was based on this technology. Aerial photographs and satellite images are being combined with GIS and visualization technology to create three-dimensional maps, such as the 2002 map of lower Manhattan.

EXTRATERRESTRIAL MAPPING HAS EXTENDED THE REACH OF CARTOGRAPHERS TO THE MOON AND BEYOND, creating new challenges similar to those faced by Renaissance cartographers during the first age of discovery and exploration. Although lunar mapping by the Soviet Union and the United States from unmanned spacecraft dates from the mid-1960s, it was not until the last of the Apollo moon missions that high-quality aerial photographs were obtained for detailed topographic maps, as illustrated by the Defense Mapping Agency's 1979 topographic orthophotomap of the moon's crater Macrobius.

Following the success of the lunar mapping programs, NASA launched unmanned mapping missions to Mars, Venus, and the moons of Jupiter and Saturn. From 1990 to 1993, the Magellan spacecraft mapped nearly the entire planet of Venus, part of it in 3-D, using imaging radar. The mapping of Mars, which began with Mariner 9 in 1971, was followed by the Mars Global Surveyor space probe in 1997, which provided the first 3-D global map of the red planet. It was generated from 27 million measurements gathered by the Mars Orbiter Laser Altimeter over a two-year period. The opening image in this section was based on data from this mission. NASA returned in 2001 with Mars Odyssey, equipped with a new imaging system based on thermal, or heat emissions, which produced the 2002 THEMIS mosaic map of Mars's Melas Canyon.

Terra Satellite

IN LATE 1999, THE NATIONAL AIR AND SPACE Administration launched its latest satellite into Earth orbit. Named Terra, it is the size of a small schoolbus. Terra carries five state-of-the art sensors; each is designed to collect specific data related to Earth's atmosphere, lands, oceans, and radiant heat. It orbits the Earth from Pole to Pole, capturing wide bands of images in the process that aid scientists in mapping and studying global climate change.

Images from ASTER, which stands for Advanced Spaceborne Thermal Emission and Reflection Radiometer, are used to create detailed maps of landforms in three dimensions and of surface temperatures. These high-resolution images of Earth are obtained in 14 different wavelengths that range from visible to thermal infrared light. This sensor was built in Japan and is monitored jointly by United States and Japanese scientists.

CERES, or Clouds and Earth's Radiant Energy System, uses two instruments to record and map the Earth's radiant heat patterns. Multi-angle Imaging Spectro-Radiometer (MISR) monitors climate change by recording successive images in four wavelengths from nine different cameras, each pointed at a different angle. Moderate-resolution Imaging Spectroradiometer (MODIS), with a 1,440-mile-wide viewing band, maps cloud cover, the extent of snow and ice, and the distribution of phytoplankton—which helps predict the onset of El Niño. MOPIT, which stands for Measurement of Pollution in the Troposphere, measures carbon monoxide and methane gases that provide data for mapping urban and industrial emissions over wide areas.

Terra is one of a series of space-based Earth sensors that have been placed in orbit by NASA's Earth Observing System (EOS) for observing and mapping Earth's land surface, biosphere, atmosphere, and oceans. Known primarily for its manned space flights, NASA also has been heavily involved in satellite mapping since launching its first successful meteorological satellite in 1960. Twelve years later, NASA initiated its Earth resource mapping mission, sending five remote-sensing Earth resource satellites (later called Landsat) into orbit between 1972 and 1984.

In early 2000, NASA joined with the National Geospatial-Intelligence Agency (NGA) to collect topographic data for much of Earth's land surfaces using reflected radar signals obtained aboard the shuttle *Endeavour*. The Shuttle Radar Topography Mission produced "the most detailed, near global topographic map of Earth ever made."

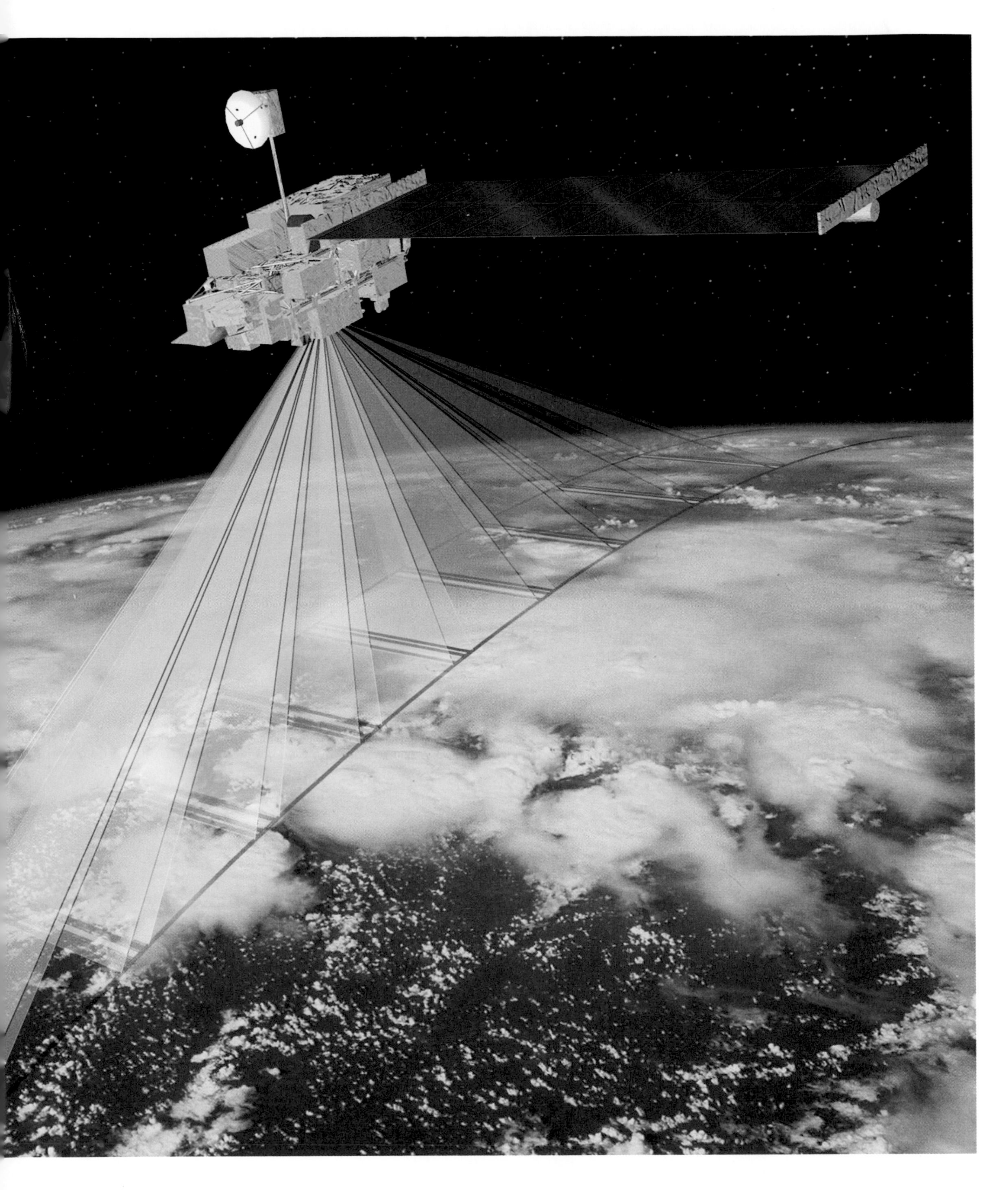

1999 Terra Satellite with Multi-Angle Imaging Spectro-Radiometer (MISR) Instrument by Shigeru Suzuki and Eric M. De Jong
Computer-Generated Image
NASA/JPL

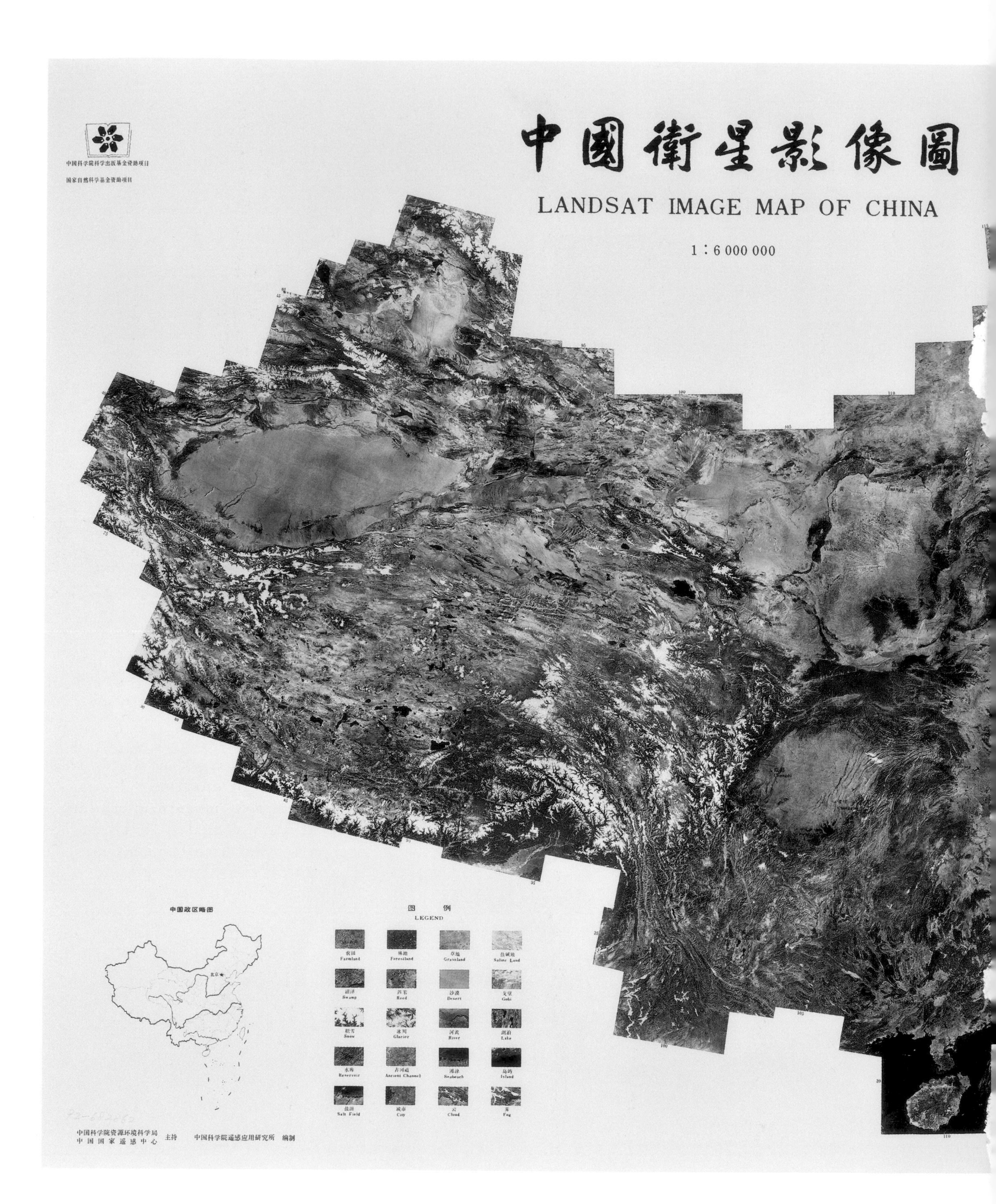

1991 LANDSAT IMAGE MAP OF CHINA BY ZHANG SHENGKAI, XIA MINGBAO, SHI JUNME
PHOTOMAP, 28 X 40 IN.
LIBRARY OF CONGRESS, WASHINGTON, D.C.

Satellite Map of China

MAPPING EARTH FROM HIGHER AND HIGHER altitudes was the dream of the first aerial photographers who took to the air in the second decade of the 20th century. Sixty years later it became a reality with the launching of NASA's first remote-sensing satellite, which was designed to record mappable data with a multispectral scanner that measured waves of light and heat energy reflected or emitted from Earth's surface.

Initially named the Earth Resources Technology Satellite (ERTS), it was later designated Landsat 1. With Landsats 4 and 5, launched in 1982 and 1984 respectively, NASA introduced the thematic mapper, an electromechanical sensor that recorded data using seven spectral bands of the electromagnetic spectrum. The thematic mapper was enhanced in 1999 for Landsat 7. Of the six Landsats successfully launched from Florida's Cape Canaveral, Landsat 5 and 7 still operate. They continue to provide complete global coverage every 16 days as they circle Earth, scanning 113-mile-wide swaths from a height of 437 miles.

During Landsat's 33-year history, its scanners have provided the basic data for satellite maps of Earth. The recorded data are beamed to NASA receiving centers, where they are processed and enhanced by sophisticated computers and converted into digital files or photographic images. These later are often transformed into maps.

This false-color infrared mosaic map of China is representative. It was prepared in 1991 from 584 separate images recorded from Landsat's thematic mapper. As with most Landsat map images, it was printed to simulate infrared photography as an aid to scientific study. With false-color infrared images, red rather than green indicates healthy farmland and forests, green-brown denotes grasslands, green designates deserts, and dark blue or black signifies water. The color white is used to indicate either snow, glaciers, clouds, or fog. Despite the obvious scientific purpose of this map, the Chinese cartographers, perhaps reflecting an earlier tradition, commented in the margins on China's "magnificent and majestic macroscopic natural scenery" and "the subtle structure of the brocade-like rivers and mountains."

Landsat's unparalleled 33-year historical archives of Earth images is maintained by the U.S. Geological Survey's Earth Resources Observation Systems Data Center in Sioux Falls, South Dakota. Scientists use these images to track changes on Earth's surface over time and across regions.

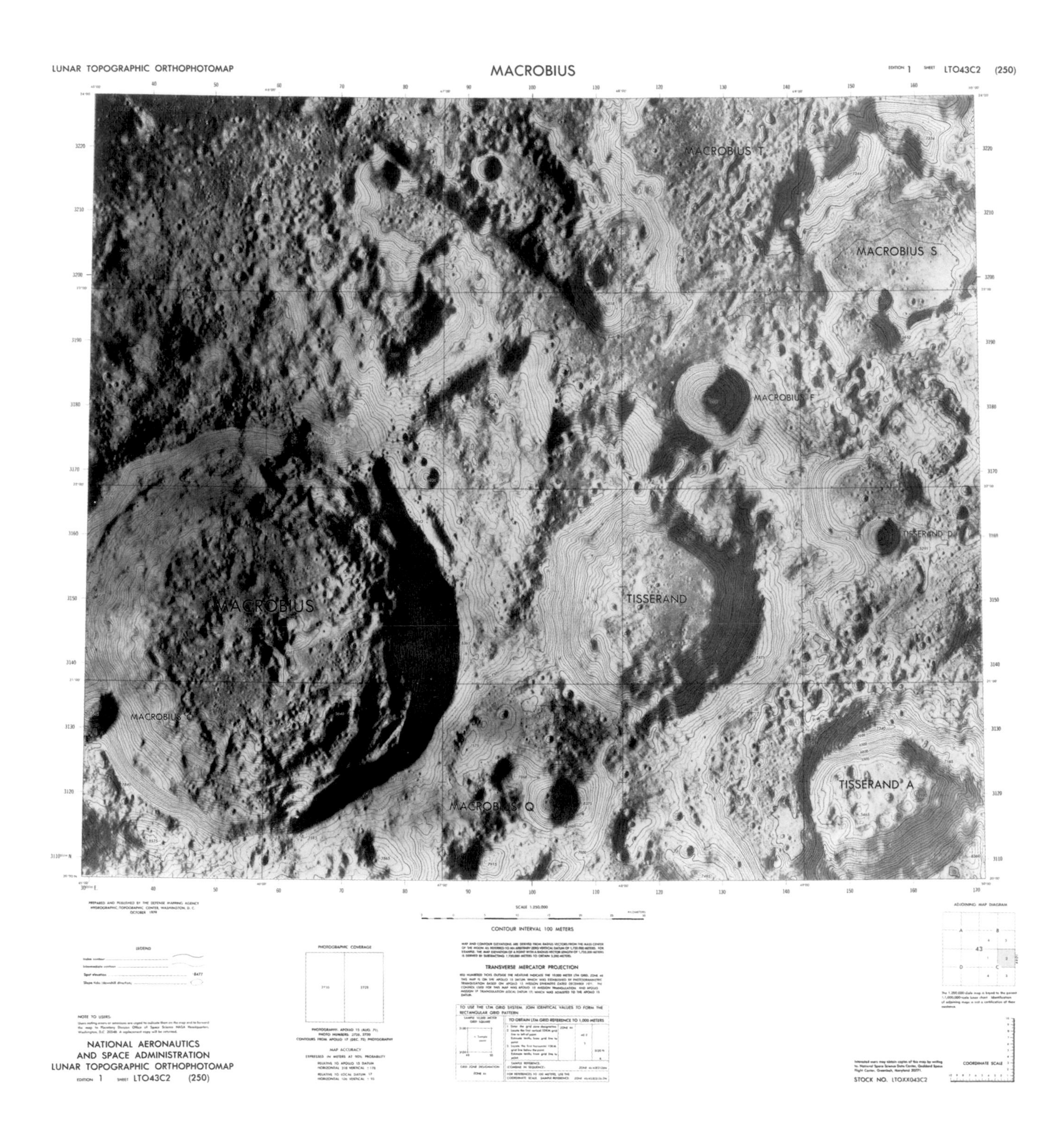

1979 Lunar Topographic Orthophoto Map, Macrobius by Defense Mapping Agency
Chromolithograph, Scale 1:250,000, 26.5 x 25.5 in.
Library of Congress, Washington, D.C.

Mapping Macrobius

WHEN GALILEO TURNED HIS 20-POWER REFRACTING telescope to the night sky in 1609 and sketched the moon, he inaugurated lunar mapping. Through the use of telescopes, then photography, lunar mapping improved through time; it found ultimate expression in the 1961 *Orthographic Atlas of the Moon,* by the University of Arizona's Lunar and Planetary Laboratory.

The next phase of lunar mapping was motivated by the Soviet-American race to the moon. Each country wanted better maps for proposed landing sites; this led each to photograph the moon from unmanned spacecraft. In 1960, the Soviet Union produced the first map of the moon's far side, based on photographs from their Luna 3. A larger scale, nine-sheet map followed in 1967, with images from Zond 3. With its Lunar Orbiter missions from 1966 to 1967, the U.S. filmed most of the moon's surface. Some 2,000 images taken by five spacecraft equipped with medium- and high-resolution cameras and film-developing labs were converted to electronic signals, then beamed by radio to Earth, where they were reconstructed at the Jet Propulsion Laboratory at Pasadena, California. Using stereogrammetric plotters, cartographers from the Air Force Aeronautical Chart and Information Center made detailed photomaps of potential landing sites. Then they produced smaller-scale maps, which guided Apollo 11 astronauts Neil Armstrong and Edwin Aldrin, Jr., to land in the Sea of Tranquility on July 20, 1969.

The last three Apollo moon missions in 1972 and 1973 carried cameras designed by aerial photographer-cartographer Frederick Doyle, who directed Apollo's Orbital Science Photo Team, a joint NASA-U.S. Geological Survey effort. More than 30 percent of the moon's surface was photographed in detail.

The Apollo images were overprinted with contours at 100-meter vertical intervals and spot elevations. Nearly 200 lunar topographic orthophoto maps were published at a scale of 1:250,000, including this map of Macrobius, a large impact crater named for Ambrosius Theodosius Macrobius, a 5th-century Roman philosopher. Also visible is the smaller Tisserand Crater, named for 19th-century French astronomer François Félix Tisserand. Considered by many "the absolutely best lunar maps ever made," several were also used by Apollo 17 astronauts Harrison Schmitt and Eugene Cernan as a substitute for a damaged fender on their lunar rover. The maps were held in place with duct tape and clamps. "Most valuable set of maps made," Schmitt told Doyle.

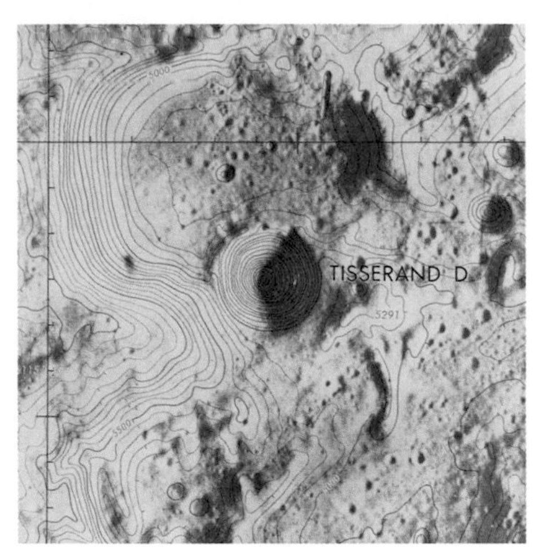

1979 CLOSE-UP OF MOON CRATER TISSERAND D

GIS Maps HIV

In 2000, when analysts at the Central Intelligence Agency's Office of Transnational Issues sought information on the distribution of the HIV pandemic sweeping Africa, they turned to Geographic Information Systems (GIS) specialists. GIS is a powerful new technology that employs computers, databases, and robust mapping software. GIS pioneer and president of Environmental Systems Research Institute (ESRI) Jack Dangermond has defined it as "an organized collection of computer hardware, software, and geographic data designed to efficiently capture, store, update, manipulate, and display all forms of geographically referenced information."

This map of sub-Saharan Africa was one of several produced. Christoper Price, an analytic methodologist with training in GIS, created a traditional statistical distribution map, called a choropleth map by cartographers, but then created an isolinear map based on interpolated values of HIV statistics, and used a linear color scale that is not restricted by administrative and political boundaries. This color scale, ranging from zero percent (light blue) to 54 percent (deep red), indicates the intensity of HIV among the adult population, defined as 15- to 49-years-olds. It creates "a more natural, detailed, accurate" map, according to Price.

Preparation involved a number of steps, beginning with the creation of a geospatial database. Statistical data collected on pregnant adult women from more than a thousand clinics was obtained from the U.S. Census Bureau and the Joint United Nations Program on HIV/AIDS data, which was then referenced to geographic coordinates. These data sets were integrated with other geographic information and manipulated using several advanced geoprocessing and data conversion software packages to produce the map of interpolated HIV prevalence in Africa. Price's map took first prize for "Best Analytical Presentation" at ESRI's 2001 User Conference, an international event promoting GIS through technical workshops, presentations, and awards. Price and his colleagues continue to refine the GIS methodology and to expand the data file. Their ultimate goal, a year or two away, Price notes, is a predictive model that would allow analysts to determine the prevalence of HIV for countries lacking the vital data, such as China, Russia, and India.

Automated mapmaking is 40 years old, but only in the last 15 years has the electronic map flourished in government and private sectors, spurred by the Defense Mapping Agency's efforts in the 1980s to develop "a complete end-to-end digital mapping capability," and by the work of Dangermond at ESRI. Contributing to this has been the widespread availability of large files of geo-referenced statistical data, particularly from the U.S. government; the pervasive adoption of the personal home computer, with its wide market of users; and a new generation of cartographers, geographers, and scientists introduced at a young age to computer language and graphics.

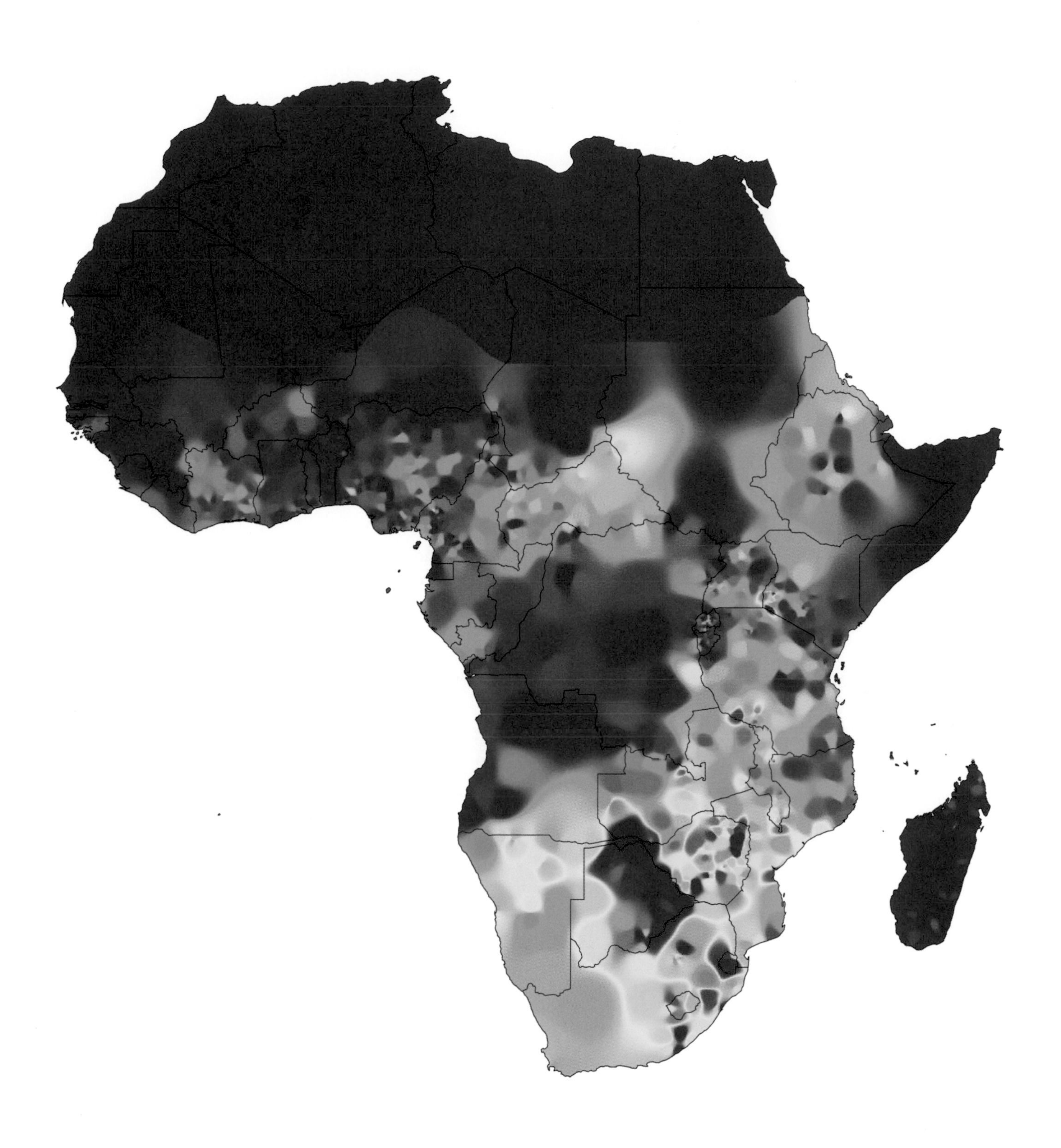

2005 Map of Interpolated HIV Prevalence in Africa by United States Central Intelligence Agency (CIA)
Computer-Generated Map by Christopher Price
CIA, Washington, D.C.

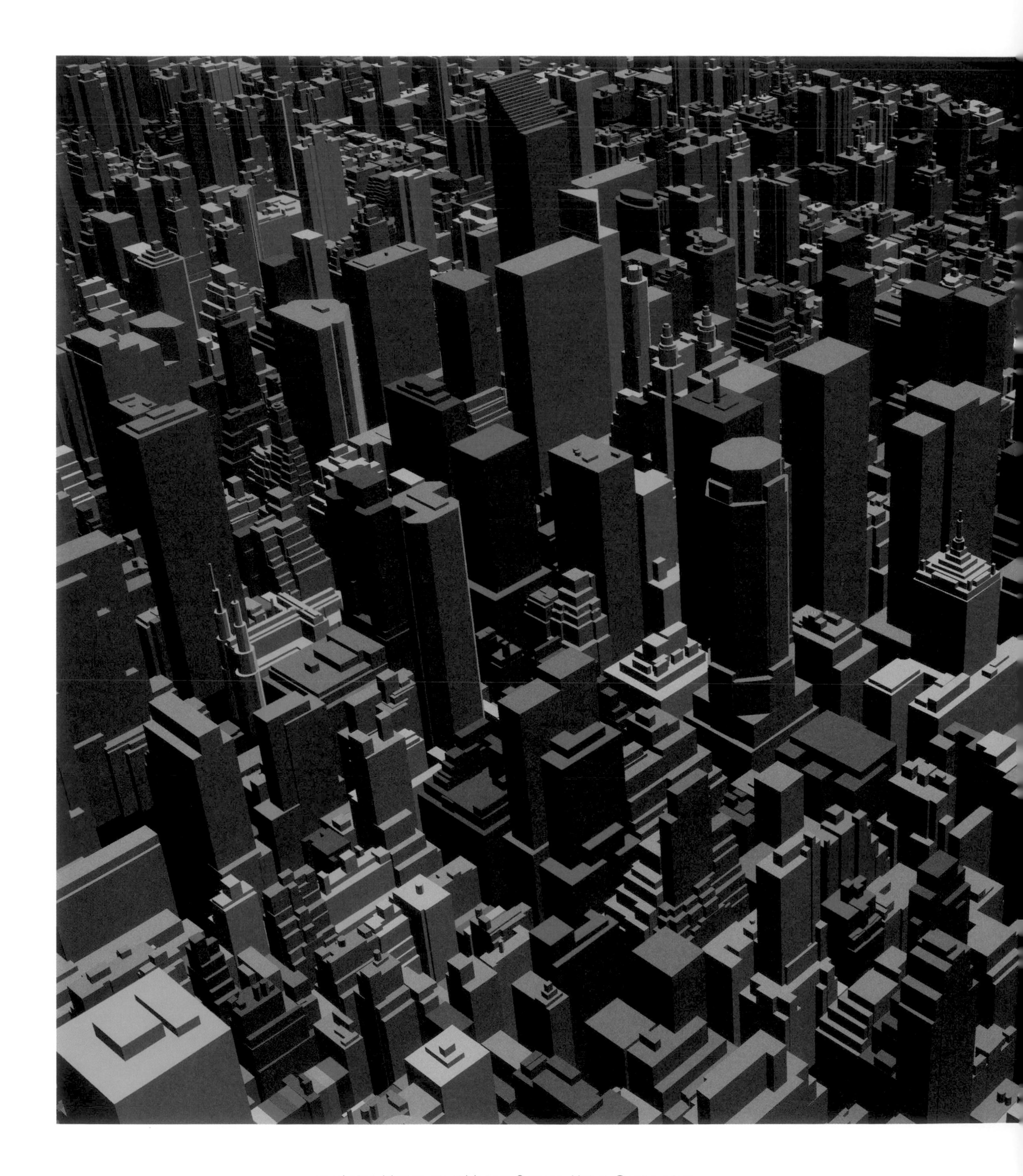

2002 Lower Manhattan by Mapping Services, Vexcel Corporation
Computer-Generated, Three-Dimensional Map
Vexcel Corporation, Boulder, Colorado

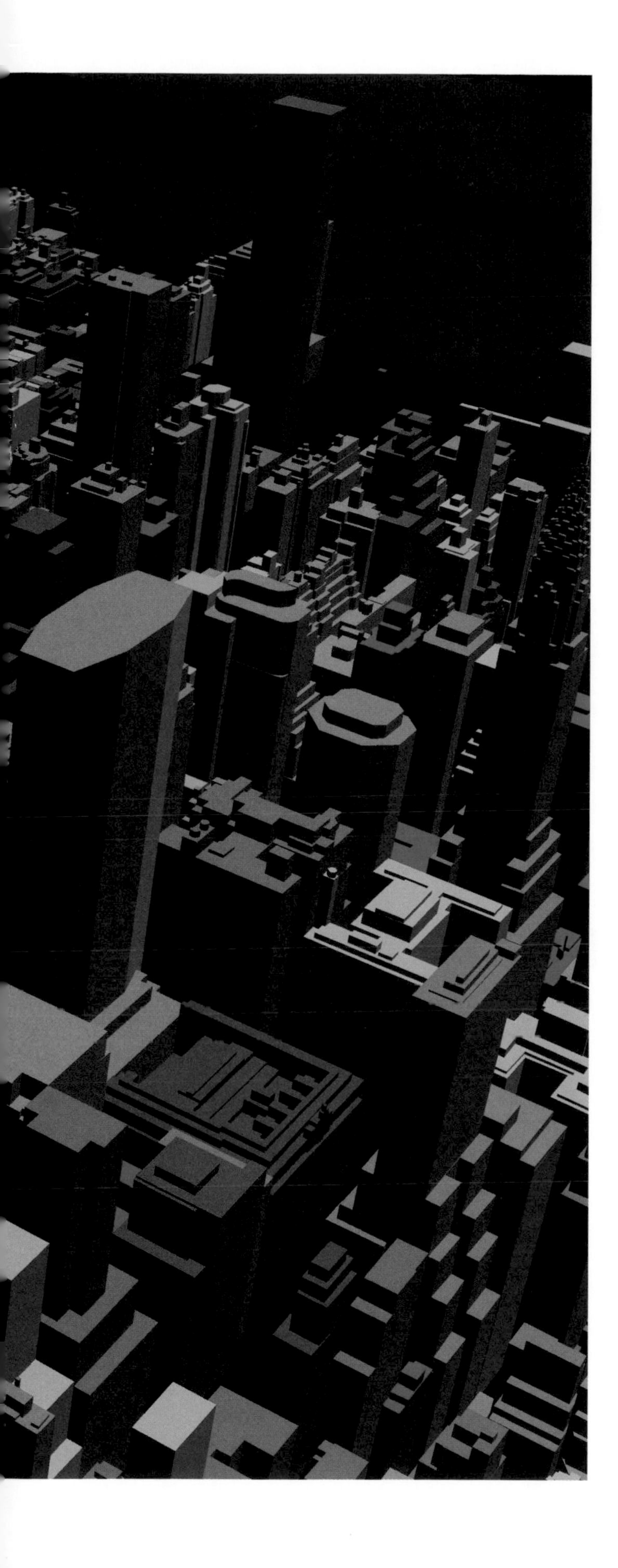

Mapping in 3-D

THE DAWNING OF THE 21ST CENTURY MADE unprecedented demands on urban planners. Wireless devices such as cell phones and personal digital assistants (PDAs) forever altered the way people communicate, and the expectation that information, from stock quotes to sports scores, could be accessed anytime, anywhere, forced new ways of visualizing urban landscapes. Initially developed for military use, remote sensing and digital imaging technologies took on a decisively civilian role, aiding in the development of wireless communications systems. Three-dimensional urban models, such as this one of lower Manhattan, enabled planners to determine clear lines of sight between geographic locations, ensuring optimum network performance and reducing maintenance costs.

The 3-D map is the linear descendant of 16th-century bird's-eye-view maps of European cities and of 19th-century panoramic maps of America. Instead of mapmakers sketching perspective views of buildings and landscapes along survey routes, 3-D cartographers work with powerful computers and work stations, using computer graphics software to store and manipulate models of reality in three dimensions. Their data are derived from a combination of aerial photographs and satellite images, orthophoto maps, and digital representations of Earth's surface elevations, dubbed Digital Elevation Models (DEM). DEMs are derived from stereographic images, digitized topographic maps, and laser scans. The data sets then combine to generate 3-D surface maps of an urban location, including buildings, trees, power lines, even cars on the street. When overlaid with aerial or satellite photography, the final product becomes an astonishingly detailed model that can be rotated and viewed from all angles and edited to assess the impact of new designs on the overall landscape.

While the bird's-eye views and the panoramic maps of the past were drawn and distributed primarily for pictorial value, contemporary three-dimensional map models are produced mainly for military, commercial, and industrial purposes. Following the events of September 11, 2001, these models assisted emergency officials to develop evacuation plans. They could also track movement of airborne toxins by simulating airflow and wind patterns within these 3-D surface models.

Moving Maps

THREE-DIMENSIONAL IMAGES OF THE LANDSCAPE, similar to bird's-eye views or panoramic maps of the past, have been transformed into moving maps, similar to 3-D animations, for the aviation industry. Synthetic vision systems, as the aviation community calls them, are currently being tested by NASA. Designed to improve safety, these terrain maps guide pilots along their projected flight paths in all weather conditions, day or night. They are generated and viewed on display screens mounted in aircraft cockpits, either on the instrument panel or in front of the pilot's eyes. Displayed also are ground obstacles, landing and approach patterns, and air data such as altitude, attitude, and air speed. "I think it's awesome," a United Airlines 767 captain told NASA. "To explain the difference in the situational awareness that you gain, it's just a complete leap from today's technology."

The synthetic visions systems being tested combine Global Positioning System (GPS) satellite signals and onboard digital navigational terrain databases to provide real-time computer-generated terrain images. GPS, also called NAVSTAR GPS, is a worldwide space-based, radio-positioning system that senses and transmits nearly precise coordinates for any place on Earth's surface. It consists of 24 orbiting satellites, a satellite control network, and receiver units that translate the satellite signals into readable position information.

Developed and maintained by the U.S. Air Force, the system became fully operational in December 1993 and is now available to anyone with a GPS receiver. Russia developed a similar system, named Global'naya Navigatsionnay Sputnikovaya Sistema or GLONASS, and the European Union plans to launch one in 2008, called Galileo. NAVSTAR GPS satellites emit continuous navigational signals as they orbit Earth, providing extremely accurate location information in three dimensions: latitude, longitude, and elevation. These signals also provide precise timing information and a worldwide geographical grid that can be converted to a local grid.

Onboard databases include digital elevation data obtained primarily from the U.S. Geological Survey (USGS) and the Defense Department's National Geospatial-Intelligence Agency (NGA). Elevation models provide values in meters and feet for the continental United States and Alaska, data collected by scanning the existing U.S. Geological Survey archive of topographic quadrangle maps. For coverage outside the United States, the Defense Department's National Geospatial-Intelligence Agency has made available its digital terrain elevation data, originally developed to support weapons and training systems. Much of this data, obtained from radar sensors carried aboard the space shuttle *Endeavour* in 2000, is considered "the most complete high-resolution digital topographic database of Earth."

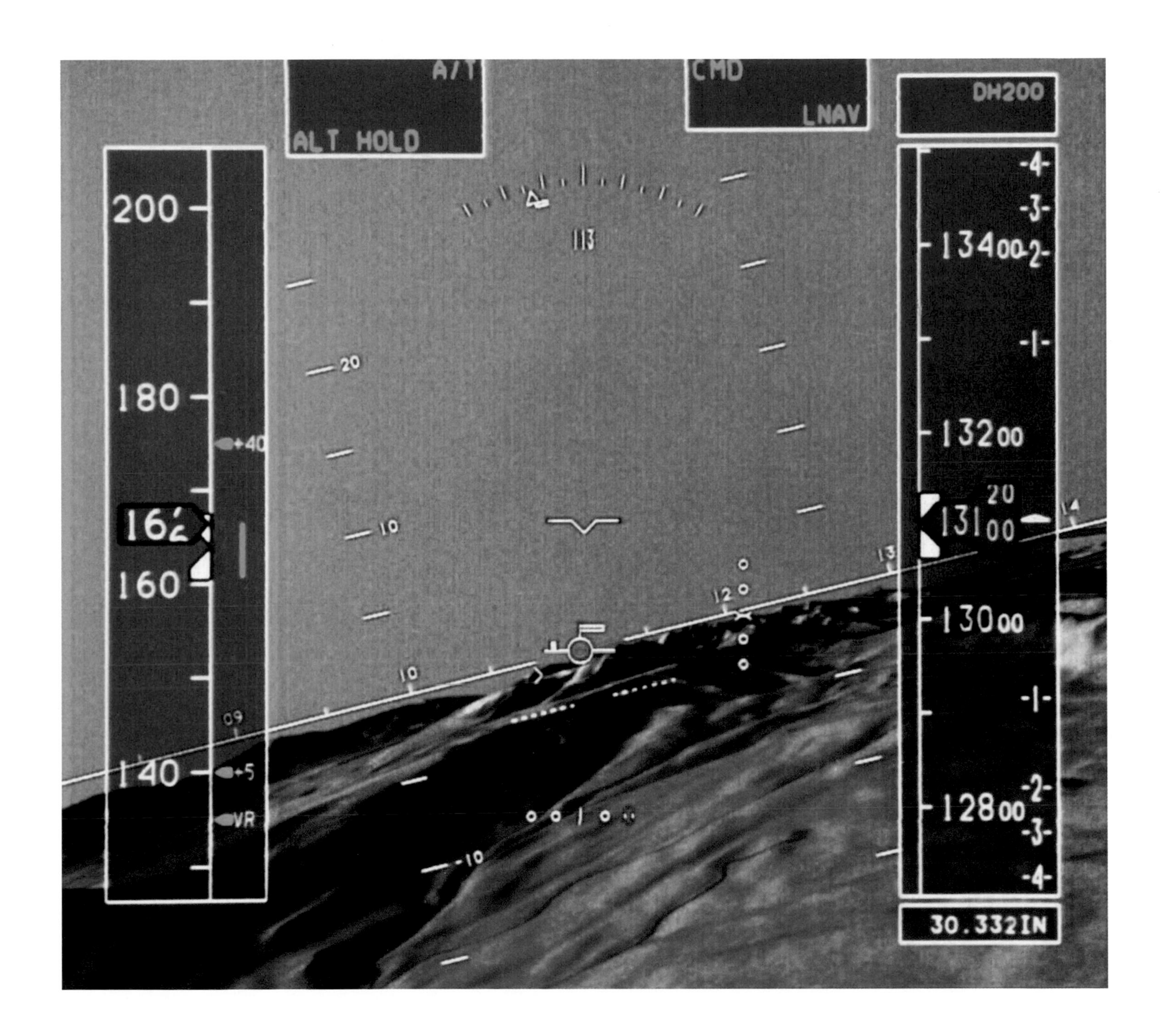

2004 Synthetic Vision Chart by NASA
Virtual Reality Display
NASA

Elevation Model of Willapa Bay

CARTOGRAPHERS AND SCIENTISTS ARE APPLYING NEW mapping technologies to old problems. An interesting example is this intertidal elevation model of Willapa Bay estuary, one of America's most productive and pristine coastal ecosystems. Located on Washington State's southern coast, just north of the mouth of the Columbia River, this vibrant habitat of more than 680,000 acres supports a wide range of marine species. Migratory birds find food and refuge. Its tidal flats support a thriving, century-old oyster industry. Coho, chinook, and steelhead salmon breed in the Willapa Bay watershed. But this rich ecosystem is threatened by *Spartina alternaflora,* an invasive saltmarsh grass. This native of the Atlantic coast was introduced into the bay with an oyster shipment more than a century ago, in 1894. Since then it has infested a third of the intertidal flats. *Spartina* is destroying the habitat, converting flats into marshes, and clogging estuaries.

Accurate topographic maps are basic to understanding and solving the infestation of Willapa Bay, but adequate traditional maps of the vast intertidal flats did not exist. These mud flats, which are alternately exposed or covered by water, are too level to be detailed in the standard topographic quadrangles by the USGS, and the water level is too shallow and intermittent to allow standard hydrographic charting by the National Oceanic and Atmospheric Administration (NOAA).

The solution was found in Light Detection and Ranging (LIDAR). LIDAR is an airborne remote-sensing system that uses laser pulses of infrared light to measure the distance from an aircraft to the ground precisely. When used with a GPS receiver and an inertial navigation system that measures an aircraft's position and altitude, LIDAR data can contribute to 3-D modeling. Because some of the energy of a laser light beam passes through water, LIDAR has an added advantage. It can be used for mapping the bottom topography or bathymetry—land shape or water depth—of shallow water.

Since 1996 NOAA has used airborne LIDAR to collect data along American coasts to assess erosion. In the spring of 2002, NOAA contracted for airborne LIDAR imaging of Willapa Bay as well.

NOAA's intertidal elevation model was prepared from this data. A detail is shown here. To reveal the subtle changes in elevation, NOAA cartographers used false-color infrared hues, ranging from dark blue to red, rather than more traditional methods of terrain representation, such as contours or shaded relief. The LIDAR survey showed that the nearly level tidal flats vary in elevation from 3.6 feet below sea level, shown in dark blue, to 5.2 feet above sea level, indicated in red. NOAA's elevation model of Willapa Bay provides the first accurate baseline—a big step toward the control and eradication of *Spartina alternaflora.*

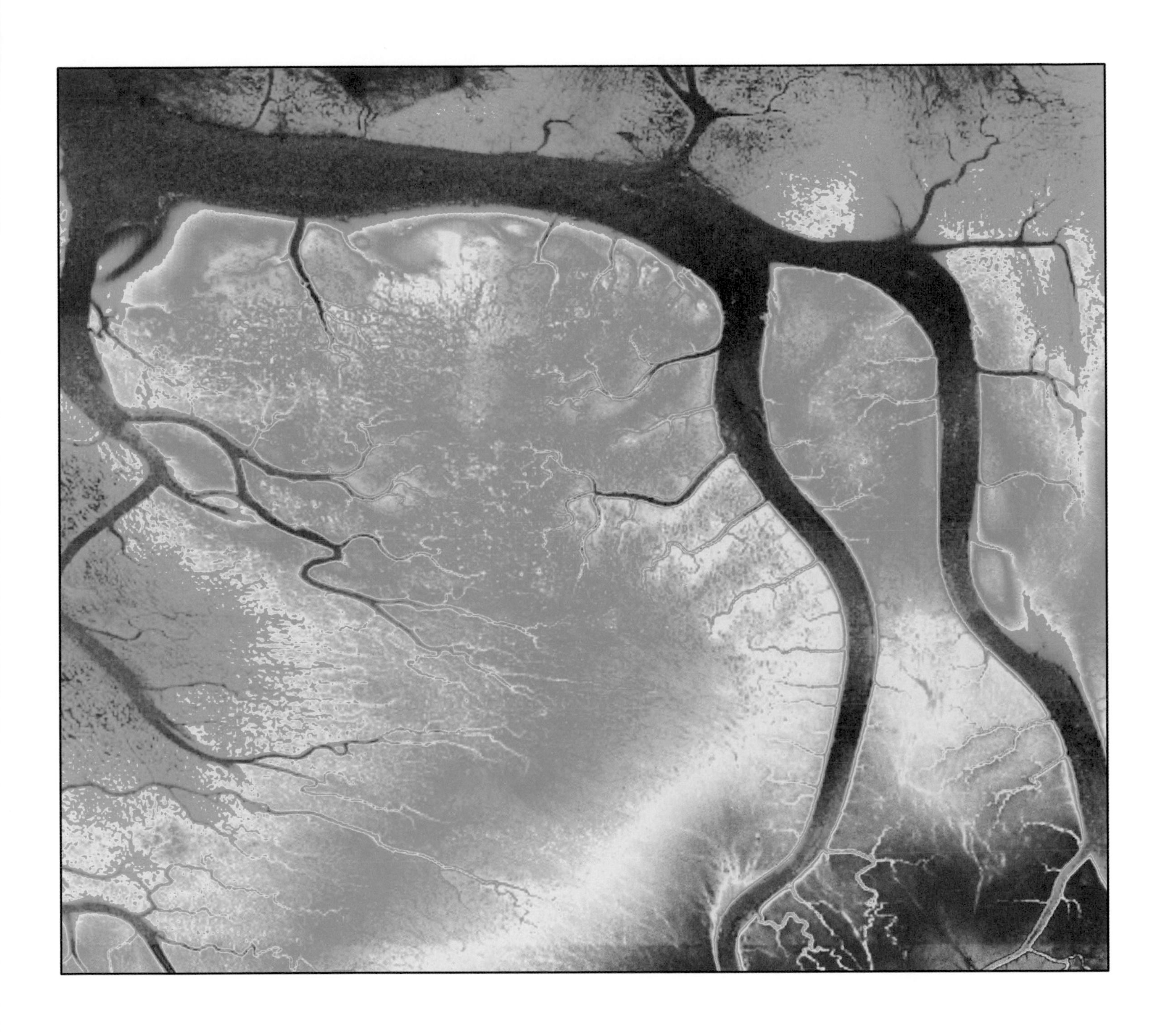

2002 Intertidal Elevation Model of Willapa Bay, Washington State, by NOAA Coastal Services Center
Computer-Generated From GPS and LIDAR Data
NOAA Coastal Services Center

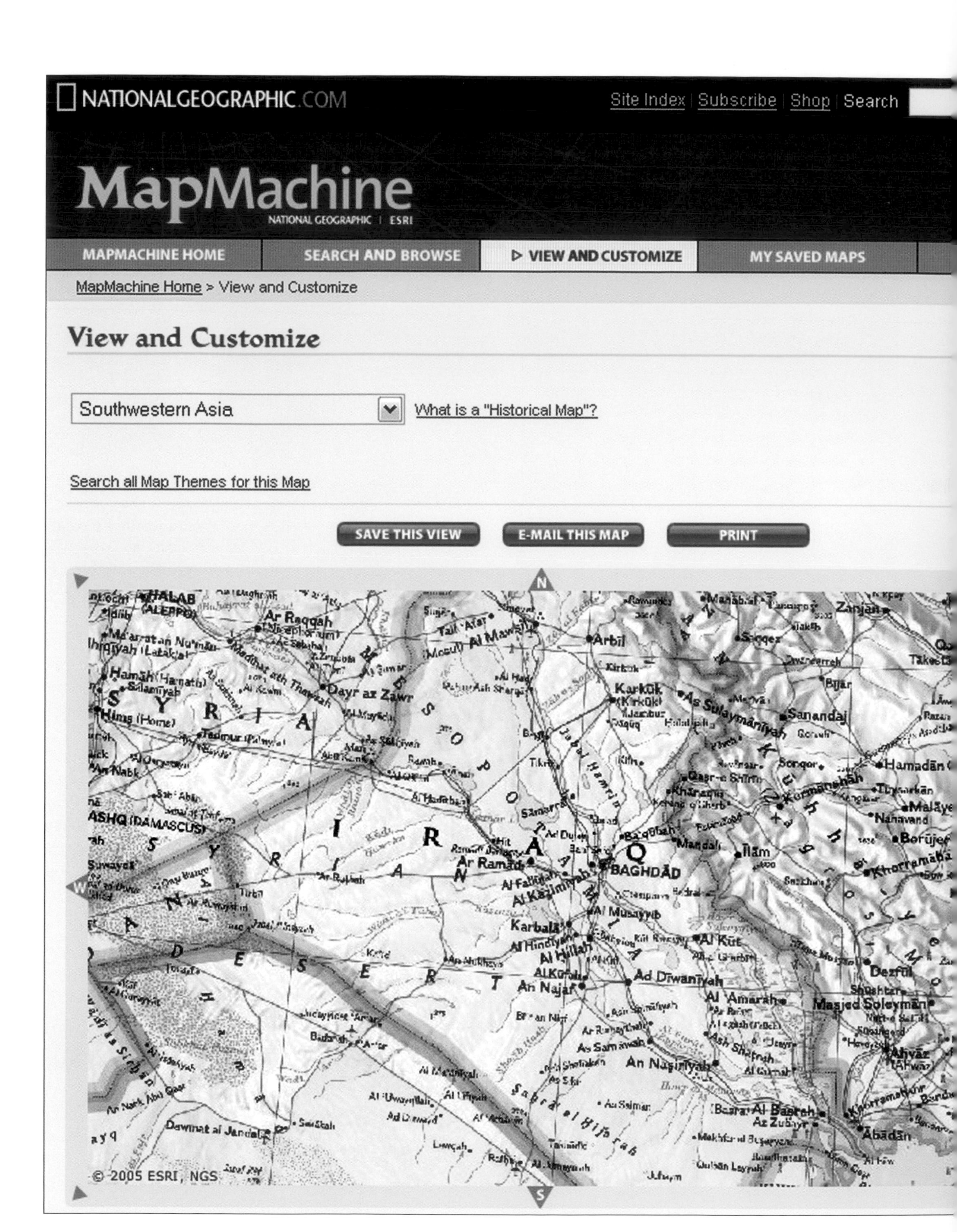

2004 National Geographic MapMachine by National Geographic and ESRI
Internet Digital Interactive Image
NationalGeographic.com

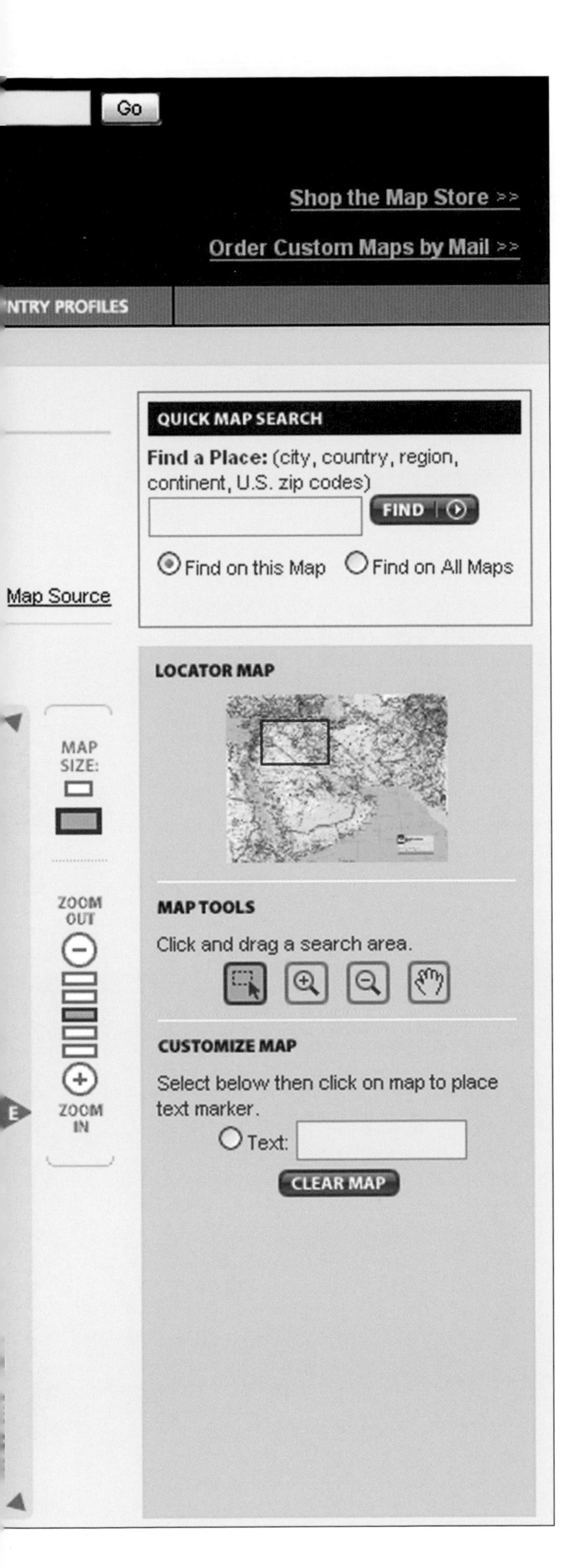

MapMachine

NATIONAL GEOGRAPHIC SOCIETY'S *MAPMACHINE* IS a multimedia online atlas, a new type of atlas comprised of virtual maps made possible by computer technology. The virtual maps that make up the *MapMachine* and other Internet map sites are nonpermanent and transitory, displayed briefly on a computer screen with the simple click of a mouse. They are computer-generated from databases of georeferenced information using mapping software. The databases are housed and maintained on computer hard drives or portable discs. Each view of a map is unique. When the computer is turned off, the displayed map is lost forever unless someone pressed the keys to make a hard copy.

The Society's *MapMachine,* and other virtual multimedia atlases, are at once similar to traditional atlases—and yet amazingly different. Like a standard atlas, maps and related images are available in various formats, but in the electronic world these graphic images are generated seamlessly through links to other virtual maps. This linkage provides an almost unlimited number and variety of maps that cannot be matched by printed atlases.

The *MapMachine*'s country profile of Iraq, for example, also leads the viewer to an entry from the CIA's *World Factbook,* an atlas plate of the Middle East from the National Geographic's own *Atlas of the World* (2004), and a dynamic map by ESRI and the Earth Satellite Corporation that offers three different layers of information—roads and railroads, political boundaries, and place-names of towns and cities. But that is just the beginning! Twenty-eight thematic maps are linked directly to this map. AirPhoto USA and Earth Satellite Corporation provide "hundreds of terabytes" of aerial and satellite imagery.

Earthquake events collected and maintained by the USGS's National Earthquake Information Service are available for the previous day, week, and month. A hazard map from ESRI's *ArcAtlas* portrays threats to the environment from agriculture practices. Hundreds of historic maps from the Geography and Map Division at the Library of Congress help place current events in an historical context. Other links include maps showing traditional themes such as elevation, land use, mineral resources, soils, population density, and annual and monthly precipitation.

Viewers can customize these maps to meet their own needs by changing scales, by reducing or enlarging map segments, and by portraying different layers of information separately or together.

Mapping the Red Planet

MARS WAS THE FIRST PLANET TO BE MAPPED. Wilhelm Wolff Beer, a Berlin banker and amateur astronomer, and his tutor, astronomer Johann Heinrich von Mädler, made a map of the red planet in the 1830s. Beer, who had built a private observatory with a four-and-one-half-foot refracting telescope, began studying Mars in 1828 with Mädler. They made a map that lacked detail, but later cartographers adopted its grid system of latitude and longitude.

Italian astronomer Giovanni Schiaparelli prepared the first detailed Earth-based map of Mars in 1877, when Mars swung close to Earth. He identified more than 60 Martian features and gave them Latin names, after biblical and classical figures, a convention still followed by mapmakers. His map introduced Mars's network of *canali,* meaning "channels," later mistranslated into English as "canals." Maps by American businessman and amateur astronomer Percival Lowell, who constructed Lowell Observatory near Flagstaff, Arizona, perpetuated the misnomer.

Unmanned lunar mapping missions led to similar voyages to Mars, Venus, asteroids, and the larger moons of Jupiter and Saturn. Extensive planetary mapping began with the Mariner 9 mission, which orbited Mars for three months in 1971. Equipped with two high-resolution TV cameras and remote-sensing instruments for determining distance and elevations, Mariner radioed 1,500 photographs to Earth. These were used to produce a topographic and a shaded relief map of the Mars surface at a scale of 1:25,000,000. More detailed maps came from two Viking spacecraft orbits starting in 1976: a series of 140 photomosaics at a 1:2,000,000 scale and 126 photomosaics on a Mars Transverse Mercator (MTM) projection system at a scale of 1:500,000.

In 2001 NASA returned to Mars to study climate and geology. Mission 2001: Mars Odyssey, named in honor of Arthur C. Clarke's book and Stanley Kubrick's movie, was equipped with THEMIS, a thermal emission imaging system. THEMIS produced this detailed 3-D infrared map of the surface of Mars; the map portrays temperature variations of a section of Melas Chasma, a 326-mile-wide canyon.

These thermal images were superimposed on a map of the canyon to study the influence of Martian terrain on day and night temperatures. Daytime temperatures, ranging from -6°F to 54°F, are portrayed in black and white, with black as coldest and white as warmest. Nighttime temperatures are shown in false-color, with cold areas represented by blue and warm areas by red. THEMIS scientists note that temperature differences in the daytime map reflect exposure to sunlight, with white areas receiving more sun. Nighttime temperature variations, they believe,"are due to differences in the abundance of rocky materials that retain heat and stay relatively warm (red)" while "fine grained dust and sand (blue) cool more rapidly."

2002 THEMIS Mosaic Map of Melas Chasma in Perspective by Arizona State University and NASA
Computer-Generated Interactive Map
NASA/JPL/Arizona State University: http://themis.asu.edu/zoom-20021216A.html

2004 The Blue Marble by Reto Stöckli, Nazmi El Saleous, and Marit Jentoft-Nilsen
Internet Digital Image
NASA

The Blue Marble

THE BABYLONIAN WORLD MAP OF 600 B.C., THE earliest visual image of the world, portrayed a flat Earth encircled by water. Twenty-six hundred years later, this image presented a similar full-disk view of Earth—a landmass surrounded by water. But the similarities stop here. This true-color composite image of the Western Hemisphere portrays Earth as if it were viewed from a distance of 22,000 miles. The image was obtained from digital data produced by satellites, rather than from the vision of a scribe in ancient Babylon.

This is one of several "Blue Marble" images produced by NASA's scientific visualizer, Reto Christian Stöckli, a scientist at the Swiss Federal Institute of Technology in Zurich, with Nazmi El Saleous and Marit Jentoft-Nilsen. Stöckli's first composite mosaic image of the Western Hemisphere was generated in 2000 using information from three different Earth-observing satellites to illustrate Hurricane Linda, one of the strongest recorded hurricanes in the Eastern Pacific.

The false-color image proved so popular with teachers, scientists, museums, and the public that NASA's Goddard Space Flight Center produced a second series of similar Earth images in 2002, this time in true color at an even sharper resolution. The Moderate Resolution Imaging Spectroradiometer (MODIS) instrument aboard NASA's Terra satellite collected much of the digital data for this more detailed image over a two-month period. MODIS is a unique instrument that captures and integrates a variety of terrestrial, oceanic, and atmospheric imaging data as it circles Earth. Cloud coverage, added to the image, came from NOAA's Geostationary Operational Environmental Satellite (GOES), which circles Earth once a day taking snapshots of clouds. Together these images help the National Weather Service identify and track severe storms and hurricanes.

Despite such accurate and constantly updated data, scientific visualization remains imprecise. "After all the data are collected, the artistry begins," noted the project's director, Fritz Hasler. In this image, for instance, the mountains of both North America and of South America were exaggerated by a factor of 50 to improve their visibility. The final touch—the addition of the moon in the upper right-hand corner—was "purely artistic license," Hasler says.

Plans for the next generation of "Blue Marble" images are underway, according to Stöckli. The images will be produced at higher spatial resolution and on a monthly basis to illustrate the seasonal climatic cycle of Earth.

Major Sources Consulted

Ahmad, S. Maqbul. "Kharita," in C. E. Bosworth, et al (eds.), *The Encyclopedia of Islam: New Edition*, 1978; Badger, Geoffrey. *The Explorers of the Pacific*, 1996; Black, Jeremy. *Maps and History: Constructing Images of the Past*, 1997; Brown, Lloyd Arnold. *The World Encompassed: An Exhibition of the History of Maps Held at the Baltimore Museum of Art October 7 to November 23, 1952*, 1952; Burden, Philip D. *The Mapping of North America: A List of Printed Maps 1511-1670*, 1996; Campbell, Tony. *The Earliest Printed Maps 1472-1500*, 1987; Clancy, Robert. *The Mapping of Terra Australis*, 1995; Colwell, Robert N., editor. *Manual of Remote Sensing: Second Edition*, 1983; Crone, G. R. *Maps and their Makers*, 1966; Dahl, Edward H. and Jean-François Gauvin, *Sphaerae Mundi Early Globes at the Steward Museum*, 2000; Dorling, Daniel and David Fairbairn. *Mapping: Ways of Representing the World*, 1997; Ehrenberg, Ralph et al. *Library of Congress, Geography and Map Division, An Illustrated Guide*, 1996; Espenhorst, Jürgen and George R. Crossman (eds.), *Peterman's Planet: Guide to the Great Handatlases*, 2003; Fite, Emerson D. and Archibald Freeman, *A Book of Old Maps Delineating American History From the Earliest Days Down to the Close of the Revolutionary War*, 1969; Friis, Herman R., editor. *The Pacific Basin: A History of Its Geographical Exploration*, 1967; Ganong, W. F. *Crucial maps in the Early Cartography and Place-Nomenclature of the Atlantic Coast of Canada*, 1964; Harley, J.B. and David Woodward, editors. *The History of Cartography*, 1987-1998; Harvey, Paul D. A. *The History of Topographical Maps, Symbols, Pictures, and Surveys*, 1980; Hébert, John, editor. *1492: An Ongoing Voyage*, 1982; Howse, Derek and Michael Sanderson, *The Sea Chart*, 1973; Karrow, Robert. *Mapmakers of the Sixteenth Century and Their Maps*, 1993; Konvitz, Josef W. *Cartography in France 1660-1848 Science, Engineering, and Statecraft*, 1987; Levenson, Jay A. (ed.). *Circa 1492. Art in the Age of Exploration*, 1991; Luebke, Frederick C., et al, *Mapping the North American Plains*, 1987; Mollat du Jourdin, Michel and Monique de La Roncière, *Sea Charts of the Early Explorers 13th to 17th Century*; Monmonier, Mark. *How to Lie with Maps*, 1991; Nanba Matsutaro, Muroga Nobuo, and Unno Kazutaka. *Old Maps in Japan*,1973; Nebenzahl, Kenneth. *Atlas of Columbus and the Great Discoveries*, 1990; Noble, John Wilford. *The Mapmakers. The Story of the Great Pioneers in Cartography – from Antiquity to the Space Age*, 2001; Parry, J. H. *The Age of Reconnaissance*, 1963; Ristow, Walter W. *A la Carte Selected Papers on Maps and Atlases*, 1972; Ristow, Walter W. *American Maps and Mapmakers Commercial Cartography in the Nineteenth Century*, 1985; Robinson, Arthur H . *Early Thematic Mapping in the History of Cartography*, 1982; Slama, Chester, editor. *Manual of Photogrammetry. Fourth Edition*, 1980; Shirley, Rodney W. *The Mapping of the World: Early Printed World Maps 1472-1700*, 1983; Snyder, John P. *Flattening the Earth: Two Thousand Years of Map Projections*, 1993; Thrower, Norman J. W. *Maps & Civilization. Cartography in Culture and Society*, 1996; Tooley, R. V. and Charles Bricker, *Landmarks of Mapmaking An Illustrated Survey of Maps and Mapmaking*, 1976; Wallis, Helen M. and Arthur H. Robinson (ed.), *Cartographical Handbook of Mapping Terms to 1900*, 1987; Wilford, John Nobel. *The Mapmakers*, 2000; Wolter, John A. and Ronald E. Grim, editors, *Images of the World. The Atlas Through History*, 1997; Woodward, David (ed.). *Five Centuries of Map Printing*, 1975; Woodward David and G. Malcolm Lewis (ed.), *The History of Cartography. Cartography in the Traditional African, American, Arctic, Australian, and Pacific Societies*, 1998; Xiaocong, Li. *A Descriptive Catalogue of the Traditional Chinese Maps Collected in the Library of Congress*, 2004; Yee, Cordell D. K. *Space and Place Mapmaking East and West: Five Hundred Years of Western and Chinese Cartography*, 1996; Wright, John Kirtland. *The Geographical Lore of the Time of the Crusades*, 1925.

Major Journals and Periodicals Consulted

AARHMS Reviews; The American Cartographer; The Cartographic Journal; Cartographica; Cartography and Geographic Information Science; Chartered Surveyor; ESSA World; Geographical Journal; The Geographical Review; Geoworld; History Today; Imago Mundi; ImCos Journal; Jewish Art; Journal of the Washington Academy of Science; The Map Collector; Mercator's World; Mid-America; The Military Engineer; National Geographic Magazine; Photogrammetric Engineering; The Portolan Journal of the Washington Map Society; Survey Review; Surveying and Mapping; Torch Magazine.

These additional sources were used by chapter:

Emergence of Mapping Traditions. Amir D. Aczel, *The Riddle of the Compass*, 2001; Robert R. Temple, *The Genius of China*, 1958; O. A. W. Dilke, *Greek and Roman Maps*, 1985; Biblioteca Nacional de España, *Tesoros de la Cartografía España*; Bibliothèque nationale de France (http://bnf.fr/); Richard Vaughan, *Matthew Paris*, 1958; Wilma George, *Animals and Maps*, 1969; Georges Grosjean, ed., *The Catalan Atlas of the year 1375*, 1978; Robert Almagia, "Presentazione," *Il mapamondo di Fra Mauro*, 1956.

Charting the Age of Discovery and Exploration. Christopher Columbus, *The Log of Christopher Columbus*, translated by Robert H. Fuson, 1992; Silvio A. Bedini, Elizabeth Harris and Helena Wright, *The Naming of America: An Exhibition at the National Museum of American History, Smithsonian Institution, Washington, D.C. October 1983*, 1983); Sebastian Münster, ed., *Claudius Ptolemaeus Geography*, 1965; Münster, *Cosmographie*, 1968; Tom Conley, *The Self-Made Map: Cartographic Writing in Early Modern France*, 1996; Henry R. Wagner, "The Manuscript Atlases of Battista Agnese," *The Papers of the Bibliographical Society of New York* 25, 1931; Samuel Eliot Morrison, *The European Discovery of America: The Northern Voyages*, 1971; Edward L. Stevenson, *Terrestrial and Celestial Globes*, 1949; James A. Welu, "Cartographic Self-Portraits," in *Imago et Mensura Mundi: Atti del IX Congresso Internazionale di Storia della Cartografia*, edited by C.C. Marzoli, 1985; John Hébert, *The 1562 Map of America by Diego Gutiérrez*, LOC brocure, 1999.

Maps for Royalty, Nobility, and Merchant Princes. John Leighly, *California as an Island*, 1972; Raleigh Ashley Skelton, "Introduction," in *Civitates orbis terrarum, 1572-1618* by Braun & Hogenberg, 1965; Abraham J. Karp, *From the Ends of the Earth: Judaic Treasures of the Library of Congress*, 1991; Johannes Dörflinger, Robert Wagner, Franz Wawrik, *Descriptio Austriae: Österreich u. seine Nachbarn im Kartenbild v.d. Spätantike bis ins 19*, 1977; Christian Sandler, *Johann Baptista Homann, Matthäus Seutter und ihre Landkarten: ein Beitrag zur Geschichte der Kartographie*; Edmund Berkeley and Dorothy Smith Berkeley, *Dr. John Mitchell, The Man Who Made the Map*, 1974

National Surveys and Thematic Cartography. Rollie Shafer, *Finding the Way and Fixing the Boundary*; Benjamin Franklin to Alphonsus le Roy, *Transactions, American Philosophical Society* 2, 1786; J. G. MacGregor, *Peter Fidler: Canada's Forgotten Surveyor 1769-1822*, 1966; Sait Maden, "Turkish Graphic Arts"; James Flatness and Christopher Murphy, "Artifacts from the 'New Order,'" *LC Information Bulletin*, 1998; Gary Moulton, *Atlas of the Lewis and Clark Expedition*, 1983; Andrew M. Modelski, *Railroad Maps of North America, the First Hundred Years*, 1984; Simon Winchester, *The Map that Changed the World: William Smith and the Birth of Modern Geology*, 2001; Matthew H. Edney, *Mapping an Empire* , 1997; Herman R. Friis, *United States Scientific Geographical Exploration of the Pacific Basin 1783-1899*, 1961; Frances Leigh Williams, *Matthew Fontaine Maury, Scientist of the Sea*, 1963; Peter Jackson, "John Tallis 1818-1876," in *John Tallis's London street views, 1838-1840*, 1969; Jonathan Potter, "Introduction," in *Antique Maps of the 19th Century*, edited by Montgomery Martin, 1989; Peter J. Guthorn, *United States Coastal Charts 1783-1861*, 1984; A. Joseph Wraight and Elliot B. Roberts, *The Coast and Geodetic Survey 1807-1957: 150 Years of History*, 1957; Carl I. Wheat, *Mapping the Transmississippi West*, 1960; Michael Friendly, "Re-Visions of Minard," *Statistical Computing and Graphics Newsletter*, 1999; Robinson, *Thematic Mapping*; Earl B. McElfresh, *Maps and Mapmakers of the Civil War*, 1999; Richard W. Stephenson, *Civil War Maps*, 1989; Thomas B. Van Horne, *History of the Army of the Cumberland*, 1885; Alan Berube and Mark Muro, "Red and Blue States Not Black-and-White: Sharp Demarcations on Electoral Map Don't Match Reality," *San Francisco Chronicle*, Aug. 15, 2004.

Maps for Everyone. John R. Hébert and Patrick E. Dempsey, *Panoramic Maps of Cities of the United States and Canada*, 1985; Eugene A. Sloane, *The Complete Book of Bicycling*, 1970; Ehrenberg, " 'Up in the air in more ways than one': The Emergence of Aviation Cartography in the United States," in *Maps on the Move: Cartography for Transportation and Travel*, edited by James Akerman, forthcoming; Isaiah Bowman, *The New World Problems in Political Geography*, 1928; Geoffrey Martin, *Mark Jefferson, Geographer*, 1968; C. H. Birdseye, *Topographic Instructions of the United States Geological Survey*, 1928; General Drafting, *Of Maps and Mapping*, 1959; Douglas A. Yorke, Jr., John Margolies, and Eric Baker, *Hitting the Road: The Art of the American Road Map*, 1996; *Fine Manuscript and Printed Americana*, Sotheby's, May 3, 1994; Donald M. Goldstein and Katherine V. Dillon (ed.), *The Pearl Harbor Papers: Inside the Japanese Plans*, 1993; Gordon W. Prange, *God's Samurai: Lead Pilot at Pearl Harbor*, 1990; Richard Edes Harrison, *Look at the World I*, 1941; Lloyd R. Shoemaker, *The Escape Factory. The Story of MIS- X*, 1990; Ehrenberg, *The Earth Revealed: Aspects of Geologic Mapping*, 1990; Russell Miller, *Planet Earth: Continents in Collision*, 1983; Tom Patterson, "A View from on High: Heinrich Berann's Panoramas and Landscape Visualization Techniques for the US National Park Service"; Matthias Troyer, "The world of H. C. Berann"; National Geographic Society, *Historical Atlas of the United States*, 1988; Brad Washburn with Donald Smith, *On High*, 2002.

Satellite Imaging, Digital Mapping, and Virtual Reality. Clare Averill, Michael Wilson, Larry DiGirolamo, Rebecca Lindsey, "Terra The EOS Flagship"; United States Geological Survey, "Shuttle Radar Topography Mission: Mapping the world in 3 Dimensions"; NASA, "NASA Develops Revolutionary Cockpit Display Technology"; Spencer B. Gross, Inc. Mapping & Aerial Photography, "LIDAR (Light Detection and Ranging"; NASA, "2001 Mars Odyssey"; Arizona State Univ. and NASA, "Mars 2001 Odyssey Thermal Emission Imaging System (THEMIS)"; NASA, Goddard Space Flight Center, "Blue Marble 2000"; NASA Earth Observatory News, "The Blue Marble"; Reto Stöckli, "The Blue Marble Next Generation."

Boldface indicates illustrations.

I.C.C. LIBRARY

ABOUT THE AUTHOR

RALPH E. EHRENBERG

Ralph E. Ehrenberg is a former chief of the Geography and Map Division of the Library of Congress and a former director of the Center for Cartographic and Architectural Archives, U. S. National Archives and Records Administration. He has also worked as a cartographer with the Department of Defense Aeronautical Chart and Information Center, St. Louis (now part of the National Geospatial-Intelligence Agency), and served as an aerial photographer with the U. S. Navy. He represented the Library of Congress on the United States Board on Geographic Names, serving as chairman for the 1989–1991 term.

Mr. Ehrenberg is founder of and an advisor to the Philip Lee Phillips Society, the Geography and Map Division's public outreach and support group; a founding member and past president of the Washington Map Society; and President of the Society for History of Discoveries.

He has lectured and consulted widely on cartographic and geographic resources, the history of cartography, and management of cartographic collections. He was an Honorary Scholar in Residence at the Mitchell Library, State Library of New South Wales, Sydney, Australia in 1997.

Major publications include *The Mapping of America*, with Seymour Schwartz (New York: Harry N. Abrams, Inc., 1980; Wellfleet Press, 2001); *Library of Congress Geography and Maps: An Illustrated Guide* (Washington, D.C.: Library of Congress, 1996); and *Scholars' Guide to Washington, D.C. for Cartography and Remote Sensing Imagery* (Washington, D.C.: Smithsonian Institution Press, 1987).

Mr. Ehrenberg graduated from the University of Minnesota with degrees in history (B. A.) and geography (M.A.).

ILLUSTRATION CREDITS

ABBREVIATIONS:

G&M: *Geography & Map Division, Library of Congress*
NARA: *U.S. National Archives & Records Administration*
NE: *Near East Section, Library of Congress*
NMM: *National Maritime Museum, London*
RB&SC: *Rare Books & Special Collections Division, Library of Congress*

Cover, National Geographic Society Collection. 4-5, G&M. 6, NE. 14-15, Horyuji Temple, Nara, Japan. 19, NMM. 20, © The Trustees of the British Museum. 22, G&M. 24-25, Bildarchiv d.ONB, Wien. 26, The Walters Art Museum, Baltimore. 27, RB&SC. 28-29, By permission of the British Library, Add.11695 f.39v-40. 30, NE. 33, G&M. 34-35, Bodleian Library, University of Oxford, MS.Greaves 42 f.119v-120r Map. 36 & 37, By permission of the British Library, Cotton Claudius D vi f.12v. 38, NE. 40-41, G&M. 43, Horyuji Temple, Nara, Japan. 44-46 (all), Bibliothèque nationale de France. 48-49, Biblioteca Estense Universitaria, Modena. 51, Ryukoku University Library, Kyoto. 52 & 53, Mappamondo di fra Mauro (Venezia, Biblioteca Nazionale Marciana), ca. 1450. 54-55, G&M. 58. NMM. 60-61, By permission of the British Library, Add.15760 f.68v-69r. 62, The Royal Library, Copenhagen, Department of Maps, Prints & Photographs. 64-71 (all), G&M. 73 Topkapi Sarayi Muzesi, Istanbul. 74-75, Bibliothèque nationale de France. 77-81 (all), G&M. 82-83, By permission of Huntington Library, San Marino, California, #HM 29(9). 84, Archivo General de Indias, Seville. 86-91 (all), G&M. 92-93, William L. Clements Library, University of Michigan. 97, RB&SC. 98-109 (all), G&M. 110-111, Courtesy Edward E. Ayer Collection, The Newberry Library, Chicago. 113, G&M. 114, Kobe City Museum. 117 & 118-119, G&M. 120-121, By permission of the British Library, 48 f.7. 122-123, Hebraic Section, Library of Congress. 124-133 (all), G&M. 134-135, By permission of the Houghton Library, Harvard University. 136-137 & 138-139, G&M. 142-143, NARA. 144-145, G&M. 146, NMM. 147 & 148, G&M. 150-151, Hudson's Bay Company Archives, Archives of Manitoba. 152-161 (all), G&M. 162, Freer Gallery of Art, Smithsonian Institution, Washington, D.C.: Gift of Charles Lang Freer, F1905.62ab. 163, NARA. 164-179 (all), G&M. 181, NARA. 182-183, G&M. 184-185 & 188-189, NARA. 190-191 & 192, G&M. 195, Private Collection. 196-203 (all), G&M. 204-205, Private Collection. 207 & 208-209, G&M. 210-211, Joe Vaghi Collection. 212-217 (all), G&M. 220-221, U.S. National Park Service. 222-223 & 225, G&M. 226-227, Created by Michael Caplinger & Michael Malin, MSSS: Data from NASA/JPL Mars Global Surveyor Mission, MOC & MOLA Science Teams. 230-231, Courtesy of Shigeru Suzuki & Eric M. De Jong, Solar System Visualization Project, NASA/JPL. 232-233, G&M. 234 & 235, Lunar & Planetary Institute, Universities Space Research Association. 237, Central Intelligence Agency. 238-239, Vexcel Corporation. 241, NASA. 243, NOAA Coastal Services Center. 244-245, NationalGeographic.com. 247, NASA/JPL/Arizona State University. 248, Image created by Reto Stöckli, Nazmi El Saleous & Marit Jentoft-Nilsen, NASA GSFC.

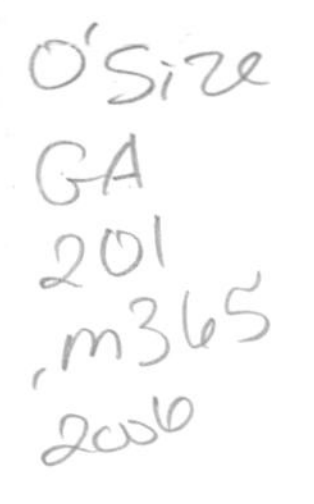
O'Size
GA
201
.M365
2006

MAPPING THE WORLD
Edited by Ralph E. Ehrenberg

PUBLISHED BY THE NATIONAL GEOGRAPHIC SOCIETY
John M. Fahey, Jr. *President and Chief Executive Officer*
Gilbert M. Grosvenor *Chairman of the Board*
Nina D. Hoffman *Executive Vice President*

PREPARED BY THE BOOK DIVISION
Kevin Mulroy *Vice President and Editor-in-Chief*
Kristin Hanneman *Illustrations Director*
Marianne R. Koszorus *Design Director*
Carl Mehler *Director of Maps*
Barbara Brownell Grogan *Executive Editor*

STAFF FOR THIS BOOK
Suzanne Crawford *Text Editor*
Michael Heffner *Illustrations Editor*
Carol Farrar Norton *Art Director*
Meredith Wilcox *Illustrations Specialist*
Barbara Seeber, Susan Hitchcock,
Scott Mahler *Contributing Editors*
Gary Colbert *Production Director*
Ric Wain *Production Project Manager*
Emily McCarthy, Lauren Pruneski *Editorial Assistants*

MANUFACTURING AND QUALITY CONTROL
Christopher A. Liedel *Chief Financial Officer*
Phillip L. Schlosser *Managing Director*
John T. Dunn *Technical Director*
Vincent P. Ryan *Manager*
Clifton M. Brown *Manager*

One of the world's largest nonprofit scientific and educational organizations, the National Geographic Society was founded in 1888 "for the increase and diffusion of geographic knowledge." Fulfilling this mission, the Society educates and inspires millions every day through its magazines, books, television programs, videos, maps and atlases, research grants, the National Geographic Bee, teacher workshops, and innovative classroom materials. The Society is supported through membership dues, charitable gifts, and income from the sale of its educational products. This support is vital to National Geographic's mission to increase global understanding and promote conservation of our planet through exploration, research, and education.

For more information, please call
1-800-NGS LINE (647-5463)
or write to the following address:

National Geographic Society
1145 17th Street N.W.
Washington, D.C. 20036-4688 U.S.A.

Visit the Society's Web site at www.nationalgeographic.com.

Copyright © National Geographic Society 2006. All rights reserved. Reproduction of the whole or any part of the contents without permission is prohibited.

Library of Congress Cataloging-in-Publication Information available upon request. ISBN 0-7922-6525-4

This book is for Phil, Klare, Diane, Erich, Lisa and Danny, and the next generation of map enthusiasts, Nichole, Dahlia, Ike, Jessica, and Jackson.

The production of an illustrated book of historic maps is not unlike the production of a map. Each requires the contributions of many persons from conception to the final printed copy. I am especially grateful to Senior Vice President Kevin J. Mulroy, who conceived the book, and Executive Editor Barbara Brownell Grogan, who brought it to completion. I also want to thank the project editor Scott Mahler, who helped the concept take form; Suzanne Crawford for her careful, insightful, and timely editing of the manuscript; Carol Norton for a book design and layout that matched the elegance and art of the historic maps that are reproduced; and Michael Heffner, for his relentless efforts to acquire digital files and transparencies of maps from repositories throughout the world.

I am indebted to my close friend and colleague of 38 years, Herman J. Viola, curator emeritus of the Smithsonian Institution, for recommending and encouraging my participation in this project.

The staff of the Library of Congress was unfailing with their support, especially John R. Hebert, Chief of the Geography and Map Division, James Flatness, and Edward Redmond; Beverly Ann Gray, Chief of the African and Middle Eastern Division; Michael Grunberger, Head, Hebraic Section; Christopher Murphy, Specialist, Middle Eastern Section; and Heather Wanser, Senior Paper Conservator.

I also wish to acknowledge the assistance of Richard Pflederer, independent scholar; Stephen M. Webb, Staff Officer, National Geospatial-Intelligence Agency; Kathryn Engstrom, Cynthia Cook, Charlotte Houtz, Michael Kline, Stephen Paczolt, Diane Schug O'Neill, and Colleen Cahill, Library of Congress; Barbara Seeber, Kris Hanneman, Susan Hitchcock, Lauren Pruneski and Meredith Wilcox, National Geographic.

My wife Tessa has been an invaluable member of this team, critically reading the entire manuscript and providing direct and indirect support in innumerable ways.

—RALPH E. EHRENBERG